Exploratory Multivariate Analysis in Archaeology

M. J. BAXTER

EDINBURGH UNIVERSITY PRESS

For Hilary

© M. J. Baxter, 1994
Edinburgh University Press Ltd
22 George Square, Edinburgh

Typeset in Lasercomp Times Roman
by Alden Multimedia and
printed in Great Britain by
Short Run Press Ltd, Exeter

A CIP record for this book is available from the British Library
ISBN 0 7486 0423 5

Contents

List of Tables	ix
List of Figures	xi
Acknowledgements	xiii

1 Multivariate Statistics in Archaeology   1
   1.1 Introduction   1
   1.2 Multivariate analysis in archaeology   2
   1.3 Aspects of computer package use in archaeology   4
   1.4 What this book is not about   7
   1.5 Literature review   10
      1.5.1 Introduction   10
      1.5.2 Statistical texts   11
      1.5.3 Quantitative archaeology texts   11
      1.5.4 Journal and conference papers   11
      1.5.5 Analysis of chemical compositions of artefacts   12
      1.5.6 Typological/morphological studies   15
      1.5.7 Assemblage comparisons   20
      1.5.8 Inter- and intra-site spatial studies   23
   1.6 The structure of the book   24

2 Univariate and Bivariate Approaches and Preliminary Data Analysis   27
   2.1 Introduction   27
   2.2 Data   27
   2.3 Dot-plots   28
   2.4 Box-plots   30
   2.5 Comparing groups   31
   2.6 Labelled scatterplots   32
   2.7 Looking at correlation matrices   33
   2.8 Assessing normality   38
   2.9 Transformations and linear combinations   40
   2.10 Tabular presentation   42
   2.11 Direct representation of multivariate data   43
   2.12 Notation, transformation and standardisation   45

3 Principal Component Analysis – The Main Ideas — 48
- 3.1 Introduction — 48
- 3.2 PCA as an exploratory method — 48
  - 3.2.1 The aims of PCA — 48
  - 3.2.2 A problem with PCA — 49
  - 3.2.3 PCA as a linear combination of variables — 49
  - 3.2.4 The geometry of PCA – a simplified view — 50
- 3.3 Examples of PCA and some further considerations — 52

4 Principal Component Analysis – Specialised Topics — 63
- 4.1 Introduction — 63
- 4.2 Distance and standardisation — 63
- 4.3 Covariance and correlation matrices — 65
- 4.4 Interpretation; biplots and related techniques — 66
- 4.5 Size and shape — 71
- 4.6 Compositional data and related problems — 72
- 4.7 Comparing configurations — 77
- 4.8 Outliers — 79
- 4.9 Mahalanobis distance — 80
- 4.10 Rotation of principal components — 83
- 4.11 Principal components and factor analysis — 85
- 4.12 An example — 90
- 4.13 Uses of PCA in archaeology — 94

5 Correspondence Analysis – The Main Ideas — 100
- 5.1 Introduction — 100
- 5.2 Some examples of correspondence analysis in archaeology — 101
- 5.3 Correspondence analysis as principal component analysis — 107

6 Correspondence Analysis – Extensions — 110
- 6.1 Introduction — 110
- 6.2 Correspondence analysis and chi-square distance — 110
- 6.3 A digression on the chi-squared test — 112
- 6.4 Correspondence analysis and chi-squared — 113
- 6.5 Decompositions of the inertia — 114
- 6.6 Correspondence analysis and seriation — 118
- 6.7 Multiple correspondence analysis — 123
- 6.8 Correspondence analysis of chemical compositions — 127
- 6.9 Examples — 128
- 6.10 Correspondence analysis in the archaeological literature — 133

7 Cluster Analysis – The Main Ideas — 140
- 7.1 Introduction — 140
- 7.2 Some applications of cluster analysis in archaeology — 141
  - 7.2.1 Chemical compositions and hierarchical agglomerative methods — 141

	7.2.2	Relocation methods and k-means – general ideas and an application	147
	7.2.3	Spatial analysis and k-means clustering	148
	7.2.4	Methods for binary data	149
	7.2.5	Mixed data and Gower's coefficient	152

8 Cluster Analysis – Some Problems 154
  8.1 Introduction 154
  8.2 Characteristics of clustering algorithms 154
    8.2.1 What types of cluster are to be expected? 154
    8.2.2 Choice of similarity measure 156
    8.2.3 Properties of clustering algorithms 157
    8.2.4 Conclusions regarding clustering algorithms 159
  8.3 Assessing cluster validity 161
  8.4 Comparing clusterings 165
  8.5 Problems with correlated variables 167
  8.6 Spatial clustering and related topics 170
  8.7 Mortuary studies and monothetic methods 172
  8.8 Other approaches 173
  8.9 An example 174
  8.10 Cluster analysis in archaeology 179

9 Discriminant Analysis – The Main Ideas 185
  9.1 Introduction 185
  9.2 Discriminant analysis 185
  9.3 The main ideas 188

10 Further Aspects of Discriminant Analysis 193
  10.1 Introduction 193
  10.2 The mathematics of discriminant analysis 193
  10.3 Statistical inference 194
  10.4 The normality assumption 196
  10.5 The equal covariance matrices assumption 197
  10.6 Sample size 200
  10.7 Assessing how good discrimination is 201
  10.8 Use for validating clusters 204
  10.9 Variable selection 206
  10.10 Examples 209
  10.11 Discriminant analysis in the archaeological literature 214

11 The Final Chapter 219
  11.1 Introduction 219
  11.2 Developments in the statistical literature and archaeology 220
    11.2.1 Exploratory data analysis 220
    11.2.2 Miscellaneous multivariate methods 221

11.3	Statistics in the archaeological literature	223
	11.3.1 Some generalities	223
	11.3.2 Statistics in archaeological publications	224

Appendix A	*Data Sets*	227
Appendix B	*Matrix Algebra – An Informal Guide*	240
Appendix C	*Computational Considerations*	252
Appendix D	MV-ARCH *and* MV-NUTSHELLX	261

Bibliography 262

Index 304

List of Tables

2.1	Correlation matrix for elements in Table A1	35
2.2	The assemblage data of Table A3 presented in percentage form and ordered by type A	43
3.1	MINITAB output of a PCA of the data in Table A1 omitting observation 12	55
3.2	Leading components for an analysis of the covariance matrix of the data in Table A1 omitting observation 12	57
4.1	Context by tool type from excavations at Ksar Akil	74
4.2	Sibson's coefficient used to compare four analyses from Chapter 3	78
4.3	Correlation matrix for the data in Table A4	91
4.4	Results of different PCAs of the data in Table A4	92
5.1	Re-ordered version of Table A6 after correspondence analysis	106
6.1	Inertia 'explained' by the correspondence analysis of Table A3	115
6.2	Statistics summarising the quality of fit of the correspondence analysis of Table A3 to the columns	115
6.3	Contributions to inertia of the columns of Table A3	116
6.4	Statistics summarising the quality of fit of the correspondence analysis of Table A3 to the rows	116
6.5	Contributions to inertia of the rows of Table A3	117
6.6	An incidence matrix in the form of a two-way Petrie matrix	118
6.7	An artificial abundance matrix	121
6.8	Three way cross-tabulation of variables 2, 3, 4 in Table A7	124
7.1	Chemical compositions of Roman glass from Norway	142
7.2	General form of table showing attributes possessed by two individuals	150
8.1	The co-phenetic correlation coefficient for different cluster analyses of the data in Table 7.1	166
8.2	Results of different classifications obtained by cluster analyses of the Mancetter data from Tables A1 and A2	177

8.3	Cross-tabulations of selected clusterings from Table 8.2 to show the relationships between methods	178
9.1	Element means for three variables from the two sites in Table A1	190
9.2	Discriminant analysis output from three variables from Table A1	190

List of Figures

2.1	Dot-plots for selected elements in Table A1	29
2.2	Box-plots for elements in Table A1	31
2.3	A comparison of the distribution of selected elements of the two sites in Table A1 using box-plots	32
2.4	A labelled scatterplot of Fe against Na for the data in Table A1	33
2.5	Scatterplots of different kinds of correlation effects	36
2.6	Examples of normal probability plots for selected elements in Table A1	39
2.7	Star plots of the assemblage data in Table A3	44
3.1	Illustrative example of a PCA using two variables	51
3.2	Component plot of correlation matrix of data in Table A1	52
3.3	Dot-plots of leading components in Figure 3.2	53
3.4	Biplot for the analysis given in Table 3.1	54
3.5	Biplot for the analysis of the covariance matrix of the data in Table A1 omitting observation 12	56
3.6	Plot of the leading components of Figures 3.4 and 3.5	58
3.7	Component plot for the PCA of the correlation matrix of the trace elements in Table A2 omitting observation 12	58
3.8	Biplot for the PCA of the correlation matrix of all elements in Tables A1 and A2 omitting observation 12	59
3.9	Plots to identify the number of 'significant' components for the analysis in Figure 3.8	60
3.10	Biplot of the assemblage data in Table A3	61
4.1	Biplot of the data in Table A1	69
4.2	Three-dimensional plots of the data in Table 4.1	75
4.3	Plot to illustrate the effect of transforming to Mahalanobis distance	82
4.4	Dot-plots of Bronze Age cup dimensions from Table 4	91
4.5	Principal component analyses of the Bronze Age cup data in Table A4	93
4.6	Principal component analysis of transformed data from Table A4	94

5.1	Correspondence analysis of the data in Table A5	102
5.2	Correspondence analysis of the data in Table A3	104
5.3	Correspondence analysis of the data in Table A6	105
6.1	Correspondence analysis showing the seriation of the data in Table 6.7	119
6.2	Multiple correspondence analysis of the data in Table A8	125
6.3	Principal component analysis of Neolithic pottery data	129
6.4	An example to show 'natural' grouping in the dimensions of Neolithic pottery data	129
6.5	Multiple correspondence analysis of the data analysed by PCA in Figure 6.3 using the first and second axes	130
6.6	Multiple correspondence analysis of the data analysed by PCA in Figure 6.3 using the first and third axes	130
6.7	Three-dimensional plots of principal component and correspondence analyses of Neolithic pottery data	131
6.8	Multiple correspondence analysis of the data in Table A1	133
7.1	Principal component plots for the analysis of the Roman glass found in Norway shown in Table 7.1	143
7.2	Dendrograms for four different cluster analyses of the data in Table 7.1	145
8.1	Clustering of random data on two variables	161
8.2	An artificial example to illustrate problems that can arise when clustering highly correlated variables	167
8.3	PCAs of the Mancetter data from Tables A1 and A2	175
8.4	Cluster analyses of the Mancetter data from Tables A1 and A2	176
9.1	SPSS-X discriminant analysis of the data in Table A1 – two groups	186
9.2	SPSS-X discriminant analysis of the data in Table A1 – three groups	187
10.1	Plot showing groups of different size and shape	198
10.2	Artificial data showing two groups and two outliers	202
10.3	Dot-plots of the Mahalanobis distance from the group centroids of the data shown in Figure 10.2	203
10.4	An h-plot of Winchester vessel glass compositional data	208
10.5	Discriminant analysis of the Bronze Age cup data in Table A4	210
B1	Two- and three-dimensional plots of data matrix Y	245
B2	Biplot of the data matrix Y	249

Acknowledgements

I began to entertain the idea of writing this book as a result of collaboration with a number of archaeologists, and talked myself into doing it in a number of bars surrounding conference venues in Heidelberg and Southampton in the spring of 1990. Among friends and colleagues who then, and subsequently, have provided encouragement, allowed me to pick their brains (sometimes unwittingly), let me use unpublished data, and lent me books and papers, etc., are Justine Bayley, John Baylis, Caitlin Buck, Neville Davies, François Djindjian, Nick Fieller, Julian Henderson, Mike Heyworth, Caroline Jackson, Kris Lockyear, John Naylor, Julian Richards and Andy Scott.

The staff of Edinburgh University Press have been helpful throughout, in particular, Vivian Bone, who did not throw me out when I (literally) walked in off the street, unannounced, with the proposal for the book, and Penny Clarke, who guided me through the various stages of production. Richard Bacon and Sue Walker are also thanked for their hospitality and help when I have been in Edinburgh.

My greatest debt is to those who have read and commented on drafts of the book at various stages. Ian Jolliffe and Bob Laxton read parts of the chapters on principal components analysis and correspondence analysis respectively. Hilary Cool, Morven Leese, Clive Orton and Richard Wright all read substantial portions of the entire book. Any misunderstandings and errors are mine, but I hope their influence will be obvious.

Apart from reading it Hilary Cool has provided encouragement throughout the gestation period of the book and helped in numerous other ways (such as ruling on the admissibilty or, usually, otherwise of the jokes), and to her the book is dedicated.

Mike Baxter
August 1993

1

Multivariate Statistics in Archaeology

> My position is that multivariate analysis is to be thought of as nothing more than the analysis of tables of data. If it is worth putting together a table of data then it is worth exploring it by multivariate methods
>
> (Wright, 1989)

1.1 Introduction

This book is concerned with techniques of multivariate statistical analysis that have found widespread use in archaeology and archaeological science. The emphasis is on the methods of principal component analysis (PCA), correspondence analysis, cluster analysis and discriminant analysis. These techniques are among the most popular methods in current use (see Sections 1.4, 1.5).

Other sections of this opening chapter spell out my view of the role of multivariate analysis in archaeology, but it may be as well to deflate immediately expectations that might be aroused by the term 'exploratory' in the title.

Firstly, the book is not about Exploratory Data Analysis (EDA) as expounded by Tukey (1977) and others. This topic – often accorded capital letters – has occasionally been discussed in the archaeological literature, usually in the context of exploring data in a univariate fashion and often obviating the need for more complex analyses. Such objectives are entirely admirable and some EDA techniques are mentioned briefly in Chapter 2; nevertheless, this book is about the more complex methods that are often needed to make sense of complex data.

The second point to emphasise is that the book concentrates on 'traditional' methods of multivariate analysis as implemented in many of the most widely used statistical packages. I am aware of the existence of statistically and graphically more sophisticated methods but have preferred to concentrate on those techniques that are, and I believe will continue to be, widely used by practitioners.

In the sense used in the book, 'exploratory' refers to techniques that usually do not postulate an underlying model for the data and often have the, perhaps limited, aim of reducing data to a form which may be used to inspect the data for archaeologically useful structure or else present known structure in a compact form such as a two-dimensional plot. It is assumed that the reader has acquired, or is prepared to acquire, a knowledge of elementary statistical methods at the level of Shennan (1988), for example. The intention has been to complement material in Shennan's book whilst avoiding undue duplication and to discuss, in detail, aspects of application only covered in outline, or not at all, there. Shennan's book deals with PCA and cluster analysis in particular, and a knowledge of this would be useful though not essential. The approach to PCA in Chapter 3 complements that of Shennan, and Chapter 4 (as with most even-numbered chapters) covers more specialised topics omitted in Shennan. There are several accessible accounts of cluster analysis in both the general and archaeological literature and I have tried, in Chapters 7 and 8, to avoid undue duplication by focusing on techniques that have found specific favour with archaeologists and on specific topics not always dealt with. Shennan does not deal with discriminant or correspondence analysis in any detail; the former topic is dealt with at greater length in the pioneering text of Doran and Hodson (1975) which, while still worth reading, is inevitably dated. Correspondence analysis in archaeology has developed since the publication of Doran and Hodson's book, particularly in France, and was not mentioned by them. The treatment here is more extensive than anything of which I am aware in the archaeological literature in the English language. The structure of this book is discussed at greater length in Section 1.6.

1.2 Multivariate analysis in archaeology

Some of the multivariate methods discussed here have been in common use in archaeology since the mid-1960s. The widespread use of multivariate methods in general was made possible by the development of the computer, and ease of application facilitated by the development of general purpose statistical packages such as SPSS and BMDP; packages devoted to a specific technique such as CLUSTAN; and packages designed for archaeologists such as MV-ARCH.

There are many good books on methods of multivariate analysis, some of which are discussed in Section 1.5, and my debt to some of these will be obvious to those familiar with them. Equally obvious will be the fact that the present book covers only a subset of the material available in these texts. The class of techniques dealt with largely revolves around the problem of reducing a table, or data matrix, of n rows and p columns to a state where some form of two- or three-dimensional summary of the data is possible. Often two-dimensional graphical displays are used, and these may be inspected to see if archaeologically useful information is revealed and to present revealed 'structure' or 'pattern' in a compact and interpretable way.

Thus, the n rows might correspond to specimens of pottery and the columns to their chemical composition with respect to p elements. The technique of cluster analysis (Chapters 7, 8) might be used to identify a small number of groups, specimens within a group being chemically similar to each other and distinct from specimens in other groups. When this can be done the technique of principal component analysis (Chapters 3, 4) is sometimes used to display the groups and their differences in two dimensions. Ideally, the chemically defined groups will have archaeological meaning and might, for example, represent groups of different provenance. In this case the technique of discriminant analysis (Chapters 9, 10) can be used to identify those elements most responsible for the discrimination, or to allocate unprovenanced specimens to a group on the basis of chemical compositions. These and related uses of multivariate methods are widespread (e.g., Jones, 1986).

A second, rather different, example arises when the rows correspond to contexts or assemblages and the columns to artefact types, with entries in the table reflecting the absolute or relative presence of types within a context. The technique of correspondence analysis (Chapters 5, 6) is sometimes used to seriate the data by reordering the rows (contexts) so that adjacent contexts have similar compositions. It will often be possible to interpret such a seriation as a chronological one, using external archaeological evidence, and to display the seriation clearly in a two-dimensional plot in which points, corresponding to contexts, have a clear order. Columns can be similarly reordered and displayed; an 'overlay' of the column and row plots will then identify which artefacts are particularly characteristic of different parts of the seriation. Several examples are given in the collection of papers edited by Madsen (1988a).

In some applications the correspondence analysis will reveal a distinct clustering of points rather than an ordering. It may then be possible to provide a useful functional interpretation. Some examples are given in Barclay et al.'s (1990) analyses of contextual information from excavations of Medieval sites at Winchester. In this case the different sites are associated with distinct types of object; a chronological interpretation is also sometimes possible but is manifested in terms of a 'clustered' association between artefacts and contexts on the plot.

In all these examples a complex body of data is reduced to a two-dimensional picture which is interpreted in the light of archaeological knowledge that may not itself be quantified easily. While this is by no means the only kind of use of multivariate statistics in archaeology, it is the predominant use and the one on which the book concentrates.

The focus is thus on techniques of multivariate analysis used, in an exploratory way, to obtain pictures or summaries of data tables which, in practice, are then interpreted in the light of archaeological knowledge. Little is said about matters of statistical inference and associated distribution theory that occupy much of the space in more mathematical treatments.

The emphasis reflects my background as a statistician with a longstanding interest in archaeology, coupled with practical experience, but no formal training. Previous, and widely praised, texts in quantitative archaeology such as Doran and Hodson (1975), Hodder and Orton (1976), Orton (1980) and Shennan (1988) have been written or co-authored by archaeologists and thus bring a perspective to bear on the material different from that here. Fletcher and Lock (1991) is a recent introductory text. The need for books that focus on statistical techniques has been questioned by Vitali (1991a), in a critique of Shennan's book, on the grounds that techniques are now well understood and that the statistics should be integrated into a more application-oriented development (see also Vitali and Franklin, 1986a, b). This latter point is an important one to which I will return briefly in Section 1.4; however, I am not sure that I agree that technique is now well understood. The even-numbered chapters 4, 6, 8 and 10, in particular, attempt to deal with topics that in my view (based on my reading of the journal and conference literature) are often misunderstood, ignored or neglected. In general an attempt has been made to draw attention to aspects of use and interpretation of which many users of statistical packages – to judge by their publications – may be unaware. This relative neglect of the potential of some of the techniques used, or their limitations, arguably stems from the lack of statistical involvement in much research work and a tendency either to use default options and little else within computer packages, or to try everything and drown in a sea of output. Much statistical use, underuse and misuse in archaeology is intimately related to the strengths and limitations of the widely-available packages and the way in which packages are often used. Widespread use of packages by non-statisticians is inevitable, and it is necessary to come to terms with this in a constructive way (rather than simply deploring over-reliance on packages). Views on this are discussed in detail in the next section.

1.3 Aspects of computer package use in archaeology

Many published applications rely on the use of one or more of the widely available statistical computer packages (e.g. SPSS-X, BMDP, CLUSTAN) or on more local, but tested, packages developed by others. These packages are increasingly available on personal computers (PCs) and will thus become even more accessible to a wider audience. At the same time user friendly teaching and/or graphics packages – available on PCs – such as MINITAB or STATGRAPHICS are extending the range of methods available and/or expanding the size of data set that it is possible to handle. Future developments of these or competing packages will almost certainly include even greater ease of use, and improved graphical output, for multivariate methods. The journal *Applied Statistics*, published by the Royal Statistical Society, publishes reviews of software packages that provides a useful means of keeping track of developments.

The point of this is that it is already very easy to use multivariate methods and will become easier. Unfortunately this very ease of package-use can constrain statistical analysis and/or promote what some would see as misuse. On the one hand a content analysis of papers in some areas of application suggests a quite clear tendency to use default options in the package of choice (and little else) and these may be inappropriate, less than optimal, or could with little extra effort be extended to provide a richer analysis. On the other hand several multivariate analyses may be undertaken on the same data set with reporting limited to results that are considered to be 'best' in some sense.

Some evidence for the foregoing observations is given in later chapters in the context of specific techniques, but they are not new (e.g., Bishop and Neff, 1989). It would be invidious and unproductive to single out instances of wholly inappropriate usage; what I have in mind are instances where the choice of technique(s) is not unreasonable, but the manner in which it is used or reported could be improved. It is hoped that the present book will be of assistance in suggesting how specific methods can be exploited. The matter of reporting is dealt with in more detail in the next paragraphs.

There are several problems that may arise when analysing a data set in several different ways. One is that the amount of output to digest can quickly become unmanageable. A second and major problem is that if different results are obtained then there is a strong temptation to report, or emphasise, those results that confirm prior expectations. Thus, in some provenance studies, several cluster analyses leading to different results may be carried out. That reported as best or most satisfying is usually that which most closely reflects the known archaeological origin of the objects. Since results are only accepted as satisfactory if they reproduce what is already known, it is sometimes difficult to see what can be learned from such a study. Detailed discussion of how and why results differ is relatively uncommon, as are methodological comparisons. The publications in the bibliography could be classified into those that report the application of a single technique; those where several techniques are used with the emphasis on the outcome of one only; and, less commonly, those that contain detailed discussion of all analyses undertaken. It is likely that at least some articles in the first category leave unreported work that was undertaken. The second category could be subdivided into those where all analyses lead to substantively similar conclusions and those where a preferred analysis is emphasised.

A (hypothetical) statistician asked for advice on the choice of method is likely to be very uneasy about using different techniques in the search for 'good' results. It is worth quoting Krzanowski's (1988, pp. 94–5) remarks on this topic (in the context of cluster analysis) at some length. He observes that 'too often the ease of computing is made an excuse for trying as many methods as possible and then selecting the results that 'look best' . . . there is a grave danger that by adopting such an approach one will choose the solution that best reflects one's original preconceptions about the problem . . . a more satisfactory approach is to consider carefully the desired objective of the analysis, and to select from the outset the clustering method that best meets this objective.' This advice reflects a real concern, but can be difficult to implement and is often ignored.

Other authors, concerned with archaeological problems (and specifically the use of artefactual trace element data), explicitly advocate the use of several techniques in an exploratory fashion. Thus Vitali and Franklin (1986b, p. 199) stress that it is 'most important not to settle for one predetermined method but to test a variety of methods and computational options weighing their outcome in terms of internal consistency and concordance with other existing information'. In general the use one makes of one's results, as well as the manner of their generation, is important and – given integrity of their reporting – I incline to the latter stance.

Rather than attempting to stem the rising tide of computer package use and meeting with the same success as King Canute, an alternative approach is to encourage more extensive, and honest, reporting of results. If several different methods produce substantively different and/or archaeologically meaningless results, there has to be a reason for this. If at all possible all results, including negative ones, should be reported and differences between them explained or synthesised. Several possibilities arise, examples of which are given in the next few paragraphs.

Firstly, substantively different but equally valid results may arise, and a synthesis of them is then more useful than either analysis separately. Thus, Baxter (1991) reports examples where a PCA of untransformed Medieval window glass compositions suggests groupings based on the major oxides, whereas the use of a form of logarithmic transformation suggests groupings based on the trace elements that influence the colour of the glass. The groupings differ but are, in their own terms, equally valid. Cross-tabulating the two groupings produces subgroups that prove more interpretable archaeologically than either grouping separately.

Secondly, choice of a method may seem reasonable, but the structure of a particular data set may 'interact' with the method to produce uninteresting results. Principal component and correspondence analyses in Baxter et al. (1990), for example, similar to those described in the previous paragraph produced results with no useful archaeological interpretation. The reason was that the transformation used, coupled with the variation in the data, meant that analyses were dominated by two elements, for mathematical reasons, having no particular interest in terms of the technology of the glass production.

Thirdly, all analyses used may produce negative or uninteresting results. With rare exceptions, such cases go unreported. While understandable, this is regrettable since – at least potentially – a false impression may be given in the literature as a whole about the general usefulness of different techniques. Published provenance studies often use one of Ward's method or the average linkage method of cluster analysis, and these have almost the status of an industry standard. It would be of interest to know how many similar analyses go unreported or do not work, and why.

Related to these last two cases are studies where one of several methods reported is deemed most interesting. Pollard (1986) reports examples where one method of cluster analysis, but not another, reproduces results in a pottery provenance study in accord with archaeological information on the origin of specimens. There is no clear statistical basis for preferring one set of results or method to the other, and one possible conclusion might be that there is no very strong structure in the data so that not too much can be read into the 'successful' analyses reported.

Ideally negative, as well as positive, results should be reported; where possible the results from different analyses should be drawn together in some form of synthesis (which may be formal or informal); the basis for preferring one set of results to another should be made clear and will preferably avoid circular reasoning; and features of the data that lead to poor results with an apparently reasonable method should be identified. In general I take the view that things happen, and results may differ, for a reason, and that if at all possible the reason should be identified and made explicit. Assuming that the choice of methodologies is not wholly inappropriate, the reasons will often have to do with unexpected aspects of data structure of interest in their own right.

The foregoing suggestions are made in response to what I see as the inevitable use that will be made of multivariate techniques given their availability within computer packages. Other approaches, not necessarily incompatible, are clearly possible. Some of these approaches view statistical methodology as something that should be firmly embedded within an analytical or (archaeological) theoretical framework that – in effect – leads to the need for specially developed techniques or restricts the range of techniques from which one might validly choose. In the next section some further remarks are made on this topic in the context of methods and approaches that I have not, for various reasons, included in the book.

1.4 What this book is not about

Haggett (1991), in a retrospective review of his 1965 geographical classic *Locational Analysis in Human Geography*, comments that he still has 'an occasional letter about the book but it comes now from archaeologists and prehistorians' who presumably find the ideas useful. In certain respects the development of quantitative methodology in archaeology can be, and has been, seen as imitating developments in other areas such as geography or taxonomy. Such imitation has been decried, with one major concern being the fear that methodologies thus imported embody assumptions inappropriate to archaeological data and problems. This, in turn, has led to calls for the development of methodology that is 'congruent' or 'concordant' with archaeological problems (e.g. Carr, 1985; Aldenderfer, 1987).

This concern cannot be gainsaid, but there is an attendant danger that can be highlighted by analogy with the development of the 'quantitative revolution' in geography. There, quantitative methods were associated with a particular theoretical standpoint – positivism – that engendered unrealistic expectations about what such methods might achieve. The inevitable disappointment with, and backlash against, this approach encompassed (in some quarters) quantitative methods with the result that methodologically useful babies were unfairly thrown out with the theoretical bathwater.

Orton's (1988) review of the collection of papers *Quantitative Research in Archaeology: Progress and Prospects* edited by Aldenderfer (1987) notes a similar phenomenon in respect of quantitative methods in archaeology and the "New Archaeology". He identifies in the book 'a broad recognition that the application of quantitative approaches in archaeology is not as easy as was thought 20 years ago, accompanied in some cases by a loss of nerve' and suggests that 'it is wrong to see quantitative methods simply as an adjunct to the "New Archaeology": the confusion of the two has led both to the inappropriate use and unreasonable rejection of the former'. His view that there is a need to 'uncouple a "quantitative idiom" from any one theoretical standpoint' is one that I share.

These issues are not, of course, new. Sneath's (1975) review of Doran and Hodson (1975) notes that 'the emphasis throughout is pragmatic' and that some readers 'may find the absence of a strong theoretical framework to be a drawback, but the authors have, I believe correctly, taken the view that we are still far from being able to reduce computer methods to rules of thumb'.

A different position is, I think, taken by Vitali and Franklin (1986b), in the context of analysing trace element data with statistical programs, where what might be termed an analytical framework that 'focuses . . . on applications not on statistics per se' (p. 195) is proposed. This concern with application rather than technique is at the heart of Vitali's (1991a) critique of Shennan's (1988) text. More recently it has been suggested that such a framework be encompassed within some form of expert system (Vitali, 1991b). The general approach, which is exemplified in Vitali and Franklin (1986a), has been criticised by Bishop and Neff (1989, p. 62) on the grounds that it embodies 'naive notions of uniform methodology involving multivariate data analysis'. This last paper calls for 'more informed application of multivariate techniques'.

Notwithstanding expressed differences of view and 'philosophical' stances, most of the authors cited above share a common concern that multivariate techniques be correctly applied within the framework of archaeologically determined problems. Any defensible stance probably contains as much 'truth' about how multivariate methods should be used as did the observations of the mythical blind wise men on the shape of an elephant after exposure to a single portion of its anatomy. I would like to think that the contents of this book reflect a pragmatic and eclectic view informed by a knowledge of application, but no doubt a particular stance is implicit.

Statisticians bring their own perspective to bear on quantitative archaeological problems. This is not always appreciated by archaeologists. Wright (1989), in his entertaining manual for the MV-ARCH package, observes that 'it may not be congenial to your statistician to advocate ancient and well-tried methods in place of experimental methods, the computerized implementation of which is unfortunately "still around the corner". You may well end up waiting.'

Some of the statistical literature related to archaeological problems is of limited or highly specialised interest. Such papers include Broadbent's (1980) work on the reality or otherwise of ley lines (!) and Kendall (1974), Freeman (1976) and others on the measurement of the megalithic yard (if it exists). These papers include seminal statistical material that has had limited application to published archaeological analyses. Closer to the mainstream of archaeological application are papers that advocate a modelling approach to problems, such as those of Naylor and Smith (1988), Buck and Litton (1991) and Scott et al. (1991). The first two of these advocate the use of Bayesian methods, to problems involving the reconciliation of radiocarbon dates and cluster analysis (of clay pipes) respectively. The paper by Scott et al. uses graphical modelling methods – incorporating ideas from log-linear models and path analysis – that require specification of the relations that exist between variables. As yet none of these approaches admit of easy and routine application and fall into the category of techniques identified in the quotation from Wright. Statistical modelling of archaeological data is likely to increase in scope and importance in the future but, in its nature, may involve methods that are not easy to apply routinely. Such topics are excluded from further discussion in later chapters.

The main casualty of the restriction to exploratory methods is the subject of log-linear models. This involves the formulation and testing of models for two- and higher-dimensional tables of counted data. Interesting archaeological applications are given in Leese and Needham (1986), for example, with Lewis (1986) providing an exposition of the method for archaeologists. A limitation of log-linear models is their difficulty in coping with sparse data tables; this is less of a problem for the exploratory method of correspondence analysis, at least for two-dimensional data.

Except as an example of the use of correspondence analysis, seriation methods have been similarly omitted on the basis that they constitute a specialised, model-based set of techniques. These methods are well understood, and are to be thoroughly discussed in a forthcoming text (Laxton, forthcoming).

Of techniques that might legitimately have been covered, multi-dimensional scaling is perhaps the most glaring omission. The basis for this is that, although it is an established method, I have located (surprisingly?) few applications in the recent literature. Other multivariate methods, such as canonical correlation analysis, occasionally arise in the literature, but not enough to merit detailed treatment here.

1.5 Literature review

1.5.1 Introduction

The topics covered in the book have, to a large extent, been dictated by my reading of statistical applications in the archaeological literature. Many of these fall naturally into a limited number of types of application which are reviewed below. I have made no attempt to define what I mean by an 'archaeological' application of statistics although many of the references in the bibliography come from journals with 'archaeology' or some cognate term in their name. Some of the papers noted in the text or bibliography possibly lie outside what even a generous definition of 'archaeology' would encompass. The only excuse for this is that I don't know how to define the boundaries between subject areas and found the subject matter of interest.

As far as journal and conference proceedings go, the coverage is a mixture of the systematic and serendipitous. Both scientific and non-scientific archaeological journals have been reviewed, with those that regularly include articles with a quantitative content such as *Archaeometry*, the *Journal of Archaeological Science*, the *Journal of Field Archaeology*, the *Proceedings of the Prehistoric Society*, *World Archaeology*, *American Antiquity*, etc. having been scanned since at least 1975 (if they date back that far). Journals that I have had access to which publish few or no articles with a multivariate component have usually been scanned since 1980. Some effort has been made to cover the relevant French literature in view of its importance in the use of correspondence analysis.

Coverage of conference proceedings is, inevitably, rather patchier. In the field of archaeological science, coverage of *Archaeometry* conference proceedings since 1970 (and related conferences) is good. These provide a rich, if rather repetitive, source of multivariate applications. For non-scientific archaeological conferences the BAR series (both the British and International) has proved useful, as it has for evidence of Ph.D. work. I do not claim to have looked at the complete BAR series. Conference papers are less rigorously refereed than journal papers, often appear in more than one place (including eventual journal publication), may simply be a report on work in progress, and are occasionally trivial. Nevertheless I think they provide a very good guide to what methods researchers are using, and how, that can be more evident than the journal literature.

It is in the book coverage that the bibliography is most serendipitous. Those cited range from singly authored works on specialised topics that make some use of multivariate methods, through multi-authored but specially commissioned works, to glossy texts that are essentially conference proceedings. Some have been specially pursued as a result of references located elsewhere; others have accidentally come to my notice while browsing the shelves of libraries where this is permitted.

Nearly all the references in the bibliography are to works that I have cast my eyes over at least once. It would be easy to multiply the list by including relevant references that I have not had the time and/or inclination to locate. The bibliography is neither comprehensive nor a random sample but is, I hope, sufficiently detailed to provide a representative indication of the way multivariate statistics have been used in archaeological applications.

1.5.2 Statistical texts

Of general works on multivariate analysis Krzanowski (1988) is a particularly good, and recent, text that focuses on application. Seber (1984) is a useful compendium of work done in a widely scattered literature. Manly (1986) is a good short introduction, with some archaeological data sets used for illustration.

On specific topics Jolliffe (1986) is especially recommended for its treatment of PCA. Greenacre's (1984) text, though heavy going in places for the non-specialist, is an important English language text on correspondence analysis. Everitt (1980) remains a useful and readable guide to cluster analysis, as does Gordon's (1981) treatment of the general subject of classification. Klecka (1980) provides a brief introduction to discriminant analysis particularly geared to the way the method is implemented in the SPSS-X package; the earlier text of Lachenbruch (1975) usefully summarises the approach most widely used in the literature.

1.5.3 Quantitative archaeology texts

Texts with a substantial statistical component are few in number and variable in approach but have usually been critically well received. The seminal work of Doran and Hodson (1975) is still well worth reading. Of more specialised interest, and now also somewhat dated, is Hodder and Orton's (1976) work on spatial analysis in archaeology. Orton (1980) is a good introduction to uses of quantitative methods for the non-mathematically-inclined but (inevitably) has little methodological detail. Shennan's (1988) undergraduate text is a level-headed account that provides an introduction to some of the ideas in this book and has a sensible attitude to both the uses and limitations of statistics in archaeology. Fletcher and Lock (1991) is a recent introductory text that includes material on inferential statistics (i.e. significance tests) in more detail than the other texts.

In a slightly different vein Madsen's (1988a) edited collection contains articles by archaeologists applying multivariate methods to a variety of material. Correspondence analysis is particularly favoured, and a valuable feature of the book is that many of the data sets used are published.

1.5.4 Journal and conference papers

In later chapters multivariate applications in archaeology are discussed from the perspective of the different techniques that have been used. It is also possible to classify applications in terms of the general structure of the data used and intentions of the analysis, and most uses can be classified into one of a surprisingly small number of 'problem types'. This kind of classification in terms of types of data-structure cuts across a technique-based classification and is reviewed in some detail here as a prelude and complement to the technique based perspective of later sections.

The four kinds of problem, the semantics of which will be covered in the discussion, are (a) analyses of the chemical compositions of artefacts: bone, etc.; (b) typological/morphological analysis; (c) assemblage comparisons; and (d) spatial analysis. From a purely statistical view this typology of use could be telescoped since (a) could often be viewed as typology based on chemical composition and (c) as assemblage typology. However, without getting into a discussion of precisely what is meant by a 'type', the classification adopted corresponds to distinct kinds of problem and forms a useful organising principle. These are now discussed in turn.

1.5.5 Analysis of chemical compositions of artefacts

An idealised caricature of problems that involve the analysis of chemical compositions can be drawn in the context of studies of pottery provenance. Specimens of known archaeological origin are grouped on the basis of chemical composition using cluster analysis; it is confirmed that these correspond to the archaeological grouping so that one is confident that the chemistry is usefully informative about provenance; having established the reality of the groups, a more digestible display of their difference is provided through the techniques of principal component analysis or discriminant analysis, this latter technique being used to 'validate' clusters; and finally, discriminant analysis is used to identify the important discriminating elements and to assign unprovenanced specimens to an appropriate group. This is an idealised portrait, in that many publications do not get beyond the cluster or cluster-and-display stage, and some aspects such as the 'validation' by discriminant analysis are questionable (Chapter 10). It is a caricature in that, while not untruthful, it does not do justice to the richness of statistical application in some papers, or the scientific scope (where the statistics plays a small role) of many.

Sweeping generalisations of the kind made above and later can be justified (to the extent that sweeping generalisations ever are) by a content analysis of statistical use in the papers listed in the bibliography. It is instructive to treat the contents of the journal *Archaeometry* over the period 1968–91, and of the published proceedings of the international *Archaeometry* conferences held biennially since 1976 (in Edinburgh, Bonn, Paris, Bradford, Washington, Athens, Toronto and Heidelberg) as a study 'population'.

Of 41 articles in the journal that use the techniques discussed in this book and are relevant to this sub-section only 4 were published in 1968–73, with numbers for the subsequent six-year periods being 15, 10 and 12. This pattern of use is reflected, with a time lag, in the 56 relevant conference publications with none in 1976, 5 in 1978 and 9, 10, 6, 11, 8 and 7 in subsequent years. There are some differences between the two sources – the journal is more varied in the material covered and methods used, for example – but these are not gross, and further analysis is confined to the joint population of 97 articles.

Ceramics and clays are the materials most studied (56%), followed by manufactured glass and metals (14% apiece), with the remaining materials including a few papers on flint and related materials and obsidian and isolated appearances of jade, lacquer and garnets. A few papers focus on aspects of statistical methodology (e.g., Bieber et al., 1976) but the primary focus is usually on a scientific archaeological problem with statistics in a supporting role. There are good reasons for the predominance of pottery in these studies; simplistically, pottery compositions reflect the compositions of the clays from which they are made and these latter are sufficiently distinct between different locations that there is hope that chemical analysis of the pottery will reveal its origin (this has been termed the provenience postulate (e.g. Harbottle, 1991)). Another reason is the large amount of pottery available for study by comparison with other types of artefact. The materials of glass and metal manufacture, by contrast, often include reused glass (cullet) or metal that will blur any distinctive 'fingerprint' that the raw materials (e.g. silica in the case of glass) might have. Thus multivariate studies of glass and metal may often focus on broader questions of technology (e.g. are compositions, and hence the presumed technology, similar across time and/or space) rather than specific questions of provenance. The nature of the material studied has potential statistical implications to which we will return shortly.

Of our study population of 97 articles, 62% use a single statistical technique with most of the remainder using just two techniques. This 62% is composed of 39% cluster analysis, 15% discriminant analysis, 6% PCA, and a single appearance of correspondence analysis. Of the 38% using two or more techniques, nearly all (there are four exceptions) use cluster analysis, either in conjunction with discriminant analysis (14%) or PCA (19%). Overall, cluster analysis is used in 74% of the papers, discriminant analysis in 31%, PCA in 28%, and correspondence analysis in just 2%. The impression created by these figures is, I believe, reflective of applications in the literature at large.

There are good (and not so good) reasons for the overall patterns of use. Correspondence analysis is a more recent technique yet to be widely used within the Anglo-American tradition but, in any case, is not especially suited to tables of compositional data of continuous variables (Section 6.8). Principal component analysis is used primarily as a display technique, once clusters have been established using cluster analysis. As a technique for revealing clusters in its own right, the most usual criticisms are that it can blur distinctions between clusters and (as often used) displays results in two dimensions whereas higher dimensions are needed to see group differences. Some authors prefer to use discriminant analysis as an adjunct to cluster analysis since, unlike PCA, a knowledge of group membership is used to obtain a sharper picture of the differences (Section 9.2). It is also often used to 'validate' the groups determined by cluster analysis, though such use is often questionable (Section 10.8).

Discriminant analysis is also sometimes the method of choice, and whether this or cluster analysis is employed depends on any prior knowledge of provenance or other group structure and how one uses it. If there is no prior knowledge of group membership then discriminant analysis is not an option. If group membership (e.g., provenance) is absolutely secure and one is not prepared to believe cluster analyses that cut across this, then there seems little point in using the latter method (some published studies are possibly open to criticism on this count). Where group membership is uncertain, or if negative results or results that cut across the prior grouping are possible and of interest, then cluster analysis is clearly of potential value. Given the popularity of cluster analysis, much of the literature clearly (implicitly?) assumes this last posture.

A simplistic explanation for the predominance of pottery provenance studies was given earlier. In practice there are many factors that complicate the simple equation between composition and source. Thus the original composition may be modified by firing and tempering (e.g., Bishop and Neff, 1989). The focus of some research is on the problems for scientific analysis and archaeological interpretation that such considerations pose, and there can also be statistical implications. Thus, if ceramic artefacts are made with raw materials from the same source but differentially tempered then standard statistical procedures may be inappropriate or misleading (Bishop and Neff, 1989). Possible responses include transforming the data before analysis to eliminate the problem, and this in turn requires a precise and explicit model of how the tempering affects the composition. A different, though related, approach is adopted by Bishop and Neff (1989) who postulate a model for the data that involves information on which elements are likely to predominate in the temper, and then interpret the output of their PCA in the light of this model. These examples are considered in more detail in later chapters.

Another example where the properties of the material studied raise issues of statistical practice arises in the context of the cluster analysis of pottery compositions. The particular methods of cluster analysis used in practice are potentially problematic if many elements are highly correlated with each other (Section 8.5). Many pottery and glass compositions contain such highly correlated elements. Although this difficulty has been known for a long time (e.g., Cormack, 1970) it is only infrequently acknowledged and addressed in application. Where the difficulty is admitted, two common responses are to argue that it is not serious for the data set under analysis, or to use a subset of variables that have low correlations. An alternative view is that cluster analysis should not be used at all.

The foregoing review is both brief and partial. A good, recent and critical review of some of the statistical problems involved in compositional analysis is given by Bishop and Neff (1989). Pollard (1986) provides a lucid account of the statistical procedures most commonly used in pottery provenance studies, as well as of their routine implementation within a laboratory. Both these papers owe something to that of Harbottle (1976) where many still current difficulties of statistical analysis were discussed. Harbottle (1982) is another useful review that notes the statistical difficulties involved with correlated data and also the materials that are, or are not, likely to be affected. Bieber et al. (1976) discuss statistical aspects of compositional studies and have been influential. The difficulties that may arise in provenance studies and their interpretation have only been touched upon; the catalogue of considerations in Rice's (1987) handbook on pottery analysis is formidable, particularly as a deterrent to such study.

1.5.6 Typological/morphological studies

In this and subsequent subsections a convenient 'population' of application articles is less easy to define than for the analysis of chemical compositions, and a different approach is adopted. The literature is scattered and the material on which any generalisations are based is the sample of applications I had to hand at the time of writing. This provides a basis for commenting on what archaeologists have tried to do and with what techniques; I have tried to compensate for potential (statistical) bias by citing more individual papers than previously so that at least the source of any misconceptions will be clear. Where the technique used within a paper is noted, it is the technique-specific part of the bibliography that should be consulted for the reference, otherwise the reference should be located in the general bibliography.

Typology is a topic on which entire collections of articles have been published, with limited agreement (e.g., Whallon and Brown, 1982), and my use of the term – which is simple-minded – had best be clarified. It is assumed that a set of entities (or 'things') is available for study, sufficiently alike in their 'thingness' that it is reasonable to study them together. (If the entities can be subdivided into different types on inspection this should be done – do not use statistical methods as a substitute for expert common sense. Different observers will, of course vary in their ability to differentiate types visually; expert artefact researchers of my acquaintance get by quite happily without the use of multivariate methods.) Observations and measurements are made on each entity; these may be qualitative, such as colour or the presence or absence of a feature, or quantitative, such as weight, length, the angle of a retouch, or a count of the number of occurrences of some feature. Quantitative variables are usually, though not inevitably, chosen to assist in the description of the shape, or morphology, of an entity.

The set of entities may be divided into subgroups on some basis before a multivariate analysis, but often is not. In this latter case the object of analysis may be to define subgroups within which entities are similar to each other. Cluster analysis has been used for this purpose, sometimes augmented by a PCA display in a manner similar to usage in the analysis of chemical compositions. Less ambitiously, a low dimensional display may be sought in which points representing very similar entities are close to each other, and points representing dissimilar entities are distant. Correspondence analysis or PCA are both suited to this purpose, depending on the kind of variables used. If subgroups are defined prior to analysis, then discriminant analysis may be used to confirm that it is a valid one in terms of the measurements used; to select those variables most important in defining such groups; or to provide a basis for classifying entities of unknown kind into a group.

The general aim of subdividing a set of entities into subgroups containing closely similar entities within each group, or of investigating a proposed subdivision, will be called 'typological', with 'morphological' used as an alternative term when only variables descriptive of shape are involved. There is no implication in this that the typology produced will necessarily have archaeological meaning or be of use (though one naturally hopes that it will), or that I have any strong views on what is an archaeologically useful 'type'. That a method such as cluster analysis will produce subgroups of similar entities is no guarantee that subgroups differ much from other subgroups. Thinking in terms of two-dimensional plots of points, subgroups may be contiguous or disjoint clusters. In the former case the division is essentially arbitrary, and may be useful as a descriptive classification but of limited application. Disjoint clusters are more likely to represent types that are in some sense 'real' and are, I think, what many applications hope to find.

The description of 'typology' that has been given is in terms of 'object clustering'. It has been strongly argued by some that archaeological types are more properly defined in terms of 'attribute clustering' – that is, the co-occurrence of variables or attributes (e.g., Spaulding, 1982). Techniques can be adapted to this end if needed. Indeed, both PCA and correspondence analysis naturally allow the investigation of relationships between variables as well as objects, and their interrelationship, so that the need for a strong point of view on the 'correct' approach is, perhaps, not necessary (Shennan, 1988, p. 284).

All this is a roundabout way of saying that producing a 'typology' – in my sense – can be accomplished using statistical methods, but that interpretation is much more difficult and the province of the archaeologist, not the statistician. In some cases automatic classification may simply provide a quick and convenient way of defining a starting-point to be inspected and rearranged at leisure or rejected if need be. The question of selecting appropriate variables and qualities for use in an analysis is both highly important, difficult in general, and also very much an archaeological problem. Selecting a finite set of variables that encompasses all the information that an archaeologist may use when classifying material can border on the impossible. This is quite apart from the problem of accounting for the differential weight that may be attached to different qualities. The rejection of methods of automatic classification for such reasons is a perfectly defensible position.

Nevertheless, many scholars have thought it worthwhile to use multivariate methods in typological study at some point in their careers. About half of the references before me as I write (excluding biological applications) date from 1985 and later, so the subject is not dead.

The abstract entities frequently referred to above have manifested themselves in a multitude of forms in applications. To give examples almost at random, studies using cluster analysis, unless otherwise indicated, include archaic Greek sculptured figures (Guralnick, 1976); prehistoric and Romano-British settlements (Smith, 1974); prehistoric southern Italian polished stone artefacts (O'Hare, 1990: using PCA); room function (Ciolek-Torello, 1984, 1985); Neolithic chambered cairns (Perry and Davidson, 1987); Maori fish hooks (Law, 1984); Maori rock drawings (Bain, 1985: discriminant analysis); Post-Medieval clay pipes (Alvey and Laxton, 1974, 1977; Buck and Litton, 1991); Post-Medieval wine bottles (Baxter, 1988: discriminant analysis); and Stone Age pit houses in Arctic Norway (Englestad, 1988: correspondence analysis) have all attracted some form of attention. This list largely excludes references to ceramics, of various periods, and lithic material, often prehistoric, which are the most studied artefacts/materials. Biological applications are also common but, in their variety, dull by comparison. Studies of skeletal remains, often crania and often human, are readily found and the *Journal of Archaeological Science* and *Archaeology in Oceania* good sources among those searched.

Generalisation about these applications is less easy than in the previous subsection. Of the archaeological (non-biological) applications in my sample about 80% make no suppositions about group structure and use one, or a combination, of cluster analysis, PCA and correspondence analysis to investigate structure. The other 20% assume some form of grouping and almost all use just discriminant analysis. Cluster analysis is the most commonly used method, occurring in about 55% of the papers; PCA occurs in about 30% of the papers, about equally split between use in its own right as the main statistical tool and use as an adjunct to cluster analysis. Though correspondence analysis has so far been used less than other methods (10%), its use is interesting; applications are mostly recent and published by scholars on the European continental mainland; it is also possibly more suited to certain kinds of data than the other methods used. This will be discussed in more detail below and in Section 6.7. With few exceptions biological applications seem to use discriminant analysis and little else; Habgood's (1986) analysis of cranial measurements is an interesting exception combining PCA and cluster analysis in an unusual manner.

There are interesting variations in, and problems posed by, the choice of variables to use and manner, if any, of their transformation. At one extreme studies may use purely metric variables. Examples include Richards' (1982) PCA analysis of Anglo-Saxon pots defined by the distances from their central axis to each of a large number of points on the profile (equally spaced); and Madsen's (1988b) PCA analysis of Neolithic pots in which landmark data in the form of the horizontal and vertical co-ordinates of 'significant' points on the profile are used. If such data are used in untransformed form then any classification is likely to be based on the size of the entity (Section 4.5); this may be a perfectly useful way of classifying the data and Richards' PCAS provide examples. More usually, however, a classification based on shape is sought, and ratios of variables (possibly logarithmically transformed) are then used. Madsen's (1988b) example is of this kind, though other possibilities are also discussed. Studies of lithic material tend to use metric variables such as lengths, breadths and widths of flint flakes, but other variables such as weight or the angle of a retouch may also be used. Green's (1980) PCA study of flint arrowheads uses weight in addition to metric variables, for instance.

At the other extreme to studies that are based exclusively on metric variables (or, more generally, continuous variables) are those which characterise an entity solely in terms of the presence or absence of a set of features (i.e. binary, dichotomous or 0–1 data). The study of rock art by Bain (1985), using discriminant analysis, in which pictures are characterised by the presence or absence of each of a set of design features is of this kind. The structure of such data is formally similar to data that arise, more commonly, in assemblage comparisons.

Binary data provide a particular example of qualitative data. More generally a quality may be able to take one of several states (e.g., colour might be one of red/green/blue). Such data can be converted to three binary variables (e.g., red/not red; blue/not blue, etc.) and analysed with techniques appropriate for binary data. This kind of approach is not always satisfactory in that qualities with different numbers of states will often implicitly receive differential weighting in an analysis.

In general the techniques discussed in the book work best when all data are measured naturally in the same units (i.e. variables are commensurate). There are problems in dealing with mixed data (e.g., binary/multi-state; binary/metric; non-commensurate metric) that await a satisfactory resolution acceptable to all. Different approaches have been tried. In a cluster analysis study of Cypriot hooked-tang weapons Phillip and Ottoway (1983) combine five qualitative and three metric variables in a measure of similarity between objects, and use this as the basis for a cluster analysis (Section 7.2). Their study was exploratory in nature and has been little emulated in published applications.

A general impression is that many applications avoid using mixed data of the qualitative/quantitative kind and, where both kind of variables naturally characterise an entity, this is clearly a potential limitation of the methodologies used. Where an attempt is made to combine variables, the most common approach is to reduce quantitative data (the higher state of measurement) to multistate qualitative data (e.g., a set of lengths between 10 and 18cm. might be coded as three categories, 10–12, 13–15 and 16–18) and apply standard methods to the latter (e.g., Smith's (1974) cluster analysis of settlement morphology). The usual methods of cluster analysis used, and PCA, are not always well suited to data coded in this form; the coding is, however, intrinsic to some applications of correspondence analysis and some protagonists of correspondence analysis believe that the method resolves problems arising in the analysis of mixed data (e.g., Djindjian, 1990b, p. 59; Dive, 1986). This is discussed at greater length in Section 6.7; however, it must be noted that the method at issue is that of multiple correspondence analysis of three- and higher dimensional cross-tabulations which is not as satisfactorily understood as correspondence analysis for two-dimensional tables (Greenacre and Hastie, 1987; Greenacre, 1988, 1990).

One final problem, not yet addressed, that should be mentioned is what has been termed variable 'redundancy' (e.g., Read, 1982). If quantitative variables are highly correlated, as is often the case, and techniques such as cluster analysis applied in a routine fashion, then problems similar to those with correlated compositional data arise (Section 8.5). If binary variables are highly correlated, then this can also cause problems. In effect certain variables are redundant and not necessary for a full description of variation within a set of entities. Discarding redundant variables can lead to the more satisfactory performance of some statistical methods and this has been advocated. One problem is that different variables can be discarded in a non-unique way so that the remaining variables used to characterise a type must not be regarded as uniquely appropriate descriptors. A second problem is that discarded variables may, in fact, be useful for distinguishing between types that do not happen to be in the sample studied.

If the impression is given that typological studies pose difficulties from both the archaeological and statistical standpoint that are not fully resolved, then this is what is intended. The problems of variable selection and weighting, and of combining variables of mixed kind, militate against an over-optimistic view of the role of statistical methods in general. In specific cases, given certain types of measurement and material, multivariate methods may be helpful; given imperfect methodology, the usefulness of different methods may have to be judged on substantive rather than statistical grounds.

Whallon and Brown (1982) provide a compendium of different positions on the subject of typology from the archaeological standpoint – dating from seminars held in the mid-1970s. Whallon's own contribution provides a pessimistic view of the potential of multivariate methods, arguing for a more sensitive data-exploratory approach in which intelligent selection of a small number of suitably transformed variables will achieve as much or more. A counter to this view is provided by Madsen (1988b) who provides a useful discussion of several of the issues involved. More recently Whallon (1990) has given an interesting discussion on the problem of selecting typologically useful clusters from a cluster analysis.

The statistical methods discussed in this book are, in the context of typological study, rather old-fashioned; the statistical study of 'shape' and pattern recognition generally is a boom area. I think it unlikely, however, that the methodology and – more importantly – means of easily implementing it will have a great influence on archaeological practice in the immediate future; a glance at the contents, and discussion, of Goodall's (1991) paper to the Royal Statistical Society will make clear the reasons for this belief.

1.5.7 Assemblage comparisons

The previous subsection deals with the classification of entities of a generically similar kind. An assemblage will be treated here as a collection of entities that will usually be of different types, characterised either by the counts of each type, by their percentage occurrence in relation to the total count across all entities, or by the presence or absence of each type. The major distinguishing feature here is the emphasis on counted or discrete data compared to the previous two subsections.

In a sense an assemblage is a super-entity, and differentiating between it and an entity is, mathematically, possibly artificial. A grave, considered as an assemblage of grave goods that are either present or absent, is mathematically identical to a piece of rock art characterised by the presence or absence of a set of motifs. Assemblages are often analysed using the same techniques as are used in typological studies, and often in the same way. Similar statistical considerations to those already discussed therefore apply. The archaeological aims in assemblage comparison are, however, different, and additional statistical problems are raised.

Examples of assemblages as understood here – references being located in the cluster analysis bibliography unless otherwise noted – are sites characterised by the numbers of tools in different categories (Hodson, 1969; Pitts, 1979: a PCA study); sites characterised by the percentage of flakes within each of a disjoint set of class sizes (Pitts, 1978b) or pottery sherds similarly defined in terms of classes of grain size (Leese, 1983: PCA); stratigraphic units characterised by counts of different kinds of lithic material (Mohen and Bergougnan, 1984: correspondence analysis) or animal bones (Ringrose, 1988a, b: correspondence analysis); burials characterised by the presence or absence of grave goods (Rothschild, 1979) or by their counts (Nielsen 1986, 1988: correspondence analysis); pottery collections characterised by morphological types (Højlund, 1988: correspondence analysis); regions characterised by the occurrence of different site types (Holm-Olsen, 1988:

correspondence analysis); and sites characterised by coins classified into different time intervals (Ryan, 1988: PCA). Analyses of assemblages of lithic material, and burial assemblages, occur most commonly. Possibly less obviously, pollen diagrams for different contexts can be viewed as assemblages in the sense used here (Fall et al., 1981: PCA; Grönlund et al., 1990: correspondence analysis). Other biological examples include Perry et al.'s (1985) analysis of insect assemblages found in archaeological contexts.

Compared to artefact typology, for example, the different kind of data that tends to be used places an emphasis on different statistical problems. Thus the use of binary data is more prevalent; Rothschild (1979), Pearson et al. (1989), O'Shea and Zvelebil (1984), Peebles (1972), O'Shea (1985), Jones (1980), King (1978) and Tainter (1975) have all analysed burial assemblages using cluster analysis of binary data. This poses interesting problems in the appropriate choice of clustering strategy to be discussed in Section 8.7. Annable (1987) has approached a similar problem through the use of correspondence analysis, while Palumbo's (1987) cluster analysis combines binary data with multistate categorical data. Other instances of the use of binary data include Pyszczyk's (1989) cluster analysis of sites on the basis of artefact presence/absence, and a discriminant analysis using similar data by Bettinger (1979); Hynes and Chase's (1982) analysis of sites using the presence/absence of tree and shrub species; and Bartel's (1981) factor analysis of assemblages of Neolithic figurines.

McClellan's (1979) analysis of burial assemblages, by contrast, used data in the form of percentages of types present – with the percentage sum for a burial of 100%. Such data are common in other contexts and include, for example, studies of assemblages of tool types by Gowlett (1986, 1988), Callow (1986c), Callow and Webb (1981: PCA), Pitts (1978b), Pitts and Jacobi (1979), and Myers (1987), as well as analyses of pollen data noted earlier. Most of these studies use some form of cluster analysis, often in conjunction with PCA – reflecting a pattern observed in the previous two subsections. The particular data structure involved, with assemblages characterised by variables that have a constant sum of 100%, poses potential problems for statistical analysis that have not always been recognised or, where recognised, have not always been adequately dealt with. These problems are discussed in various places in the book (e.g., Section 4.6); here it may be noted that some scholars such as Gob (1988) have concluded that correspondence analysis, rather that PCA or cluster analysis as typically applied, is the more appropriate research tool, and this view is certainly apparent in recent research emanating from the European mainland.

One of two main aims can be identified in most assemblage studies that use multivariate methods. The first might be termed 'assemblage typology', where general considerations of use are similar to those of the previous subsection. The second aim, which raises additional issues, is that of seriation in which assemblages are ordered in a hopefully unambiguous fashion with a view to providing a chronological or other archaeologically useful interpretation. Burial or mortuary studies can be of either type, for example, with some studies seeking a classification or typology that might then be interpreted as reflecting groups of different status, while others seek to sequence graves chronologically. Brown (1987) gives a discussion of the former kind of use and Nielsen (1986, 1988) has examples of the latter. Cluster analysis tends to be the method of choice in the first case and, increasingly, correspondence analysis in the latter.

Petrie's (1899) formulation of the seriation problem – for Egyptian grave assemblages – was the inspiration for important work by Kendall (1971) that in turn has influenced other mathematicians and statisticians to work on the problem (e.g., Wilkinson, 1971; Laxton and Restorick, 1989; Laxton, 1990). Part of the attraction is that the seriation problem is readily formulated in a mathematical fashion. Here we shall only be concerned with the use of correspondence analysis for seriation (Sections 5.2, 6.6), considered as a mechanical sorting procedure. Hodson (1990) provides a useful discussion of archaeological practicalities in which any automatic seriation is treated simply as a starting-point on which to base a useful archaeological ordering.

Much assemblage data arises, in its raw form, as a table of counts which can be (and often is) converted to either binary form or standardised to percentage form before analysis. It is, of course, possible to analyse the counts directly, and several examples are given in the edited collection of Madsen (1988a). The archaeological examples in Madsen's book are very varied; the statistical technique used for the analysis of assemblage data is invariably correspondence analysis. This is not surprising since correspondence analysis is a natural exploratory method to use with tables of discrete data (i.e. counts) in the same way that PCA is a natural exploratory approach to unstructured tables of continuous data.

Several problems exist in the use of counts in their raw form; the fact that such data usually have a non-normal distribution has occasioned some cause for concern leading to the use of logarithmic transformations, (e.g., Bard's (1989) use of cluster analysis). More importantly, if assemblages are of widely different sizes, then the size rather than 'shape' of the assemblage is likely to dominate a PCA or cluster analysis of the counts. Usually it is 'shape' that is of interest, and correspondence analysis is then a reasonable tool to use because of its emphasis on shape (Section 6.2).

As with other areas of application, cluster analysis has been the method most commonly used in about half of the references, and is particularly dominant in the earlier literature. Correspondence analysis has grown in popularity in recent years, particularly on the European mainland, and is used in about one-third of the references. This is about the same level of use as PCA, which it has to some extent replaced; discriminant analysis, occurring in about one-sixth of the references, has been the least used method.

1.5.8 Inter- and intra-site spatial studies

Studies in each of the preceding three categories may sometimes have a spatial component in the sense that each may result in a classification of material or assemblages that can then be examined in relation to geographical distribution. Here we will be mainly concerned with a subset of papers that have used multivariate methods in the context of inter- and intra-site spatial analyses. Unless otherwise indicated, methods of cluster analysis have been used.

One strand of investigation derives from the work of Kintigh and Ammerman (1982) who presented a method for the cluster analysis of the co-ordinates of settlement or artefact locations that respects the spatial relationship between the objects of study. The aim is to define spatially compact, possibly distinct, concentrations of sites or objects. The approach, in the terminology of the original paper, is an 'heuristic' one that can provide a starting-point for archaeologically-based interpretations of site usage. Examples, particularly concerned with artefact scatters on French prehistoric sites, are to be found in several papers by Simek (1984a, b, 1987, 1989), Simek et al. (1985) and Rigaud and Simek (1991), as well as Ammerman et al. (1987). Van Waarden (1989) has used the method to examine the intra-site distribution of a class of structures rather than artefacts.

An alternative approach, developed originally as a 'competing' and improved methodology, is that of Whallon (1984). Observed artefact densities are used to define smoothed maps of densities across a site, and the set of densities at different locations are used as the basis for a cluster analysis of locations. Unlike Kintigh and Ammerman's method, spatial relations are not built into the clustering – hence the terminology 'unconstrained clustering'. The clusters obtained are subsequently inspected to see if a natural spatial interpretation is possible. Ridings and Sampson (1990) have used the approach in an inter-site spatial analysis of Bushman pottery decorations; while Cribb and Minnegal (1989) have used it in the spatial analysis of a dugong consumption site.

Whallon's (1984) technique can be thought of as generating observations at points on a map, or for spatial units across a map, that have a similar formal structure to assemblage data as understood in the previous section. Other techniques, such as PCA, might be applied to such data (e.g., Cribb and Minnegal, 1989) and used as a basis for generating scores to be plotted in the hope of revealing spatial structure.

Whallon's (1984) paper was intended, in part, to improve on this kind of approach, and an illustrative example to show its limitations was provided. More recently Whallon and Kintigh have co-authored an interesting paper (Gregg et al., 1991) that compares their approaches. They are shown to be complementary in the spatial information that can be recovered from 'perfect' anthropological data. The performance of the two approaches is also examined using such data, degraded to simulate archaeological sites.

The structure of the data generated in Whallon's approach is such that technical problems can arise in its analysis by PCA or cluster analysis because of its compositional nature. This difficulty was recognised by Ridings and Sampson (1990). Alternative approaches that may be worth exploring, as with assemblage data that are subject to similar difficulties, include correspondence analysis (Blankholm, 1991; Johnson, 1984; Djindjian, 1988b). This last paper is, in many respects, in the same spirit as Whallon's work, with correspondence analysis considered to be more appropriate than PCA (and capable of providing information on variables as well as objects) and a less 'archaic' method of cluster analysis used (Djindjian, 1988b, p. 98). A few other applications of correspondence analysis to intra-site spatial analysis are referenced in Djindjian (1990b).

Kintigh (1990) reviews the general area of intra-site spatial analysis including approaches not dealt with here. Blankholm (1991), similarly, is a comprehensive review of different techniques of analysis illustrated on several data sets using methods of cluster and correspondence analysis.

1.6 The structure of the book

Although this book is not about 'exploratory data analysis' as popularised by Tukey (1977), some simple univariate techniques that are covered by his approach are discussed in Chapter 2. Chapter 2 is of a rather miscellaneous nature and serves, in part, to remind readers of material that would be acquired in a reading of Shennan (1988), for example, and in part to introduce ideas and notation used elsewhere in the book.

The four main topics discussed – PCA, correspondence analysis, cluster analysis and discriminant analysis – each have two chapters. The odd-numbered chapters introduce the methods in what is intended to be a reasonably accessible fashion with examples based, for the most part, on real data. Even-numbered chapters deal with specific points of detail and special topics that may be important in practice but are not always discussed, particularly in introductory treatments. The penultimate section in each even-numbered chapter consists of a detailed example or examples designed to draw together material from the other sections. The final sections of the even-numbered chapters review the use that has been made of the technique in archaeology, noting aspects of the history of usage and attempting, in some cases, to identify trends in use.

The mathematical level of different sections is highly variable, though none of it would tax a mathematician. The non-mathematician may be unfamiliar with the matrix algebra that is used in places; the (informal) guide for such readers in Appendix B treats a matrix as a table of numbers and matrix algebra as a set of rules for converting such tables to other tables that have useful properties. Where they have occurred to me, non-mathematical analogies of what the mathematics is doing have been provided (and like all such analogies should be treated with caution). Passages that look forbidding can usually be omitted by the reader without seriously impairing understanding of a lot that follows later – though I would prefer the reader to skim rather than skip such sections.

The material covered, including the examples, is intended to reflect both how methods have been used in practice and what methods I think are of current interest. For example, the chapters on PCA use examples from scientific archaeology because it is widely used in this area and appropriate to the kind of data often available; the correspondence analysis chapters, by contrast, use mainly non-scientific examples and the method is arguably more appropriate to certain kinds of data that PCA has been used for in the past.

A word about my methodological prejudices may be in order. Wright's (1989) discussion of PCA includes statements to the effect that 'we should welcome the eigenanalysis concept for the routine analysis of archaeological tables of data' and that 'nothing of the richness of eigenanalysis is to be harvested by using the alternative multivariate method of cluster analysis'. One reason cited for this is the incapacity of cluster analysis to 'provide us with the dual characterisation of objects in terms of variables and variables in terms of objects'. These prejudices, if that is what they are, are ones that I share and undoubtedly colour my treatment of different topics. Also echoing Wright, I think of correspondence analysis as simply a non-linear form of PCA. Correspondence analysis has been unjustly neglected on the one hand, and oversold by its protagonists on the other. It is hoped that the treatment here avoids either extreme. In writing on cluster analysis, what emerged sometimes came out more pessimistically than I had originally intended; given its extremely widespread use in archaeology there are a lot of unappreciated (by some users) and/or unresolved difficulties with many commonly-used approaches. The proof of the statistical pudding has, I think, to be in the archaeological eating but it is clear that many cooks and consumers find much of what is commonly done indigestible. Discriminant analysis is the technique discussed here that has been least used in my own research, and it probably shows. It sits uneasily with the other topics in that, properly used, it should not really be considered as an exploratory technique. It is, however, often used simply as a display technique or to accompany an exploratory method (usually cluster analysis) and could not be omitted.

The analyses in the book present output in the form that, apart from minor modifications, is obtained with the packages I normally use. In the course of writing the book the packages available to me have changed and I might do things differently now if starting from scratch. The original intention was to base examples on MINITAB where possible, and SPSS-X or CLUSTAN where not. Subsequent exposure to the STATGRAPHICS package, which is used in several places, and MV-ARCH modified this intention. While I find MINITAB an easy and flexible package to use, readers will have their own favourites and, given the pace of development of packages, may well have access to software capable of producing high-quality graphics in a painless fashion. The packages exploited here were simply those available while writing the book and do not reflect any views about packages, such as BMDP, that I have not used. A more detailed discussion of computational considerations is given in Appendix C.

2

Univariate and Bivariate Approaches and Preliminary Data Analysis

2.1 Introduction

Preliminary data inspection is essential before undertaking any form of multivariate data analysis and is sometimes all that is needed (Whallon, 1987; Lock, 1991). The importance of such inspection, and a willingness to let commonsense override any itch to use multivariate methods because they are there, cannot be overemphasised. This book is, nevertheless, about multivariate methods, and has been written with the aim of encouraging an appreciation of both the scope and limitations of widely used methods. In Chapters 3–10, therefore, the analyses undertaken are presented at least partly for illustrative purposes and do not always include the kind of preliminary analyses that – in practice – ought to be used. The present chapter illustrates a miscellany of methods that are useful for looking at data. In addition the opportunity is taken to introduce ideas that are of importance in later chapters.

A data set that will be used for illustrative purposes is introduced in the next section. Sections 2.3–2.5 examine some methods suitable for the rapid univariate inspection of data. The aim may be to acquire some 'feel' for the data; to identify obvious structure that may obviate the need for more complex analyses; or to identify features that more complex analyses may be sensitive to.

Bivariate data inspection using scatterplots and correlations is covered in Sections 2.6–2.7. Sections 2.8–2.10 cover a range of topics – assessing normality, data transformation and tabular presentation – that lead into issues discussed in more detail in later chapters. Some approaches to the direct display of multivariate data are mentioned in Section 2.11, and the chapter closes in Section 2.12 with a discussion of notational conventions used in the book.

All the methods presented are easy to understand and implement. The illustrative examples in this chapter are based mainly on MINITAB output. Computational aspects of application are discussed generally in Appendix C.

2.2 Data

Appendix A contains some of the data sets used for illustrative examples in the book. The data used for most of the examples in this chapter are given in Tables A1 and A2 which show the concentrations of twenty-two elements in 105 specimens of waste Romano-British glass found in excavations at Leicester and Mancetter. Table A1 shows the concentration in percentages of the major and minor elements (in their oxide forms, but referred to as 'elements' throughout); Table A2 shows the concentrations in parts per million (ppm) of the trace elements. The major component of these glasses is silica, typically present at levels between 60 and 70% and conventionally obtained as 100% less the sum of the major and minor oxide concentrations. The final column in each table identifies the site.

Tables A1 and A2, separately or together, are typical of measurements that arise in the scientific study of archaeological artefacts of pottery, glass or metal (Section 1.5). One common question of interest is whether or not there exist chemically distinct groups within the data and, if so, whether they are associated with archaeological groupings.

These data have been discussed in Jackson et al. (1991). They are used here and elsewhere in the book as an example of a real (and typical) data set for which some statistical analyses will lead to no clear conclusions. It is evident, using simple techniques, that the composition of glasses from the two sites is different on average, but that there is a subset of specimens that cannot, in terms of their chemical composition, be readily identified with the correct site of origin. The use of multivariate methods represents an attempt to see whether more complex statistical techniques can provide a sharper definition between specimens from different sites. The answer, ultimately, is 'no'.

Real data sets are often like this; they have stories to tell, which statistical techniques can help to elucidate, but have no clear ending or 'moral'. Several of the illustrative examples used in the book are of this kind. Published uses of statistical methods, by contrast, are often based on examples where the methods 'work'. My preference is for a presentation of techniques that does not restrict examples to successful illustrations of a method. Users of statistics (potential and actual) need to steer between the Scylla of undue faith in what statistics can achieve (followed by acute withdrawal symptoms when the methodology fails to deliver) and the Charybdis of total abstinence occasioned either by fear of the unknown or an untested belief in its likely impotence.

2.3 Dot-plots

'Dot-plots' is the term used in MINITAB for what some may know as linegraphs; with large amounts of data they can be thought of as similar to histograms. They provide a quick way of getting some 'feel' for distinctive or unusual features of a set of data. The plots for some of the elements of Table A1 are shown in Figure 2.1 (omitted elements had no particularly distinctive features). Several of the plots provide some signs of multimodality, or distinct grouping (e.g., Fe, Ca, Mn, Sb). If these variables are

Univariate and Bivariate Approaches

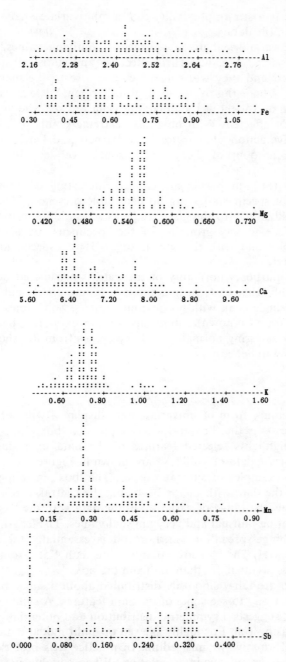

Figure 2.1: Dot-plots for selected elements in Table A1

related there is a strong possibility that a multivariate analysis will reveal groups within the data.

For some variables there is evidence of outlying values, singly or in small groups (e.g., Mg, K). Their effect on any statistical analysis needs to be assessed and they will sometimes give rise to the main features of an analysis. Where this is the case it will often be sensible to conduct analyses with and without the offending values to see whether conclusions are materially affected. With obviously extreme points it is often best to take corrective action prior to analysis. Barnett and Lewis (1984) provide an extensive account of the general problem of detecting and treating outliers.

Antimony (Sb), in particular, has an interesting distribution, with a sharp peak of specimens having no detectable presence. This element can be added deliberately to the glass recipe as a decolourant and is of potential importance in any grouping of the specimens. Its association with other variables and with the site of origin is of especial interest and is examined later.

For some purposes normality of the data is considered desirable. This means, roughly, that plots such as those in Figure 2.1 should be symmetrical about a single peak with no obvious outliers. Sometimes this property is desirable for (unknown) sub-groups of the data. Dot-plots provide a rapid way of assessing normality and departures from it; other approaches are considered in Section 2.8.

2.4 Box-plots

A complementary form of univariate data display – with advantages and disadvantages – is the 'box-and-whisker plot' or 'box-plot'. It shows less detail but highlights selected features of the data in a starker fashion. Examples for the data of Table A1 are shown in Figure 2.2.

These are examples of MINITAB output. The 'box' encompasses the central 50% of the data with the '+' showing the central, median value. The whiskers show the range of variation in the data with the '*' and 'O' symbols highlighting unusual and very unusual values. The length of the box is called the 'hinge-spread' in MINITAB and is essentially the inter-quartile range – call it H. The '*' shows points more than 1.5H from the box and the 'O' shows points more than 3H from the box.

If the data are fairly smoothly distributed about a single peak, the box-plot is a good way to see some of its main features. A symmetric diagram, for example, suggests a symmetric distribution and potentially troublesome values are easily seen. For highly skewed or multi-modal data the box-plot can be uninformative or misleading. In practice it is best examined with the dot-plot to avoid mis-interpretation. When suitable it is useful for a compact display of the data and, particularly, for comparing groups.

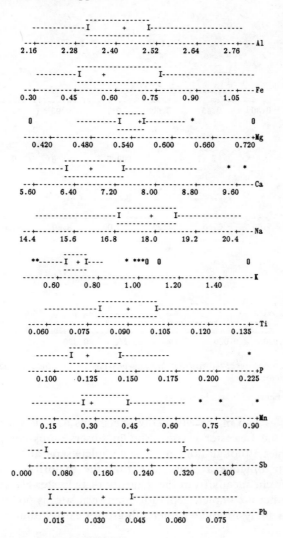

Figure 2.2: Box-plots for elements in Table A1

2.5 Comparing groups

Where distinct archaeological groups exist independently of the chemistry, or where statistical analysis suggests distinct chemical groups, both dot- and box-plots are useful for comparing distributions between groups. Figure 2.3 uses box-plots to compare distributions of three of the elements for the two sites.

About 75% of the Leicester specimens have a higher Fe content than about 75% of the Mancetter specimens, while for Ca the reverse is true.

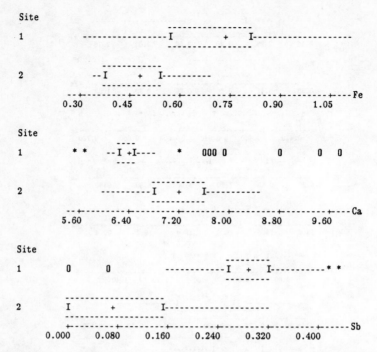

Figure 2.3: A comparison of the distribution of selected elements for the two sites in Table A1 using box-plots

Note: Site 1 is Leicester, site 2 is Mancetter.

For Sb, 75% or so of the Mancetter specimens have lower values than all but two of the Leicester glasses. Similar observations could be made for other variables although none provide complete separation between sites.

It is fairly obvious at this stage that the majority of the Mancetter glasses differ chemically from the majority of Leicester glasses. What is less clear is whether a complete separation between sites is possible, based on a multivariate analysis; and whether sub-groups exist within sites.

2.6 Labelled scatterplots

Figures 2.1 to 2.3 look at one variable at a time but raise questions about possible relationships between variables. Such relationships can be investigated using scatterplots or correlations (Section 2.7). If additional information on the objects is available this can usefully be included by labelling the plot.

An example is shown in Figure 2.4, which shows a plot of Fe against Na with points labelled according to the site of origin. The Mancetter specimens tend to have lower values on both variables than the Leicester specimens but with several exceptions. Some outlying Leicester specimens are

Univariate and Bivariate Approaches

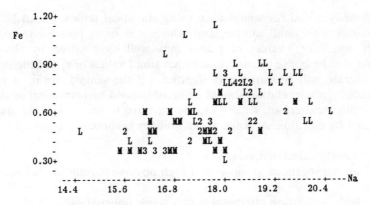

Figure 2.4: A labelled scatterplot of Fe against Na for the data in Table A1

Note: An L identifies specimens from Leicester, an M specimens from Mancetter; numbers show multiple occurrences.

also evident. Similar plots will be used in later chapters to display the output of different multivariate analyses.

2.7 Looking at correlation matrices

In Figure 2.4 there is some suggestion of a linear relationship between the levels of Fe and Ca, albeit with a lot of scatter involved and some unusual values. The strength of this relationship can be formally measured using a (product-moment) correlation coefficient. Such coefficients play an important role in certain forms of multivariate analysis and are discussed at some length here. Shennan (1988, pp. 126–131) discusses correlation (and the summation notation to be used below) and this, or similar material, should be reviewed if necessary before progressing with this section.

The product-moment correlation coefficient, r, is a measure of the strength of *linear* relationship between two variables. It varies between -1 and $+1$, the extremes indicating a perfect negative or positive relationship. Values close to zero suggest little or no linear relationship.

For two variables, X and Y say, the estimated covariance based on a sample of n is

$$c_{xy} = \sum_1 (x_i - \bar{x})(y_i - \bar{y})/(n-1) \qquad (2.1)$$

and the estimated variance – of X for example – is

$$s_x^2 = \sum_i (x_i - \bar{x})^2/(n-1) \qquad (2.2)$$

where $\bar{x}$ is the mean of X etc. The correlation between X and Y is then defined as

$$r = c_{xy}/s_x s_y \qquad (2.3).$$

Although often used in a descriptive fashion the correlation coefficient can be formally tested for significance using statistical tables. It can be useful to do this with small sample sizes where it is quite possible to read too much into 'large' values of r that may well have arisen by chance. It should also be borne in mind that rather small values of r, of little substantive interest, will be statistically significant if the sample size is very large. In practice, interpretation should be accompanied by inspection of the corresponding scatterplot, since r can be sensitive to features of the data not reflected by its value. A range of possibilities is shown in Figure 2.5 as follows:

(a) weakly correlated variables;

(b) an homogenous group with a high positive correlation between variables;

(c) high correlation attributable to a single unusual value;

(d) high correlation arising because there are two distinct subgroups within the data – within subgroups the variables are weakly correlated;

(e) negative correlation arising from two subgroups within which variables are positively correlated;

(f) low correlation arising from two perfectly, but non-linearly, related variables.

Assuming the variables in the plots to be element concentrations, a plot such as (b) would identify elements whose high correlation helped characterise that group. In (d) or (e), by contrast, the high correlation identifies elements whose high or low presence distinguishes between groups but says nothing about the relationships within groups. A plot such as (c) identifies unusual data which might predictably emerge as the main feature of any subsequent analysis and whose removal from a data set (for separate consideration) is often sensible. Non-linear plots such as (f) would caution against the use of methods that assume a linear structure such as principal component and discriminant analyses as often implemented.

If two variables are strongly correlated, then much of the 'information' in a plot may be deduced by inspection of a single variable only. This is demonstrated explicitly in Section 3.2. More generally, if several variables are highly intercorrelated, it may be possible to capture much of the information in the data set by using a smaller number of variables. This is the basic idea underlying the method of principal component analysis discussed in the next two chapters.

Preliminary analysis of the correlations for a set of variables will often suffice to establish whether or not a multivariate analysis is worthwhile and may sometimes be all that is needed. For example, if all variables are very weakly correlated there is little point in undertaking principal component analysis. Table 2.1(a) shows the correlations produced by MINITAB for the elements of Table A1 in the form of one-half of the correlation matrix. This is typical of computer-generated output and is not in an immediately comprehensible form. Chatfield and Collins (1980, p. 41), following

Table 2.1: Correlation matrix for elements in Table A1

(a)	Al	Fe	Mg	Ca	Na	K	Ti	P
Fe	−.115							
Mg	.010	.433						
Ca	.385	−.553	−.108					
Na	−.446	.509	.252	−.785				
K	.113	.195	.365	−.202	.171			
Ti	−.130	.773	.543	−.674	.620	.340		
P	.346	−.346	.126	.484	−.558	.103	−.347	
Mn	.151	−.247	.118	.100	−.166	.184	−.172	.371
Sb	−.468	.727	.390	−.818	.829	.209	.813	−.638
Pb	−.234	.451	.386	−.634	.606	.294	.635	−.275

	Mn	Sb
Sb	−.247	
Pb	.124	.704

Note: The above correlation matrix was produced, in the form shown, using the MINITAB package.

(b)	Sb	Na	Ti	Fe	Pb	Mg	Ca	Al	P	Mn	K
Sb		.8	.8	.7	.7	.4	−.8	−.5	−.6	−.2	.2
Na	.8		.6	.5	.6	.3	−.8	−.4	−.6	−.2	.2
Ti	.8	.6		.8	.6	.5	−.7	−.1	−.4	−.2	.3
Fe	.7	.5	.8		.5	.4	−.6	−.1	−.3	−.2	.2
Pb	.7	.6	.6	.5		.4	−.6	−.2	−.3	.1	.3
Mg	.4	.3	.5	.4	.4		−.1	.0	.1	.1	.4
Ca	−.8	−.8	−.7	−.6	−.6	−.1		.4	.5	.1	−.2
Al	−.5	−.4	−.1	−.1	−.2	.0	.4		.3	.0	.1
P	−.6	−.6	−.4	−.3	−.3	.1	.5	.3		.4	.1
Mn	−.2	−.2	−.2	−.2	.1	.1	.1	.0	.4		.2
K	.2	.2	.3	.2	.3	.4	−.2	.1	.1	.2	

Note: The same correlations as Table 2.1 (a) in reordered and modified form.

Ehrenberg (1975, 1977), give some useful advice for simplifying such output. This includes (a) rounding correlations to one or two decimal places and suppressing the preceding zeroes; (b) presenting the full correlation matrix apart from the diagonal of 1's that should be suppressed; and (c) rearranging the ordering so that large values cluster round the diagonal. In the presentation of Table 2.1(b), a reordered version of Table 2.1(a), variables are grouped according to the sign as well as strength of association with the spacing designed to identify groups of intercorrelated variables.

On looking at Table 2.1(b) there is a clear group of positively correlated variables (Sb, Na, Ti, Fe, Pb, Mg) negatively correlated with another group (Ca, P, Al, Mn); with K showing generally weak correlation with other elements. Correlations are, apart from Mg, generally stronger in the first group than the second. This suggests that the information in the

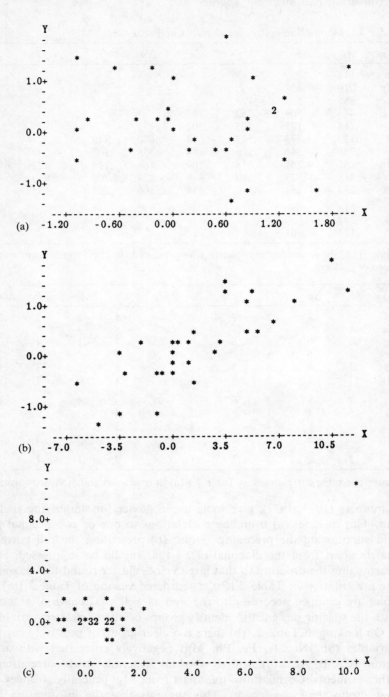

Figure 2.5: Scatterplots of different kinds of correlation effects
(a) Random scatter ($r = -0.13$)
(b) Strong and genuine linear relationship ($r = 0.82$)
(c) High correlation caused by an outlier ($r = 0.81$)

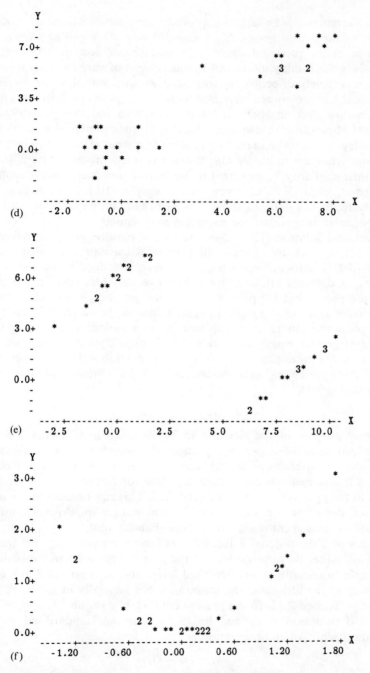

(d) High correlation caused by group structure with little correlation within groups (r = 0.94)
(e) High negative correlation arising from group structure with high positive correlation within groups (r = −0.77)
(f) Low correlation with a perfect non-linear relationship (r = 0.26)

eleven variables can be largely captured with a smaller number – perhaps three or four – in a reasonably successful way. This will be shown to be the case in the principal component analyses of Section 3.2, where the analysis is predictably influenced by the subsets of variables just identified. There is a variety of problems associated with the way that the correlation coefficient has sometimes been used in archaeological applications, and discussions are given in Speth and Johnson (1976) and Cowgill (1990a), for example. Several of these difficulties arise in comparing collections or assemblages using abundance or relative frequency data of artefact types. If many types are mutually absent from two assemblages (the 0-0 problem) their similarity as measured by the correlation coefficient is artificially inflated. Cowgill (1990a) gives several simple artificial examples, using relative frequency data, to show how r can take on virtually any value for pairs of assemblages that are essentially very similar.

Speth and Johnson (1976) also discuss the 'closure' problem, referencing unpublished work by Cowgill, that arises when data are constrained to sum to 100% across all rows (e.g., assemblages defined by the relative frequency of different artefact types). For two variables that have a significant presence, what happens is that as one gets bigger the other tends to get smaller and this produces a negative bias in the correlation coefficient. This phenomemon is quite common in archaeological data and is discussed at greater length in Section 4.6. Representation of the correlation matrix in a simple picture is discussed in Section 4.4, and other simple graphical approaches are illustrated in Hills (1969) and Baxter and Heyworth (1991).

2.8 Assessing normality

A major purpose of the methods to be discussed in Chapters 3–8 is the identification of homogeneous groups of 'individuals' where the individuals may be artefacts or assemblages, for example. Having identified such groups it is sometimes considered desirable for further analysis that variables in the groups be normally distributed. This can be assessed by inspection of dot-plots for each variable and subgroup. Another informal method of assessment is via a normal probability plot.

In essence, the ordered values of a variable are plotted against the standardised values they should have if they came from a normal distribution. The term 'standardised' is often used later; since it is not used in a consistent way in the literature, the meaning it has generally in the book is discussed in Section 2.12. In the present context, if a variable has mean $\bar{x}$ and standard deviation s, defined by (2.2), then the standardised value or 'standard score' of an observation is

$$y_i = (x_i - \bar{x})/s \tag{2.4}$$

This simply converts the original variable into a new variable having zero

Univariate and Bivariate Approaches

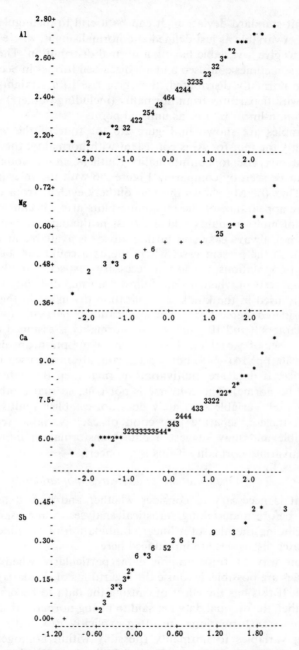

Figure 2.6: Examples of normal probability plots for selected elements in Table A1

Note: Plots are of element values against expected values of the normal order statistic. Numbers show multiple occurrences with + showing ten or more occurrences.

mean and unit standard deviation. It can be useful to distinguish between standardising a variable as just defined and normalising it, where the data is transformed to give a variable having a normal distribution. These distinct processes are sometimes confused and are discussed further in Section 2.12.

If data are normally distributed, they give rise to a straight-line probability plot, with departures from normality (including outliers) showing in the form of a non-linear plot or as unusual points.

Some examples are shown in Figure 2.6 for four of the variables of Table A1. Only the plot for Al is consistent with normality; the features of the data that give rise to non-normality (outliers; multi-modality; skewness, etc.) can be seen by comparing Figure 2.6 with the relevant parts of Figure 2.1. That for Mg shows the two outliers evident from the dot- or box-plots; the appearance of the remaining plots arises because of the differing forms of multi-modality and skewness in the data. Interpretation of such plots is not always easy and further advice is available in Daniel and Wood (1980). In the present case, where the data could well be a mixture from different populations, we do not necessarily expect normality.

More formal tests of whether data follow a normal distribution are possible but rarely used in the kinds of application discussed in the book and are not considered here. For some applications, particularly those discussed in Chapters 9 and 10, multivariate normality is assumed for the distribution of a set of variables. Tests of this assumption are discussed in Aitchison (1986, pp. 143–148) but, once again, are rarely used in archaeological practice. If data are multivariate normal then the individual variables should be normal. The converse is not true, so that establishing the normality of each variable separately does not establish multivariate normality. Nevertheless, separate inspection of each variable separately is usually sensible and may suggest a data transformation more likely to result in multivariate normality if this is a concern.

2.9 Transformations and linear combinations

In practice it is necessary to consider whether and how data might be transformed before undertaking statistical analyses. Some aspects of this problem, including the idea of a linear combination of variables, that are relevant to later discussion are introduced here.

A common way of transforming data, particularly when only non-negative values are possible, is to use the logarithms of the data rather than the raw data. If this has the effect of causing the data to become normally distributed, then the original data are said to be log-normally distributed.

Logarithmic transformations are often explicitly used with the aim of transforming variables to normality (possibly within homogeneous subgroups of the data). Harbottle (1976) and Pollard (1986) discuss why this may be desirable in pottery provenancing studies, based on the chemical composition of the pottery. One argument is that the trace elements found in the clays used to make pots have a naturally occurring log-normal distribution.

Another possible motive for transformation concerns the relative magnitude of observations. Two numbers differ by an order of magnitude if one is 10 times bigger than the other; by two orders of magnitude, if one is 100 times bigger; and so on. With data that are skew, or have outliers, and where numbers differ by (possibly) orders of magnitude graphical representation of the data can be very unsatisfactory with much of the detail obscured. Logarithmic transformation can then provide a clearer picture of the data as well as possibly achieving normality.

In some cases the base of the logarithm used – usually either natural logarithms or to base 10 – is important; in others not. In analysing chemical compositions data expressed in parts per million (ppm) range from about 10 to 1000 and convert to 1 to 3 if logarithms to base 10 are used. Figures expressed as percentages from .01 to 100 convert to -2 to $+2$ if logarithms to base 10 are used. Thus, with data in mixed %/ppm form, use of logarithms to base 10 as is often recommended will convert the data to a similar order of magnitude.

The logarithm is undefined for zero values. In these cases it is conventional to add a small and arbitrary value to all the data. This practice can be misleading since it is possible for subsequent statistical analysis to be dominated by those variables with an original value of 0; this should be checked carefully. With many zero values for a variable, as with Sb in Table A1, it will sometimes be necessary to omit the variable from an analysis if logarithmic transforms are to be used.

Another form of transformation is to use ratios or the logarithms of ratios. For certain types of data the use of ratios is suggested by statistical considerations, and this will be discussed in Section 4.6.

Scientific and archaeological considerations can also suggest the use of ratios. A good example occurs in the statistical analysis of pottery compositions, where the use of a reasonably pure temper can differentially affect the raw elemental composition but leaves element ratios unchanged. Discussion of this and/or examples occur in Harbottle (1976), Wilson (1978), Mommsen, Krauser and Weber (1988) and Topping and Mackenzie (1988).

Leese et al. (1989) use the logarithms of ratios in a paper that is considered in more detail in Section 4.6. Henderson (1989, p. 50), in a study of compositional categories of Iron Age glass, plots the logarithms of ratios of tin and antimony oxides to manganese oxide. This reveals interesting compositional patterns in the data, the transformation being necessary because the untransformed ratios differ by about four orders of magnitude. In morphological studies of vessel shape, for example, ratio transformations, possibly logarithmically transformed, can be used to eliminate comparisons dominated by the size of a vessel (Sections 4.5, 4.6).

A different kind of example of the use of ratios is given by Robertson-Mackay (1980) who, for a sample of early cattle skulls, plots log(Y/X) against log(YX) where Y is basal length and X is least frontal breadth. The plot provides clear evidence of differences between the sexes, and between wild and domesticated cattle.

Log(Y/X) has an interpretation as skull shape and log(YX) as size. The logarithms of ratios and products can be written alternatively as

$$\log(Y/X) = \log(Y) - \log(X) \tag{2.5}$$

and

$$\log(YX) = \log(Y) + \log(X). \tag{2.6}$$

These are very simple examples of linear combinations (of logarithmically transformed variables) having the general form

$$a_1 \log(Y) + a_2 \log(X).$$

In (2.5) and (2.6) (a_1, a_2) is $(1, -1)$ and $(1, 1)$ respectively. Linear transformations of untransformed data (e.g $a_1 Y + a_2 X$) also occur. Foy (1985), for example, in a study of Medieval French glass, demonstrates compositional groupings using graphs based on Ca + Mg.

The idea of linear combinations will play an important role in the sequel, starting with Section 3.2. There, and elsewhere, ideas of multivariate analysis will be introduced using linear combinations that involve more than two variables; have non-integer values of a_1, a_2 etc.; and may lack a simple interpretation.

2.10 Tabular presentation

The data of Table A3 are a subset of Table 2 in Mellars (1976) and are in the form used by Pitts (1979) (see also Myers, 1987). They show the abundance of five artefact types in assemblages from thirty-three British mesolithic sites and are used here to make a simple point about tabular presentation. In statistical analyses of such data the decision needs to be made as to whether comparison should use the raw numbers or not. Table 2.2 shows Table A3 re-expressed with numbers showing the percentage of objects in an assemblage accounted for by each tool type and with sites ordered by the percentage of microliths in the assemblage.

A minor point is that not all rows sum to 100% (e.g., sites 11 and 20). This is attributable to rounding error and is to be expected. Later chapters look at ways of graphically displaying such tabular material. With a relatively small number of variables, often most of the main features of the data are evident on inspection. As an example, the unusually high proportion of tool type C for site 27 (the main feature of Pitts' (1979) principal component analysis) is evident on inspection.

Table 2.2: The assemblage data of Table A3 presented in percentage form and ordered by type A

Site	Tool type (%)				
	A	B	C	D	E
3	8	90	2	0	0
18	27	61	12	0	0
27	27	35	36	1	1
11	30	63	7	1	0
20	37	49	0	0	13
22	39	50	0	6	5
5	44	49	4	4	0
19	46	38	0	0	16
16	46	32	22	0	0
8	47	33	1	3	15
33	51	37	12	0	0
23	51	48	2	0	0
28	57	26	11	2	4
15	57	40	3	0	0
7	60	32	7	0	1
30	61	32	5	0	2
31	62	35	3	0	0
17	67	29	2	0	2
26	71	22	1	0	6
10	76	20	3	2	0
1	76	21	2	1	0
24	81	19	0	0	0
2	82	17	1	0	0
14	85	14	1	0	0
6	90	8	2	0	0
4	90	10	0	0	0
12	91	8	1	0	0
25	91	6	3	0	0
32	92	8	0	0	0
9	93	0	7	0	0
13	93	7	0	0	0
21	93	0	7	0	0
29	94	0	6	0	0

2.11 Direct representation of multivariate data

The bulk of this book is concerned with methods for transforming multivariate data in such a way that structure can be perceived by inspection of two- or three-dimensional plots or pictures. A number of techniques exist for the direct representation of multivariate data that are briefly mentioned here. Gower and Digby (1981) provide a convenient reference for several of them.

To represent a third dimension, three-, rather than two-dimensional plots are sometimes used (Rauret et al., 1987); alternatively the size of the plotting symbol on a two-dimensional plot may be used to represent the

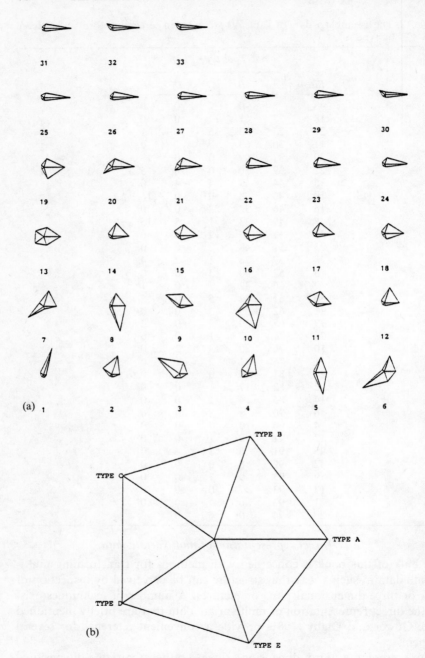

Figure 2.7: Star plots of the assemblage data in Table A3
(a) Star plots for the data in Table A3
(b) Key to the star plot of Figure 2.7(a)

third dimension. Rays emanating from the plotting symbol, whose length and direction represent magnitudes in the third and fourth dimensions, have also been used (Doran and Hodson, 1975, p. 227).

In star plots there are a number of rays corresponding to the number of variables; the length of each corresponds to the magnitude of the variable. This kind of plot is available in the STATGRAPHICS package and has been used by Williams et al. (1990) in an exploratory study of historic land use patterns. The data of Table 2.2 are displayed in such a fashion in Figure 2.7 where it can be seen, for example, that number 1 (site 3) is unusual because of the high value of tool type B and low values elsewhere, whereas 25-33, the last nine sites in Table 2, are very similar. That number 3 (site 27) is fairly unusual is also apparent, apart from a general similarity to 9 (site 16) arising from the relatively high proportion of tool type C.

Other plots based on similar ideas are discussed in Gower and Digby (1981). Faces, in which each variable is represented by some feature of a human face whose appearance depends on the value of the variable, are fun but highly dependent on the allocation of variables to features. Sustained use of this device is made, in a non-archaeological context, by Flury and Riedwyl (1988).

2.12 Notation, transformation and standardisation

In this section some notational conventions used in later chapters are developed. The observation in the i'th row and j'th column of a data table (or matrix) will be written as x_{ij} and the raw data matrix as **X**. Typically, multivariate methods are not applied to the x_{ij} but to some modification of it which may involve transformation and/or standardisation. Transformation is typically undertaken to endow the data with 'nice' properties, such as a normal distribution; standardisation is usually meant to give the variables equal weight in some sense.

The two procedures are distinct; a logarithmic transformation of the form
$$z_{ij} = \log(x_{ij}) \tag{2.7}$$
transforms the data (possibly to a normal distribution), but in the sense used here, does not standardise it. To achieve standardisation one possibility is to use (2.4) applied to the raw data or, for the transformed data use
$$y_{ij} = (z_{ij} - \bar{z}_j)/s_j \tag{2.8}$$
where $\bar{z}_j$ and s_j are the mean and standard deviation of the j'th transformed variable. The effect is that the new variables, Y_j, all have unit variance. This usage is not universal; the term 'standardisation' is sometimes used for what is referred to above as (logarithmic) transformation − the idea being that variables of similar orders of magnitude, but not necessarily equal weight, are obtained. Similarly the term 'normalisation' is sometimes used for what I would call 'standardisation'.

Unless otherwise stated y_{ij} will be used to denote the data input into a multivariate analysis with $\mathbf{Y}$ the table of such numbers. Upper case and subscripted letters such as X_j or Y_j denote variables, with lower case letters used for observed values of the variable. The y_{ij} may be derived directly from the untransformed data or after transformation; which applies should be clear from the context. The (transformed) data is said to be 'centred' if it has the form

$$y_{ij} = c(z_{ij} - \bar{z}_j) \qquad (2.9)$$

and standardised if

$$y_{ij} = c(z_{ij} - \bar{z}_j)/s_j \qquad (2.10)$$

unless some other form, such as division by the range, is explicitly noted. For most purposes c can be taken to be 1, although advantages in a different choice are discussed in Section 4.3.

For many published analyses, using different techniques, (2.10) is used (whether rightly or wrongly) and is the default in the book. Where logarithms are used, assume natural logarithms are intended unless otherwise indicated.

Apart from (2.7) it will be convenient to list other transfomations here. If the x_{ij} are proportions, a desire to make the data more nearly normal underlies the use of the arc-sin transformation

$$z_{ij} = \sin^{-1}(\sqrt{x_{ij}}) \qquad (2.11).$$

With data in the form of proportions and with a constant row sum, mathematical considerations may suggest the use of the transform

$$z_{ij} = \ln(x_{ij}) - p^{-1} \sum_{j=1}^{p} \ln(x_{ij}) \qquad (2.12)$$

which will be discussed in more detail in Section 4.6. Equation (2.12) is the logarithm of an observation divided by the geometric mean of a row. Though it is not obvious (2.12) is related to the use of ratio transformations

$$z_{ij} = x_{ij}/x_{ip} \qquad (2.13)$$

for $j = 1, \ldots, (p - 1)$; this or its logged version have been used in a variety of applications.

The form of transformation used, if any, may be dictated by practice in an area or by the problems presented by the data set to hand. A common guiding principle seems to be a desire to make data as nearly normally distributed as possible. This is not an essential requirement for most of the multivariate methods to be discussed, but may improve interpretability since the use of highly skewed data, for example, can cause problems.

Preliminary data inspection will sometimes suggest if a transformation is needed, although if there are groups in the data, variables may have a multi-modal distribution so that identifying a suitable transformation will be difficult.

While interpretation is easier if all variables are similarly treated, the use of different transforms within analyses is not uncommon. Several of the references to analyses of artefact compositions, for instance, follow Pollard's (1986) advice and transform the trace elements logarithmically but not the major or minor elements. In another context involving mixed data – the shape analysis of fibulae – Doran and Hodson (1975) use both logarithms of ratios of metric variables and logarithms of angles. In these cases a desire to achieve normality is the underlying motive.

3

Principal Component Analysis – The Main Ideas

3.1 Introduction

In Section 3.2 some alternative ways of looking at principal component analysis (PCA) are reviewed. Emphasis is on PCA as an exploratory tool for obtaining a two- or three-dimensional picture of complex multivariate data. Aspects of the methodology are illustrated in Section 3.3 using the data in Tables A1 and A2. Applications of PCA to such data are among the more common in the archaeological literature (see Section 1.5). After artefact compositional analysis, other common uses have been to inter-assemblage comparisons and to morphological (or typological) analysis. The former may be more suited to correspondence analysis (see Chapters 5, 6); the latter is illustrated in a detailed example of Shennan's (1988) and in Section 4.12.

Most textbooks on multivariate analysis contain chapters on PCA. The best single reference is Jolliffe (1986), a book devoted entirely to the topic. Jackson (1991) is also comprehensive. Shennan (1988, Chapter 13) provides a good introductory account for archaeologists. Chapter 4 will deal with aspects of PCA dealt with only briefly, or not at all, in Shennan's book.

3.2 PCA as an exploratory method

3.2.1 The aims of PCA

The first eleven columns of Table A1 provide an example of a 105 by 11 data matrix. Structure in this set of data, such as the existence of chemically distinct groups, is not readily discerned by looking at or 'eyeballing' the figures. If distinct chemical groups exist then their relationship – if any – to archaeological information such as the site of origin is also of interest. PCA provides a quick way of looking at the data such that obvious structure and other interesting features of the data can be seen. Specifically PCA may be used:

(a) to obtain a two-dimensional picture of the rows of the data matrix that allows any groups differentiated by their chemical composition, and possible associations with archaeological variables such as site of origin, to be examined;

(b) to obtain a two-dimensional picture of the columns (variables) of

Principal Component Analysis – Main Ideas

the data matrix that, in the context of Table A1, shows how elements are associated;

(c) by putting the two kinds of picture in (a) and (b) together it can be seen which elements affect the structure in the data and how.

Essentially the pictures provided by PCA are obtained by a series of mathematical operations on the data matrix (Appendix B) that can always be carried out. Consideration, therefore, will also need to be given to how well the pictures obtained really represent the data.

3.2.2 A problem with PCA

It is essential at the outset to realise that the results of a PCA depend on how the data are scaled or transformed. If an analysis is carried out on the centred data of Table A1, for example, different results will be obtained than if the data are standardised so that variables have zero mean and unit variance. Similarly, the results of an analysis of Table A3 depend on whether it or the percentage data of Table 2.2 are used, and whether or not the percentage data are standardised or otherwise transformed.

For reasons to be discussed in Section 4.2 it is common practice to standardise, and this is often the default in computer packages. It will be assumed that PCAs reported in this section are of standardised data (as in equation 2.4), unless otherwise stated.

A PCA of standardised data can be obtained by a series of mathematical operations on the correlation matrix (Appendix B). If data are unstandardised then the matrix of covariances (equation 2.1) rather than the correlation matrix is used. The terms 'analysis of the correlation matrix' and 'analysis of the covariance matrix' that are sometimes used in the sequel thus indicate whether or not data have been standardised prior to analysis.

3.2.3 PCA as a linear combination of variables

The idea of a linear combination of variables was introduced in Section 2.9. For the eleven variables in Table A1 (standardised) define two linear combinations of the form

$$P_1 = a_1 Al + a_2 Fe + \ldots + a_{11} Pb \qquad (3.1)$$

and

$$P_2 = b_1 Al + b_2 Fe + \ldots + b_{11} Pb \qquad (3.2).$$

For specific values of the coefficients $(a_1 a_2 \ldots a_{11})$ and $(b_1 b_2 \ldots b_{11})$ it is possible to calculate P_1 and P_2 for any row in the table and plot it as a point on a graph of P_2 against P_1.

In PCA the coefficients in (3.1) are determined, roughly speaking, such that the data plotted on the first axis is spread out as much as possible. The coefficients in (3.2) are determined so that P_1 and P_2, the first two principal components, are uncorrelated and, subject to this, the spread on the second axis is the second greatest possible. Further components P_3, P_4, ... are similarly defined. The general idea is that as much of the variation in the data as possible is shown in two dimensions, this variation being related to the 'spread' of the data on the two axes. Because of the way the mathematics works a successful analysis will often account for much of the correlation structure as well (Jolliffe, 1986, p. 11). The a_i and b_i can be multiplied by an arbitrary constant without affecting this general outline and it is necessary to 'normalise' them in some way (using 'normalise' in a different sense from in Chapter 2). Different choices are possible; that assumed in the book, unless otherwise stated, is

$$a_1^2 + a_2^2 + \ldots + a_p^2 = 1 \tag{3.3}$$

By plotting P_2 against P_1 for all the rows in the data matrix a picture of the relationship between the rows is obtained. A second graph, obtained by plotting the coefficients against each other (i.e. a_1 against b_1 etc.) shows the relationship between the variables. Its use is illustrated later.

3.2.4 The geometry of PCA – a simplified view

Another view of what PCA does, which may be helpful, is given here. Figure 3.1(a) shows the plot of standardised values of Cr against Mg for a sample of eighty-five specimens of Romano-British vessel glass found in excavations at Winchester. Points are labelled A or B according to the two groups identified in the original work by Heyworth and Warren (see Baxter and Heyworth, 1991). The same data are shown in Figure 3.1(b) after a PCA, with the origin now at the centre of the points scatter and with the scatter rotated so that the greatest spread is along the first axis. In PCA these new axes are defined as outlined in the previous subsection. Apart from the shift of origin and rotation, the two graphs are the same because there are only two variables.

Figure 3.1(c) shows the point scatter plotted in one dimension on the first axis only. The bi-modality of the plot suggests the possible existence of two groups. Because of the high correlation shown in Figure 3.1(a) the distance between points in Figure 3.1(c) is similar to that in Figure 3.1(b) or 3.1(a). Consequently the final figure largely preserves distances between points, and features in the two-dimensional plot of 3.1(a) can be investigated by looking at 3.1(c).

This is essentially what a PCA analysis is trying to do except that the data normally exist in p dimensions (where p is bigger than 3 and the data cannot be visualised – see Section 4.4) and are approximated by a two-dimensional picture. Figure 3.1(c) provides a good approximation because the original data are strongly correlated. In general a PCA is only worthwhile if there is a reasonably high level of intercorrelation among the variables. This is often to be expected for data of the kind in Tables A1 and A2 so

Principal Component Analysis – Main Ideas

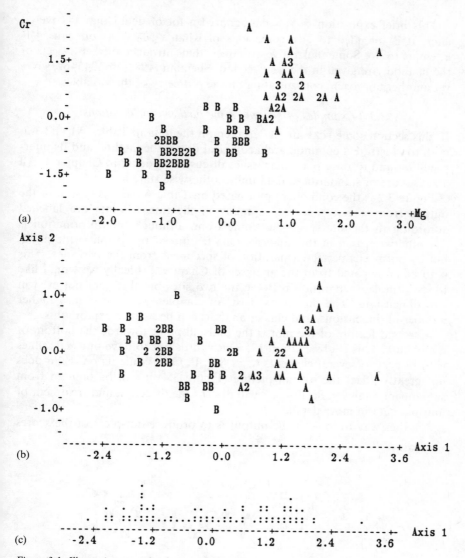

Figure 3.1: *Illustrative example of a PCA using two variables*

Note: Figure (a) shows a plot of Cr against Mg for 85 specimens of Romano-British vessel glass with letters labelling which of two groups specimens belong to. Figure (b) shows the same data after a PCA and (c) shows scores on the first component.

that PCA can be expected to be a useful tool. Vierra and Carlson (1981) advocate tests to see if the data are randomly patterned before doing a PCA, but it will often be clear from inspection of the correlation matrix that there are sufficiently high correlations to make PCA a worthwhile exercise.

This brief exposition of PCA has ignored a lot of detail that is covered later. It is intended to give some idea of what a PCA – as often used – attempts to do. Some illustrative examples, that introduce further aspects of the method, are given in the next section. Shennan (1988, pp. 245–62) gives yet another approach, concentrating on the geometry of the variables.

3.3 Examples of PCA and some further considerations

In this section some PCAS are undertaken on the data in Tables A1 and A2, both to illustrate a common archaeological use of the method and its interpretation and to raise particular issues discussed in detail in Chapter 4. All analyses are of standardised data unless otherwise stated.

Figure 3.2 is the component plot based on Table A1 and is typical of the output from a PCA reported in many applications. Points are labelled according to the site specimens come from, although this information is not otherwise used in the analysis. One feature of the graph is the good, but by no means perfect, separation of specimens from the two sites. This is to be anticipated from the analyses of Chapter 2. Ideally we would like to see complete separation between the two sites but this does not exist in two dimensions. The possibility exists of complete separation in a higher number of dimensions, and cluster analysis can be used to explore this.

A second feature of the plot is the unusual point towards the bottom of the picture. This is observation 12, which turns out to have unusual values with respect to several elements (e.g., Mg, P, Cu). While its omission does not greatly affect the overall picture (Figure 3.4) it will be omitted from subsequent analyses. Section 4.8 will discuss the detection and treatment of unusual data in more detail.

Another way to look at the output is to produce dot-plots of the scores

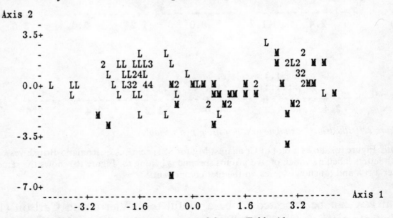

Figure 3.2: Component plot of correlation matrix of data in Table A1

Note: The letters L and M label specimens from Leicester and Mancetter; numbers show multiple occurrences of specimens.

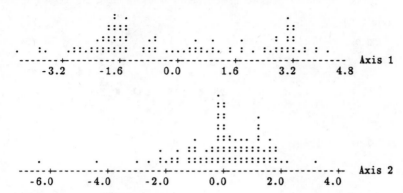

Figure 3.3: *Dot-plots of leading components in Figure 3.2*

Note: The upper graph shows the distribution of scores on the first component of Figure 3.2 and the lower graph the distribution of scores on the second component.

on the leading components as in Figure 3.3. This shows, fairly clearly, grouping on the first component and an outlier on the second. Inspection of higher-order components may also be useful.

Table 3.1 shows MINITAB output for the analysis of the data omitting observation 12 and Figure 3.4 shows the associated row and variable plots. Similar information should be available in any package that carries out PCA.

The information in Table 3.1(a) allows an answer to the question as to how good a representation of the data Figure 3.4(a) provides. Specifically, the second row in Table 3.1(a) can be interpreted as the proportion of variation in the data (on the scale used) 'explained' by a component. The third row gives the cumulative variation explained. Thus the leading two components account for 63.6% of the variation in the data, with the leading component accounting for 48.4%. This is not perfect, but is high enough to hold out hope that the component plot will be reasonably informative about structure in the data. Appendix B and Section 4.4 give a more detailed discussion.

These numbers are quite typical of PCAs of glass, metal and pottery data (e.g., see the analyses of pottery data in Jones, 1986). The structure of these materials is such that a high level of inter-elemental correlation is to be expected. This means that PCA is often a sensible exploratory technique to use and can be expected to produce informative results in many cases. Similar remarks apply to the correlations to be expected between metric variables in many morphological studies.

The output in Table 3.1(b) gives the coefficients of the components that define the axes of Figure 3.4(a) and the points in Figure 3.4(b). Thus the two leading components or axes (equations (3.1),(3.2)) are, on rounding coefficients to two decimal places,

$$P_1 = .18Al - .33Fe + \ldots - .32Pb$$
$$P_2 = -.42Al - .12Fe + \ldots - .20Pb.$$

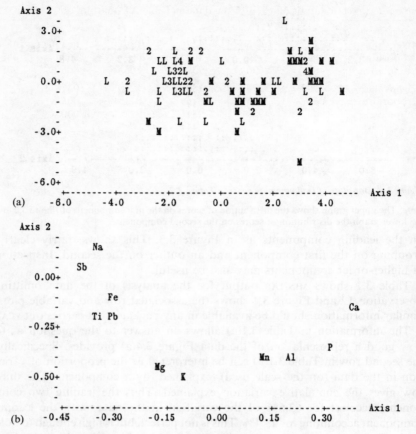

Figure 3.4: Biplot for the analysis given in Table 3.1

Note: The upper graph is a plot of the specimens on the first two components; the lower graph is a plot of the variables. Letters in (a) label sites, with numbers showing multiple occurrences. The graphs are of MINITAB screen output with minor editing to identify elements in (b).

Substituting for the (standardised) values of Al, Fe, etc. from Table A1 defines values of P_1 and P_2 for each row that are the points plotted in Figure 3.4(a).

Shennan (1988) discusses the interpretation of the coefficients in some detail. In MINITAB the components are defined as in (3.3) such that the sum of the squared coefficients for a component is unity (e.g. $.182^2 + .331^2 + \ldots + .425^2 + .323^2 = 1$). A squared coefficient for a variable multiplied by the corresponding eigenvalue can be interpreted as the variation in that variable accounted for by the component. Thus 96% of the variation in Sb (i.e. $5.3239 \times (-.425)^2$) is accounted for by the first component.

It will sometimes be the case that P_1 and P_2 can be given a simple and

Table 3.1: MINITAB output of a PCA of the data in Table A1 omitting observation 12

(a)
Eigenvalue	5.3239	1.6759	1.1276	.7391	.6639	.4576
Proportion	.484	.152	.103	.067	.060	.042
Cumulative	.484	.636	.739	.806	.866	.908
Eigenvalue	.3909	.2797	.1806	.1288	.0378	
Proportion	.036	.025	.016	.011	.003	
Cumulative	.944	.969	.985	.997	1.000	

(b)
Variable	PC1	PC2	PC3	PC4	PC5	PC6
Al	.182	−.420	−.340	−.425	−.540	−.283
Fe	−.331	−.122	−.364	−.031	−.219	.279
Mg	−.202	−.454	−.244	−.613	−.208	−.268
Ca	.366	−.126	−.229	−.309	.191	−.203
Na	−.372	.127	.175	.049	.025	−.127
K	−.122	−.495	.117	.533	.645	−.037
Ti	−.378	−.184	−.208	.039	−.141	.116
P	.302	−.327	.037	−.127	.022	.783
Mn	.102	−.379	.675	−.176	−.321	−.216
Sb	−.425	.043	.004	−.068	.017	−.010
Pb	−.323	−.198	.310	−.086	−.202	.194

Variable	PC7	PC8	PC9	PC10	PC11
Al	−.237	.178	.060	.092	−.151
Fe	.554	−.134	.525	.022	.126
Mg	−.109	.199	−.098	.334	.174
Ca	−.070	−.214	.322	−.632	−.264
Na	−.147	.721	.333	−.370	.076
K	.093	−.094	.086	−.005	.005
Ti	.056	−.088	−.634	−.561	.135
P	−.091	.361	−.044	−.019	−.173
Mn	.443	.003	−.037	−.098	−.055
Sb	.021	−.030	−.068	.129	−.889
Pb	−.617	−.445	.283	−.029	.121

Note: The output shown above is exactly in the form produced as the default using the MINITAB PCA command. This is shown in this manner for illustrative purposes but would usually be put into a more presentable form for the purpose of publication.

useful interpretation. In the present case such an interpretation is not obvious. This does not matter, as interpretability of the components is not necessary for the analysis to be useful (see Section 4.4 and 4.10 for a fuller discussion). If structure is shown in a component plot the reasons for it can usually be identified either by further examination of any groups shown or by inspection of a plot such as Figure 3.4(b). Some scholars advocate 'rotation' of the components to aid interpretation – this is the subject of some confusion and will be discussed in more detail in Sections 4.10 and 4.11.

The variable plot in 3.4(b) is based on the coefficients of the leading two components, for example Al is plotted at the (.18,−.42) position. Three groups of variables − (Sb, Na, Fe, Ti, Pb), (Ca, P, Al, Mn) and (Mg, K) − are suggested. Comparison of these groups with the correlation matrix in Table 2.1(b) shows that, broadly speaking, within any group elements are positively correlated. The first two groups listed, opposite each other on the plot, tend to identify pairs of negatively correlated elements. Thus the plot provides quite a good picture of the correlation matrix for the data. There is a sense in which plots similar to 3.4(b) can be quite precisely interpreted as two-dimensional approximations to the correlation matrix. This is considered in Section 4.4.

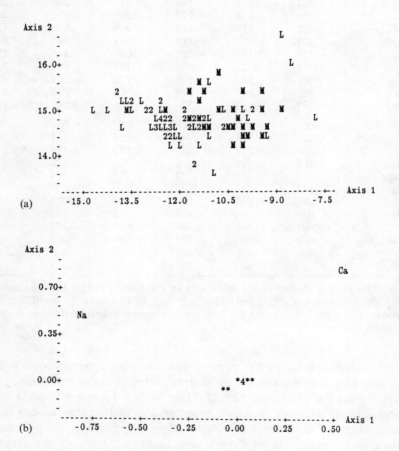

Figure 3.5: *Biplot for the analysis of the covariance matrix of the data in Table A1 omitting observation 12*

Note: Conventions are as in Figure 3.4 except that only Ca and Na are identified in (b).

Table 3.2: Leading components for an analysis of the covariance matrix of the data in Table A1 omitting observation 12

Variable	PC1	PC2
Al	.045	.010
Fe	−.073	−.075
Mg	−.006	−.021
Ca	.498	.856
Na	−.858	.508
K	−.018	−.021
Ti	−.009	−.008
P	.008	.001
Mn	.017	−.019
Sb	−.088	−.044
Pb	−.007	−.005

If the axes are suitably scaled, an informal judgement of how good the representation is can be gained by noting how near the points are to forming a circle about the origin – the closer to a circle, the better. With a large number of variables, or with many weakly correlated variables a less good representation will usually result (e.g., Figure 3.8(b)). A fuller discussion is given in Section 4.4.

The joint inspection of Figures 3.4(a) and 3.4(b) – a biplot (Section 4.4) – is also useful. Examples of such biplots are suprisingly hard to find in the archaeological literature given their potential use. Specimens plotted on the component plot and away from the origin will tend to have relatively high values on those elements that occupy the equivalent space on the variable plot. Thus the Leicester glasses which tend to the left of Figure 3.4(a) will have high values on those elements to the left of Figure 3.4(b) and low values on elements to the right, relative to the Mancetter glasses. Graphical analyses, such as those in Figure 2.3, readily confirm this. In effect Figure 3.4 presents, in a single diagram, much of the information gleaned from the analyses in Chapter 2.

The effects of the scale of measurement – the subject of Section 4.2 – can be seen by comparing Figure 3.4(a) and Table 3.1(b) with the equivalent Figure 3.5 and Table 3.2 for a PCA of centred but unstandardised data. The contrast is most stark if one compares the two tables where it can be seen that the components for the unstandardised data are almost entirely determined by Na and Ca. Section 4.2 will show that this is entirely predictable, so that Figure 3.5 could be more easily obtained by plotting Na against Ca directly.

As it happens the two figures are less obviously different. This is because most of the structure in the data is contained in the first components in each case which, as Figure 3.6 shows, are strongly related. This is not inevitable and raises the issue as to how different analyses of the same data might be compared. This is taken up in Section 4.7.

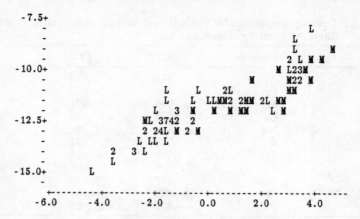

Figure 3.6: Plot of the leading components of Figures 3.4 and 3.5

Note: The plot is of scores on the first component of the analysis of the covariance matrix against those for the analysis of the correlation matrix.

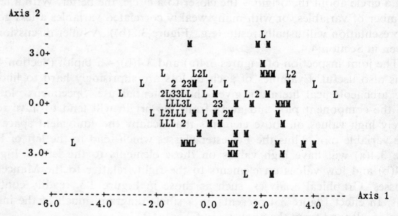

Figure 3.7: Component plot for the PCA of the correlation matrix of the trace elements in Table A2 omitting observation 12

Note: Conventions are as in Figure 3.2.

Another instance where a comparison of analyses may be useful is when different variable subsets are used to obtain results. Figure 3.7 shows the component plot for the trace elements in Table A2. This will be compared more formally with Figure 3.4(a) in Section 4.7.

Finally the biplot for all the data is shown in Figure 3.8. The first graph suggests three possible groups, that to the left mainly based on the Leicester specimens and characterised by relatively high values of the variables to the left of Figure 3.8(b). The other two groups suggest a possible subdivision of the Mancetter material. The variable plot, which is attempting to 'squeeze' information on the interrelationships between twenty-two

Principal Component Analysis – Main Ideas

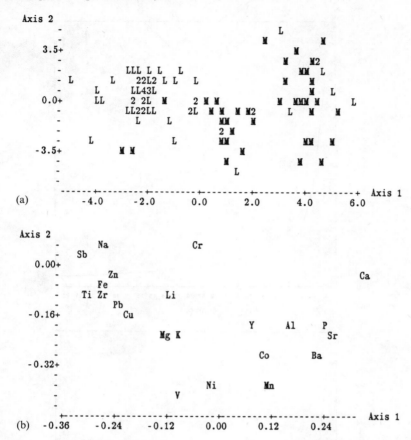

Figure 3.8: Biplot for the PCA of the correlation matrix of all elements in Tables A1 and A2 omitting observation 12

Note: Conventions are as in Figure 3.4. The position of Ti and Zr have been perturbed very slightly in order to render them distinct.

variables into a single plot, is less good than that of Figure 3.4(b), for example. In particular elements such as Cr and Li, close to the (0,0) point, are poorly represented.

It has been assumed in this chapter and others that a plot based on the first two or three components gives an informative, if not perfect, view of the data. Various rules, many *ad hoc*, have been proposed for deciding how many components are needed to describe the data well (Jolliffe, 1986, pp. 92–107). Requiring the proportion of variation explained to be greater than some arbitrary number, such as 70% or 80%, is common. For analyses based on the correlation matrix, selecting components with an eigenvalue bigger than 1 is often suggested, though Jolliffe (1986, p. 95) notes that a cut-off value of 0.7 may be better. Another somewhat subjective

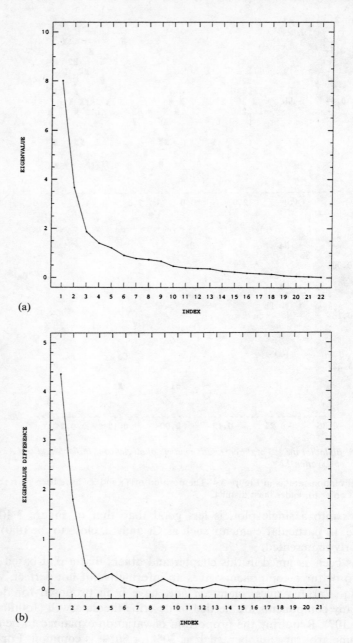

Figure 3.9: Plots to identify the number of 'significant' components for the analysis in Figure 3.8
(a) A plot of ordered eigenvalues against their index.
(b) A plot of differences in successive eigenvalues against their index.

Principal Component Analysis – Main Ideas

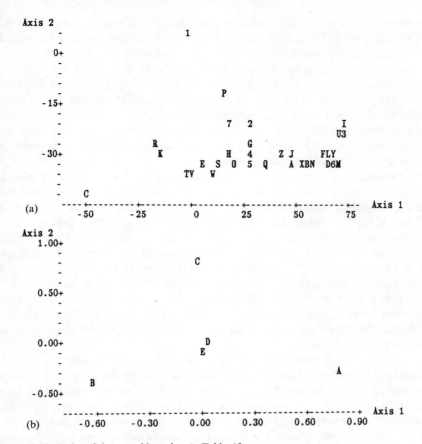

Figure 3.10: Biplot of the assemblage data in Table A3

Note: Diagram (a) is the component plot for the assemblages using the covariance matrix of the data in Table A3. The letters A to Z label assemblages 1–26 with the numbers 1–7 labelling assemblages 27–33. The graph is an edited version of MINITAB screen output with slight perturbation of positions in order to distinguish between sites. Diagram (b) is the corresponding variable (tool type) plot with identifications as in Table A3.

approach is to plot eigenvalues against their index and look for an 'elbow' in the plot – the so-called 'scree' plot – or to plot changes in the eigenvalues and look for a levelling-off of values. For the PCA of Tables A1 and A2 the eigenvalue sum is 22 and the leading eigenvalues are 8.01, 3.66, 1.87, 1.40, 1.20, 0.90, 0.77, 0.73, etc. with the cumulative proportions of variation 'explained' being 36.4%, 53.1%, 61.6%, 67.9%, 73.4%, 77.5%, 81.0%, 84.3%, etc. Requiring 70% or 80% of the variation to be explained would suggest five or seven components are adequate; using cut-off values for the eigenvalues of 1 and 0.7 leads to a choice of five and eight components respectively. The scree plot in Figure 3.9(a) suggests perhaps four to six components, as does the plot of eigenvalue differences in Figure 3.9(b). Thus, between four and eight components are suggested as adequate, according to the criterion used and the subjective interpretation of the graphs. In any event, the reduction from the original twenty-two variables is substantial. While Figure 3.8(a) accounts for only just over 50% of the variance, a quite typical value for the number of variables used, it is nevertheless suggestive of structure. The two-dimensional plot is thus useful, but it should be borne in mind that it is only an approximation to structure in what for practical purposes is between a four- and an eight-dimensional space.

As a final, and different, example to illustrate points taken up elsewhere, the percentage data of Table 2.2 have been subjected to a PCA without further standardisation, and results are shown in the form of a biplot in Figure 3.10. The upper diagram essentially reproduces Pitt's (1979) analysis and its main feature is the identification of a site (Star Carr) in the upper part of the diagram as unusual in terms of its assemblage composition. Comparison of the two plots shows that this site is associated with a high level of tool type C – a feature evident on inspection of Table 2.2. The first axis, from Figure 3.10(b), is entirely determined by the opposition between types A and B; the fact that D and E are close to the (0,0) point indicates that they have little influence on Figure 3.10(a). All these features are entirely predictable.

Firstly, types A and B are much more common than other types and, since unstandardised data are used, will tend to have the greatest influence on the analysis (Section 4.2). Secondly, the data are constrained to sum to 100% for each site and since A and B are dominant types they are necessarily highly negatively correlated. This in turn results in their opposition on the variable plot in the figure (and would occur even were the data standardised). These features of the data are potentially problematical for a PCA and are discussed at greater length in Section 4.6. An alternative approach using correspondence analysis is illustrated in Section 5.2.

4

Principal Component Analysis – Specialised Topics

4.1 Introduction

A miscellany of topics related to the interpretation of PCA and extensions to it are discussed in this chapter. Data transformation and standardisation have already been discussed in Section 2.12. Archaeological applications tend towards the use of standardised data and some of the reasons for this are suggested in Section 4.2; the choice between standardised and unstandardised data is not clear-cut and some different views are considered in Sections 4.2 and 4.3. Presentation of results from a PCA is often limited to a row plot; this is unnecessarily constraining and Section 4.4 looks at the benefits to be gained from plotting both rows and columns, in addition to other aspects of interpretation. If thought is not given to data transformation before an analysis, the first component will sometimes be interpretable as a 'size' effect which is not always what is intended; this is considered in Section 4.5 and leads into a discussion of the problems of compositional data in Section 4.6, when the rows of a data matrix sum to the same value. Chapter 3 raised the problem of comparing the results of different analyses and this is addressed in Section 4.7, while 4.8 briefly considers the problem of outliers. The possibility of using Mahalanobis rather than Euclidean distance is covered in Section 4.9; this is a more specialised topic needed for an appreciation of Section 8.5 and parts of Chapters 9 and 10. Section 4.10 looks at the subject of component rotation and this is related to the thorny topic of factor analysis discussed in Section 4.11. A detailed application to illustrate some of the ideas discussed is given in Section 4.12, and the chapter closes with a review of applications of PCA in the literature.

4.2 Distance and standardisation

One interpretation of the usual plot of component scores with the 'normalisation' of (3.3) is that it attempts to approximate the (p-dimensional) Euclidean distance between the rows of a data matrix in just two dimensions (Section 3.2.4). The definition of the (squared) Euclidean distance in p dimensions is

$$d_{ik}^2 = (y_{i1} - y_{k1})^2 + (y_{i2} - y_{k2})^2 + \ldots + (y_{ip} - y_{kp})^2$$

$$= \sum_j (y_{ij} - y_{kj})^2 \qquad (4.1).$$

This generalises the definition of distance in everyday space.

One consequence of this definition is that the main contributions to d_{ik}, and hence to the PCA analysis, arise from those variables that can make a large contribution to (4.1). Where variables differ widely in their variance only those with a large variance will contribute significantly to the PCA. If unstandardised data are used, such contributions may well come from only a small subset of the variables – effectively ignoring information contained in other variables.

To give an example: if PCA is applied directly to the centred but unstandardised data of Table A1 then Ca and Na, with the greatest magnitudes and variances, swamp other variables with their contribution. They account for 98% of the distance between rows 3 and 4, for instance, and clearly dominate the analysis (Table 3.2, Figure 3.5).

This particular problem arises because of the differing orders of magnitudes of the variables. A logarithmic transformation can partially effect a cure but can also result in other variables (often those with a low presence) becoming dominant on the transformed scale. It is to avoid such problems that standardisation is commonly used even after transformation. This gives variables equal weight (variance) so that each may potentially contribute to the distance (4.1) and the PCA (Pollard, 1986).

Whether or not variables should be standardised (as well as transformed) is not as clear-cut as the common practice of doing so might suggest. Bishop and Neff (1989) and Leese et al. (1989) both give examples where the archaeologically interesting variation in their data on artefact compositions is revealed by transformed but unstandardised data. This appears to be because the interesting variables – e.g., those associated with the degree of temper added to a ceramic product – are small in number and have a relatively high variance. Using unstandardised data highlights groups made from similar clays but differing in the extent to which the clays have been tempered. If the data are standardised the importance of these variables – relative to the objectives of the studies – is obscured. Another argument for not standardising is that if the variables are measured in the same units, then standardisation amounts to an arbitrary weighting of them.

In contrast to the above, Baxter (1989, 1991) and Baxter et al. (1990) provide examples of chemical compositional analysis where the use of unstandardised data leads to results that are entirely predictable and/or archaeologically uninformative, and examples where the use of standardised data are more interpretable than the results of using unstandardised data. Different choices of transformation without standardisation will implicitly weight variables differentially – sometimes, but by no means always, in an archaeologically useful way. Standardisation is an attempt to avoid this at the possible cost of smearing real and useful differences based on a subset of variables. It will often be useful to look at results obtained using different treatments of the data, always provided that different outcomes are both explained and fully reported.

A useful discussion of the general problem is provided by Wright (1989) who suggests that for metric variables, in morphological studies for example, standardisation will nearly always be needed. The argument here is that the morphological characteristics measured by the variables are inherently of equal interest and should be given equal weight. For counted data, by contrast, transformation to percentages (or their square roots) is suggested followed by a PCA of the covariance matrix. The argument in this case is that rare types should not be given equal weight with common types, as would be the case if the correlation matrix were used. The concept of 'rarity' applies to counted data but not metric data and hence different approaches are needed. Duncan et al. (1988, p. 9) also suggest that for the analysis of assemblage data the use of unstandardised percentages is appropriate.

The foregoing discussion assumes that variables are measured in similar units. Where this is not the case then standardisation is essential if an analysis using PCA is to be undertaken at all. Examples include Hinout's (1984) typological study of flint artefacts using angles of retouch as well as metric properties, and Doran and Hodson's (1975) discussion of La Tene fibulae using metric properties, angles and counts.

The issues raised in this section concerning standardisation often amount to a choice of a PCA based on the covariance or correlation matrix of transformed data. This is examined in more detail in the next section.

4.3 Covariance and correlation matrices

If, in (2.9) and (2.10), c is defined as

$$c = (n-1)^{-1/2} \tag{4.2}$$

then analysis is of the covariance or correlation matrix respectively. Shennan (1988, p. 262) observes that 'probably the majority of statisticians are extremely cautious about using the correlation matrix rather than the covariance matrix for principal components analysis, on the grounds that it tends to destroy the validity of available statistical distributional theory; it can make the results difficult to interpret; and it allows the dubious possibility of combining different types of measurement (Fieller, pers. comm.).' A less pessimistic (statistician's) assessment of the usefulness of the correlation matrix approach is given in Jolliffe (1986).

Archaeological and general practice leans towards the use of the correlation matrix. The issues involved are essentially those discussed in the last section.

With reference to the quotation from Shennan, the point may be made that nearly all of the references to archaeological uses of PCA in the bibliography use the technique in an exploratory or descriptive way. From this perspective, concerns about the validity of available distributional theory are largely irrelevant. The issue is whether or not reported uses, which tend to be based on the correlation matrix, are substantively informative. Similar considerations apply to the use of mixed variables, the possibility of combining which is sometimes seen as a positive merit of standardisation. In effect the issue devolves to whether or not the results of using PCA tend to be interpretable from a substantive, archaeological standpoint; and whether or not PCA, rather than some simpler approach, is the best technique for achieving a desired end. If Wright (1989) is correct, then counted data of the kind arising in assemblage studies are best left unstandardised, though possibly transformed, or analysed by some other method such as correspondence analysis (Chapters 5, 6). Metric data, by contrast, should usually be standardised. Chemical compositional data are usually standardised, and produce interpretable results, but sufficient examples of successful use of the covariance matrix exist to caution that the choice should not be an automatic one.

It is undoubtedly the case, as suggested by Bishop and Neff (1989), that PCA and other multivariate techniques are often applied because they are there, in the form made most accessible by the available software. This typically involves the use of the correlation matrix. More thought ought to be given to the use of alternative scales on which to analyse data than is sometimes evident.

Nevertheless, PCA is quick and easy to apply, and many published examples provide a concise summary of archaeologically useful information – often based on the correlation matrix. In summary, this suggests that potential statistical objections to the use of PCA in archaeological applications that use a correlation matrix may, from a practical view – and regarding PCA as an exploratory method – be overstated. Explicit modelling of the data, rather than exploratory analysis, is an alternative approach (e.g. Buck and Litton, 1991; Scott et al., 1991) and more sophisticated exploratory methods exist. These alternative methods are not widely used at present and, because of the demands made of the data and/or because of the lack of easily used and widespread software, may find little practical application in the immediate future.

4.4 Interpretation; biplots and related techniques

Some additional aspects of the interpretation of a PCA are addressed in this section along with the neglected topic of the use of biplots (or similar display methods). Some use of matrix algebra is unavoidable, and it may help to review Appendix B if this is unfamiliar, although if this is taken on trust, much of the section can be understood by the non-initiate. The relevant sections of Shennan (1988) provide a complementary treatment.

A transformed and/or standardised data matrix, **Y**, can be factorised using the singular value decomposition as

$$\mathbf{Y} = \mathbf{U\Sigma V'} \tag{4.3}$$

where $\mathbf{U'U} = \mathbf{V'V} = \mathbf{I}$, the p by p identity matrix; and $\mathbf{\Sigma}$ is a p by p diagonal matrix. From (4.3) it follows that

$$\mathbf{Y'Y} = \mathbf{V\Omega V'} \text{ with } \mathbf{\Omega} = \mathbf{\Sigma}^2 \tag{4.4}.$$

For practical purposes it suffices to know that with (2.9) and (4.2), equation (4.4) defines the covariance matrix of the data; with (2.10) and (4.2) the correlation matrix is defined; and the output of many statistics packages that use PCA is based on the columns of **V** and the diagonal elements of $\mathbf{\Omega}$ that we assume to be ordered from largest to smallest. The diagonal elements of $\mathbf{\Omega}$ are called eigenvalues and the columns of **V** are called eigenvectors (the terms 'latent' or 'characteristic' replace 'eigen' in some treatments of the subject).

The following specific points may be noted:

(i) The columns of **V** hold elements that define the coefficients of the principal components P_1, P_2 etc. in equations (3.1), (3.2) 'normalised' as in (3.3). Computer packages that allow a PCA usually report these or equivalent coefficients (see (iv) below).

(ii) The sum of the eigenvalues is equal to the sum of the diagonal elements of $\mathbf{Y'Y}$. If the correlation matrix is used this sum is just p; if the covariance matrix is used it is just the sum of the variances of the Y_i. It is usual to express the 'success' of a component plot or its 'goodness of fit' as the proportion of the sum accounted for by the first two components which is

$$100(\lambda_1 + \lambda_2)/\sum_{i=1}^{p} \lambda_i \tag{4.5}$$

where the λ's are the eigenvalues ordered by size. All good packages should report this information.

(iii) Packages which provide a component plot or, better still, allow the component scores to be saved are, in effect, using values stored in the first two columns of $\mathbf{U\Sigma}$.

(iv) As Shennan (1988) notes, and uses, some packages normalise coefficients in a slightly different way. Define the diagonal elements of $\mathbf{\Sigma}$, the singular values, to be σ_i, and note that

$$\lambda_i = \sigma_i^2 \tag{4.6}.$$

Then an alternative normalisation is to multiply the coefficients in the columns of **V** by the corresponding σ_i. The resultant coefficients are sometimes known as loadings and their squares have an interpretation as the amount of variation in a variable accounted for by a component. See Shennan for more detail; this section for another use; and Section 6.5 for equivalent uses in correspondence analysis.

These statistics and their use may be illustrated by reference back to Table 3.1 and Figure 3.4. Here we see that the component plot (based on scores (iii)) accounts for 64% of the variance in the data (using (ii) and (4.5)) which is reasonable but leaves about a third of the variation 'unexplained'. Using an arbitrary cut-off of 0.3 to identify important variables in the definition of a component (i) we see that (Fe Na Ti Sb Pb Ca Pb) define the first component – the sign of the last two differing from the rest; and (Al Mg K P Mn), with the same sign, identify the second.

It is neither easy to interpret these components although this is often attempted, nor is it essential for PCA to be useful to have such an interpretation (though see Section 4.10). In fact an alternative approach to the display of the data that is often more informative is possible. This approach – the biplot and related methods – is the subject of the rest of this section.

Biplots have already been introduced in Figures 3.4 and 3.5 where a joint display of the rows (objects) and columns (variables) of the data matrix was given. The row plot is based on co-ordinates in the first two columns of $\mathbf{G} = \mathbf{U}\Sigma$ (see (iii) above), while the column plot is based on co-ordinates held in the first two rows of $\mathbf{H} = \mathbf{V}$ (see (i) above). The joint display is a biplot in the sense of Gabriel (1971) in that the data matrix analysed $\mathbf{Y} = \mathbf{GH}'$, so that the joint plot has an interpretation as a visual approximation to **Y**. An interpretation of the plot was given in Section 3.3.

It is possible to define **G** and **H** in different ways retaining the feature that $\mathbf{Y} = \mathbf{GH}'$. A particularly useful definition is to have $\mathbf{G} = \mathbf{U}$ and $\mathbf{H} = \mathbf{V}\Sigma$. The subsequent row plot has an interpretation as approximating to Mahalanobis distances between rows (see Section 4.9 for further discussion of this). The variable plot has a particularly nice interpretation and this is now illustrated by example.

This second kind of biplot can easily be constructed from the first kind using standard output simply by dividing the scores on the i'th component by σ_i and multiplying the coefficients of the i'th component by σ_i (this then gives the loadings described in (iv) above). Thus from Table 3.1 we have σ_1 equal to 2.307 (the square root of 5.3239) and $\sigma_2 = 1.295$. The coefficients for the variable plot are then $.182 \times 2.307$, $-.331 \times 2.307$ etc. for the first component and $-.420 \times 1.295$, $-.122 \times 1.295$ etc. for the second component. The plot based on these is shown in Figure 4.1(b) with the associated row plot in 4.1(a).

In Figure 4.1(b) the variables are represented as vectors from the origin and the plot has been surrounded by a unit circle. On its own this is called

Principal Component Analysis – Specialised Topics

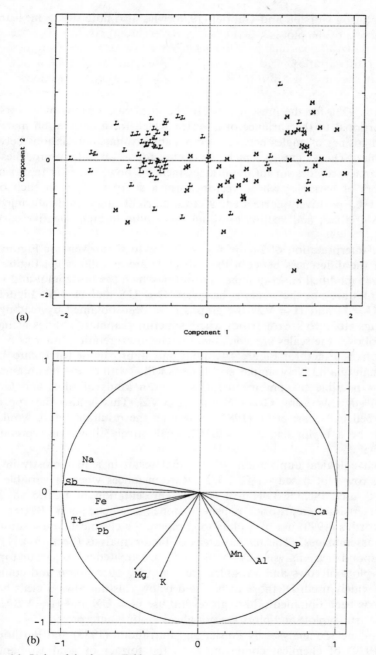

Figure 4.1: Biplot of the data in Table A1

Note: In diagram (a) L and M label the sites Leicester and Mancetter.

an h-plot by Corsten and Gabriel (1976) who also note that a measure of the success of the plot is

$$100(\lambda_1^2 + \lambda_2^2)/\sum_{i=1}^{p} \lambda_i^2 \qquad (4.7)$$

which is 92% for the present case. In the ideal case lengths of vectors are proportional to the variance of a vector which for a correlation matrix is 1, and cosines of angles between vectors approximate correlations between variables. The extent to which the vector for a variable approaches the unit circle provides an informal assessment of how well it is represented. Clusters of variables whose vectors form a sharp angle with each other should be positively correlated; vectors at right angles indicate uncorrelated variables; and vectors at about 180° indicate high negative correlation.

The interpretation of Figure 4.1 is similar to that given for Figure 3.4. where the differences between the two plots are as follows. (i) Figure 3.4, which is modified MINITAB screen output in which the horizontal and vertical scales are not the same, is based on $\mathbf{G} = \mathbf{U\Sigma}$ and $\mathbf{H} = \mathbf{V}$. Figure 4.1 uses $\mathbf{G} = \mathbf{U}$ and $\mathbf{H} = \mathbf{V\Sigma}$; the graphs have been obtained by reading the loadings etc. into STATGRAPHICS where superior graphics facilities including control over the scales are available. (ii) The interpretation in terms of correlations is 'exact' for Figure 4.1(b) but only approximate for Figure 3.4(b) although qualitatively similar conclusions will – with care – be obtained.

It is possible to combine features of both kinds of plot, and Gower (1984) suggests using $\mathbf{G} = \mathbf{U\Sigma}$ and $\mathbf{H} = \mathbf{V\Sigma}$. This is also the approach described in Lebart et al. (1984). Although the resultant figure would no longer be a biplot (since $\mathbf{Y} \neq \mathbf{GH'}$), qualitatively similar interpretations are likely.

Archaeological applications of PCA that result in a plot mostly involve just a row plot based on $\mathbf{G} = \mathbf{U\Sigma}$. Rare instances where a variable plot only is used include O'Hare's (1990) representation of metric variables characterising prehistoric polished stone artefacts, and Baxter and Heyworth's (1990) use of h-plots to compare the chemical compositions of glass assemblages. (The related areas of factor analysis (Section 4.11) and correspondence analysis (Chapters 5, 6) are treated as separate topics.) Joint plots of rows and variables are relatively uncommon and could be more widely used. Of those to be listed below it is not always clear how a plot has been obtained (although several use $\mathbf{G} = \mathbf{U\Sigma}$ and $\mathbf{H} = \mathbf{V\Sigma}$ rather than a true biplot) and plots are not always jointly interpreted.

They include analyses by Poirier and Barrandon (1983) and Berthoud et al. (1979) of chemical compositions – the former of gold Roman and Byzantine coins and the latter of Iranian copper ores. Typological studies include Hinout's (1984, 1985) analyses of flint artefacts; Larsen's (1988) analysis of biconical urns; and Speiser's (1989) analysis of Byzantine ceramic vases. Ringrose (1988a, b) has studied bone assemblages from different layers in this way (although his main interest was in correspondence analysis) and Tomber (1988) has similarly treated pottery assemblages. These examples, with others noted in Section 4.13, suggest that scholars on the European mainland (as opposed to Anglo-Americans) have been much more ready to exploit this particular approach to data display. We shall see in Chapter 6 that this is quite clearly the case for the related technique of correspondence analysis.

4.5 Size and shape

It will sometimes be the case in an analysis that all the coefficients of a component have the same sign and are of similar magnitude. Alternatively a subset of the coefficients may have the same sign with others close to zero. In either case the component, which is often though not inevitably the first one, may be interpreted as a 'size' component.

Such 'size' components arise, predictably, when all variables in an analysis are positively correlated. Typically, coefficients on the first component will then have the same sign and be of reasonable magnitude. This has a clear physical meaning in morphological studies based on metric properties where, for example, the measurements taken on an urn may all correlate positively with its height and with each other. A simple plot of the component scores against height would then be sufficient to demonstrate the physical interpretation. Richards (1987) and Shennan (1988, pp. 263–70) give examples that illustrate this kind of interpretation. Bishop and Neff (1989) observe that a similar phenomenon can arise in compositional studies of artefacts where all trace elements are positively correlated.

It is the fact of general positive correlation, rather than the actual magnitudes, which is important in producing this effect. Chatfield and Collins (1980, p. 72) give a small example in which two rather different matrices of positive correlations give very similar first components. This lack of sensitivity of the coefficients to the values of the correlations has been seen by some as an argument against using the correlation matrix. In archaeological practice it is sometimes the case that, having established a 'size' effect, this is associated with some variable, such as height or volume of an urn or facial length of a sculpture (Siromoney et al., 1980), regarded as a more 'fundamental' indicator of size than the other variables correlated with it.

When the first component has the interpretation envisaged above, the second component will typically contain coefficients with a mix of positive and negative signs and may be interpreted as a 'shape' component. More generally, higher-order components may reflect different aspects of 'shape'. If 'size' is of limited interest, then its dominance of the first component is sometimes considered a disadvantage of PCA which has been unfavourably contrasted with the 'shape'-oriented method of correspondence analysis for this reason.

If the 'size' effect is not of interest it can be dealt with in different ways. One way is to eliminate it by an intelligent choice of data transformation. A common approach is to transform to ratios, dividing (p-1) of the variables by the p'th. The divisor is usually chosen to be a good indicator of the size of an object. Thus Siromoney et al. (1980), in a stylistic analysis of South Indian sculptures based on facial characteristics, find that their first component in their initial analysis is a size component and transform by dividing by face length. Madsen (1988b) in similar vein, in an analysis of Neolithic pots, notes the possibility of dividing measurements by pot height. This turns out not to be entirely satisfactory and the analysis reported scaled vertical measures by dividing by height, and horizontal measures by dividing by rim width. Madsen also notes that if measurements are divided by the cube root of the volume

$$z_{ij} = x_{ij}/\sqrt[3]{V_i} \qquad (4.8)$$

where V_i is volume, then the size effect is successfully eliminated. Another approach is taken in Habgood's (1986) analysis of cranial measurements where rows as well as columns are standardised. It is also possible to retain the analysis, size effects and all, and base subsequent graphical presentation and interpretation on components other than those reflecting size.

Although some authors (e.g. Benfer and Benfer, 1981) have argued against the use of ratios, because they are rarely normally distributed, this overstates the need for strict normality with many multivariate methods used in an exploratory way. It can also sometimes be corrected for by taking logarithms of ratios (e.g., Doran and Hodson, 1975). In the next section we shall see how a similar transformation turns out to be useful in other contexts.

4.6 Compositional data and related problems

One view of the problem described in the previous section is as follows. We have n objects (say pots, to be specific) and p metric measurements. These are x_{ij} for pot i and j = 1, ... , p. The measurements can be written as $x_{ij} = c_i s_{ij}$ for all j where c_i is an unknown size component and real interest focuses on a comparison of the s_{ij} for different i. To eliminate the size effect a possible transformation is to define

$$z_{ij} = \ln(x_{ij}/x_{ip}) \qquad (4.9)$$

j = 1, ... , p − 1. This involves an asymmetric treatment of the data in the sense that the use of the p'th variable as a divisor is arbitrary and different choices will lead to different PCA results.

A similar problem was faced by Leese et al. (1989) in a totally different context and was resolved in a slightly different, symmetric, fashion. Their problem was to compare the trace element compositions of medieval tiles where x_{ij} is the measured elemental concentration, s_{ij} represents the contribution from the clays used in manufacture, and c_i is a dilution factor arising from the addition of a relatively neutral temper. The problem is to compare the clays used in manufacture in the presence of different but unknown dilution factors. The resolution of this problem is to define the geometric mean for an artefact

$$g(x_i) = (x_{i1}x_{i2}\ldots x_{ip})^{1/p} = c_i(s_{i1}s_{i2}\ldots s_{ip})^{1/p} \quad (4.10)$$

and then use the centred log-ratio transformation

$$z_{ij} = \ln[x_{ij}/g(x_i)] \quad (4.11).$$

It is easy to see that (4.11) both eliminates the dilution effect and involves the symmetric treatment of the variables. An alternative way of writing (4.11) has already been given in equation (2.7).

Mommsen, Krauser and Weber (1988) report similarly motivated work but use centred ratios rather than log-ratios. Pike and Fulford (1983) do not give (4.11) explicitly but seem to have used it in their study of black-glazed pottery from Carthage to correct for variation in the counting statistics giving rise to their neutron activation analysis results. The actual inspiration for the methodology in Leese et al. was the work of Aitchison (1986) on the analysis of compositional data. Similarly inspired, in a different context, is Baxter's (1989) note on the analysis of glass compositions where the compositional sum is 100%. Such data pose potential problems for analysis with PCA, and to such problems we now turn.

Ceramic studies using elemental concentrations typically omit the major contribution of silica and focus on the minor and trace elements. By contrast, silica is an important and deliberately controlled constituent of glass manufacture, often included in statistical analyses where the element sum is 100% apart from measurement omissions and error. Metal studies, similarly, often use the full composition. Such fully compositional data often occur in other contexts; for example, assemblages characterised by the percentage of different types present or artefact densities at a location converted to relative frequencies (Whallon, 1984).

There are, at least in principle, problems in applying 'standard' statistical methods to such compositional data. These are elaborated at length in Aitchison (1986). These difficulties can be visualised graphically using a simple example. Table 4.1 is adapted from Doran and Hodson (1975) and shows the percentage presence of three different tool types from different stratigraphic contexts in excavations at Ksar Akil. Figure 4.2 shows two different views of these data in three-dimensional plots. In the upper diagram a view is shown where the data are seen to lie in a two-dimensional plane. This arises because the compositional constraint, that the three

Table 4.1: Context by tool types from excavations at Ksar Akil

	Tool types (%)	
A	B	C
17.1	26.0	56.9
17.7	25.6	56.6
9.5	49.1	41.4
3.8	56.7	39.5
7.0	45.4	47.6
8.1	50.0	41.8
9.3	50.1	40.6
6.7	54.5	38.8
6.6	74.7	18.6
6.0	66.6	27.5
6.8	72.8	20.4
5.8	73.9	20.3
7.5	58.9	33.6
4.4	72.0	23.6

Source: Doran and Hodson (1975, p. 259)

types sum to 100%, means that only two numbers can be independently determined. A consequence is that correlations between variables will be subject to negative bias. The view in the lower diagram shows that the pattern of variation in the data is non-linear; PCA is a linear method and Aitchison (1983, 1984, 1986) has argued that its application to data that show non-linear variation, as in the figure, is inappropriate.

As an example of the effect of the negative bias, the problem addressed in Poirier and Barrandon (1983) may be cited. They examined the composition of gold coins whose purity may be degraded by the addition of ingredients other than gold. An idealised coin would be composed of 100% gold. Any addition of other elements (e.g., Ag, Pb) means that as their content goes up, the percentage of gold in a coin goes down. This leads to an inevitable negative correlation between gold and the other elements, and in their PCA analysis gold was inevitably opposed to other elements on the first component, in a manner similar to that for types A and B in Figure 3.9(b). Whether or not this is a bad thing is discussed shortly. For analyses based on the covariance matrix, Aitchison (1983, 1984, 1986) argues for transformation of the data using (4.11) before analysis, in order that PCA is capable of capturing the non-linear variation.

Baxter (1989) noted the potential for applying Aitchison's methodology to archaeological glass compositions; subsequent empirical studies have explored the practicalities involved (Baxter, 1991, 1992b; Baxter and Heyworth, 1989; Baxter et al., 1990; Heyworth et al., 1990) and produced mixed results. Examples are given in these papers where the use of (4.11) recovers useful archaeological structure not revealed by more standard

Principal Component Analysis – Specialised Topics

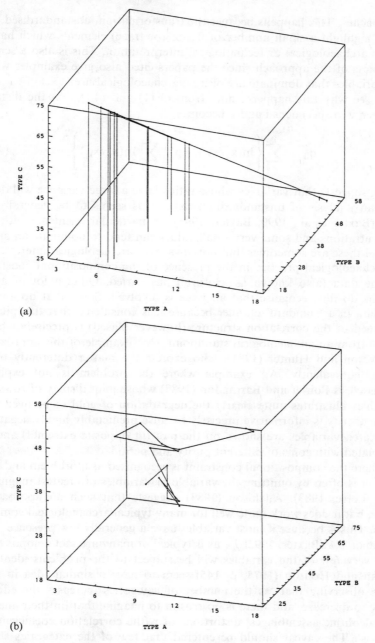

Figure 4.2: Three-dimensional plots of the data in Table 4.1

Note: In diagram (a) the angle of view is selected to show the linear relationship that exists between the variables: in (b) the view is chosen to show the 'curved' pattern of the data.

approaches. This happens because the transformation, unstandardised, will often highlight a small number of minor (or trace) elements which have a useful archaeological or technological interpretation. This is also a serious weakness of the approach since the papers cited also give examples where the variables that dominate are of little archaeological use.

To see why this happens, note from (4.11) and (4.1) that the distance between two specimens i and k becomes

$$d_{ik}^2 = \sum_{j=1}^{p} \left\{ \ln(x_{ij}/x_{kj}) - p^{-1} \sum_{j=1}^{p} \ln(x_{ij}/x_{kj}) \right\}^2 \qquad (4.12)$$

which suggests that variables whose ratios have a large variance will dominate an analysis of unstandardised data. This seems to be generally the case (Baxter et al., 1990; Baxter, 1991). In particular, elements with low concentrations and some very small values can tend to dominate an analysis and these are sometimes, but not always, of archaeological interest.

Archaeological practice in the presence of the compositional constraint on the data (also known as 'closure') has varied. Quite a lot of applications do not recognise that an issue is involved. Some that do analyse the data in a 'standard' manner because it is considered that the 'picture' obtained of the correlation structure (however biased) is precisely what is useful from an archaeological standpoint. One example of this approach is Sanderson and Hunter (1981), who express the matter differently but, I think, equivalently. An example where the problem is not explicitly discussed, is Poirier and Barrandon (1983) whose joint display of rows and variables illustrates quite clearly the degradation of gold coins over time. Their display is informative precisely because predictably highly negatively correlated variables are shown on the plot (at opposite extremes) and are associated with coins of different purity and period.

Where the compositional constraint is recognised as a problem and dealt with, it is often by omitting the variable or variables of greatest magnitude (e.g., Leese, 1983). Aitchison (1983) suggests that such an approach is naive, but it does work quite well for many typical archaeological compositions. This is because if most variables have a generally low presence – say less than 3% (Baxter, 1992b) – as is typical of many artefact compositions, then very few of the variables will be subject to the problems identified. Doran and Hodson (1975, p. 145) seem to have a similar idea in mind when observing that 'as the number of categories increases the effect is likely to decrease, and there is no reason to imagine that [in their analyses of Paleolithic assemblages] distortions of r [the correlation coefficient] are serious'. The caveat should be entered that few of the categories should contain significantly large values. Close (1977, 1978) recognises the problem and uses an arc-sin transformation (2.11) in the hope that it will help 'de-correlate' her assemblage data; in fact this particular transformation is usually used because, empirically, it is found to make such data more nearly normal than for any other reason. Gob (1988) suggests that correspondence analysis may be a more appropriate method for treating assemblage data, and this is considered in the next two chapters.

A serious practical limitation of equation (4.11) for many archaeological problems is that it is undefined for zero values of x_{ij}. This is a problem in compositional studies of artefacts if measurements are below the level of detection. Similarly, many assemblage studies contain zero elements in the data matrix. In general (4.11) is unsuitable with zero or near zero values in the data. In Section (4.13) some PCA analyses of pollen assemblages are noted and Birks and Gordon (1985) have observed that such data might be treated by Aitchison's approach. They also observe, however, that they know of no such applications and that the methodology is subject to the problems just outlined. In Chapters 5 and 6 some alternative approaches that use correspondence analysis and are not affected by zeroes are discussed.

To summarise, there are theoretical objections to the use of PCA applied to fully compositional data, but archaeological applications of standard PCA procedures nevertheless may produce useful results. It is best to show an explicit awareness of the problem, even if one chooses to use the standard approach. In the case of the analysis of chemical compositions, many of the measured elements typically have a low presence and are relatively unaffected by closure bias.

Baxter (1992b) suggests that the analysis of subsets of the data, possibly on separate scales including (4.11), may be useful; or that the full data set could be examined on different scales. Some comparison of results from different analyses is then essential and this is considered in the next section and in Section 6.8.

4.7 Comparing configurations

In Section 3.3 analyses were undertaken (a) on the same objects and variables using different scales and (b) on the same objects using different variables. A variant on the latter strategy, noted in the last section, would be to analyse the different subsets on different scales. The outcome of such analyses may be two or more component plots and it may be of interest to compare these in a formal way.

The ideas to be discussed here extend to more than two dimensions but are most simply envisaged in two. Imagine that the component plots for two analyses are plotted on transparencies. One of these is fixed; the aim then is to see how closely the second can be made to 'fit' the first given a common origin. To do this you are allowed to move the second around (rotation); turn it over (reflexion); or stretch it in a uniform manner (scale change). Some measure of the closeness of fit needs to be defined and the aim is to minimise this measure.

Sibson (1978) proposed one approach to this that encompassed earlier work of Gower (1971b). Mathematically the result is messy, requiring some use of matrix algebra. A MINITAB macro to effect computations is given in Appendix C, and an example is given below following the general result.

Let the two configurations to be compared be held in data matrices $\mathbf{X}$ and $\mathbf{Y}$ where columns are centred on their origin (for the two-dimensional case these would just be the co-ordinates of the points for the two analyses). Then the measure defined by Sibson (1978) intended to reflect the goodness of fit in a least squares sense and to be symmetric is

$$\gamma = 1 - [\{\text{tr}(\mathbf{Y'XX'Y})^{1/2}\}^2/\text{tr}(\mathbf{Y'Y})\text{tr}(\mathbf{X'X})] \qquad (4.13)$$

where $'$ is the transpose and tr( ) the trace operator (Appendix B). The coefficient varies between 0 and 1 taking the former value for a perfect fit.

In the one-dimensional case γ reduces to $(1 - r^2)$ where r is the correlation between the scores on the leading components for the two analyses. The two-dimensional case is of most interest here and will be illustrated by example.

It is not obvious what constitutes a 'good value' of γ. To get some feel for this the different analyses of Section 3.3 are compared. Four analyses are reported there in Figures 3.4–3.5 and 3.7–3.8 using the correlation and covariance matrices of the major/minor elements; the correlation matrix of the trace elements; and the correlation matrix of all elements. Table 4.2 summarises results.

From the table it can be seen that the least similar analyses, with the highest values, are those for the covariance matrix of the major/minor elements and the correlation matrix of the trace elements. The level of similarity is low. A high degree of similarity is shown between the analysis of all the data, using the correlation matrix, and the equivalent analyses using the major/minor and trace subsets. Intermediate levels of similarity are shown by the other comparisons, that between the correlation and covariance matrices of the major/minor elements arising from the similarity on the first component (Figure 3.6). These results are to be expected given the nature of the data sets analysed. A more extensive use of γ, used to compare the similarities between different methods, is given in Baxter (1991).

Table 4.2: Sibson's coefficient used to compare four analyses from Chapter 3

		Figure		
		3.5	3.7	3.8
	Figure		100γ	
	3.4	43	38	11
	3.5	*	69	51
	3.7	*	*	11

Earlier applications of these kind of ideas, based on Gower's (1971b) work, have gone under the heading of constellation analysis. Different subsets (or constellations) of variables are used to define a configuration of points using PCA, for example. The 'distance' between different configurations is measured using a coefficient that does the same job as γ and these distances are used as input in an analysis to produce a 'map' or picture in which the similarities between analyses are displayed. Published applications include Doran and Hodson (1975) and Pitts (1978b).

Another approach is implemented in the MV-ARCH package and illustrated in Wright's (1989) accompanying manual. Suppose an analysis gives rise to p components and q analyses are undertaken that need to be compared. Distances are defined between each pair of the pq components and these may be used as input to a cluster analysis, for example, in order to determine the similarities between components from different analyses.

4.8 Outliers

It has been suggested in the previous two chapters that if univariate or bivariate exploratory analysis identifies clear outliers in the data, then it may be sensible to omit such observations from a subsequent PCA. A good reason for doing this is that such observations will often predictably be identified by the PCA itself, if included, and will sometimes be the main feature of a component plot obscuring other useful structure in the data. Jolliffe (1986, p. 175) explains why such univariate or bivariate outliers are likely to be picked up by the first few components.

This is emphatically not saying that outliers are to be ignored; it is simply saying that they are so different from the bulk of a sample that no information is gained, and much may be obscured, by including them in a multivariate analysis. Heyworth et al. (1990) give an analysis of the composition of colourless Roman glass where two clear outliers, omitted from their PCA, were the most archaeologically interesting specimens in the sample.

It is possible for a multivariate outlier not to emerge from univariate inspection. In this case such outliers can sometimes be detected by plotting the last few components for reasons discussed in Jolliffe (1986, p. 176).

The foregoing discussion envisages obvious outliers being detected and omitted before a PCA. It is quite possible, given the ease of application, to use PCA as an exploratory tool to detect outliers and then to re-run an analysis omitting them. Most archaeological applications of PCA that explicitly consider outliers seem to operate in one of these two modes. An alternative to the 'subjective' use of PCA plots to identify outliers has been presented by Leese (1983) who uses probability plots based on the gamma distribution. Other more 'objective' approaches are summarised in Jolliffe (1986, Chapter 10).

While the omission of obvious outliers that will clearly affect an analysis seems sensible (so long as this is not equated with ignoring the observations) it can sometimes give rise to unease. Such unease, when manifest, seems to stem from a feeling that the data are being manipulated to get the results one wants. An interesting compromise is to omit suspected outliers from an analysis, but to include them on a component plot as supplementary points so that their relation to other data can be seen. This kind of idea, more common in correspondence analysis, has not, to the best of my knowledge, been applied in the published literature. Scores are obtained for the outliers using the components that have been calculated omitting them and this provides the co-ordinates for subsequent plotting. This is a simple idea that deserves further exploitation.

4.9 Mahalanobis distance

Earlier sections have noted that the usual PCA plot provides a two-dimensional approximation to the Euclidean distance between rows in p dimensions (4.1). In Section 4.4 it was noted that one form of biplot approximated to Mahalanobis rather than Euclidean distance and this former distance is now discussed in a little more detail. The idea is necessary for discussion in Chapters 8–10 but could be omitted on a first reading.

Before looking at the ideas mathematically, a simplistic explanation may help to understand the ideas. In Figure 4.3(a) a plot of two variables shows a reasonably high positive correlation of 0.767. This gives rise to an elliptical scatter of points which turns out to be unsuitable for the application of certain common methods of multivariate analysis (Section 8.5). Ideally the scatter should be circular, and the idea behind the use of Mahalanobis distance is that such an elliptical scatter is transformed to circularity before analysis. More generally, if data are highly correlated and normally distributed, the scatter of a set of points will form a (hyper-)ellipsoid in p-dimensional space. Imagine this to be a solid made of malleable material (such as dough), then in transforming to Mahalanobis distance we are attempting to 'squeeze' this solid into a spherical shape.

More formally let S be the covariance matrix based on an unstandardised data matrix Z, then the Mahalanobis distance between two rows, δ_{12} say, is defined by

$$\delta_{12}^2 = (z_1 - z_2)'S^{-1}(z_1 - z_2) \tag{4.14}$$

where, for the purposes of this section, z_i' represents the $(1 \times p)$ vector of values in the i'th row. This assumes that the inverse of S exists. In the same notation Euclidean distance can be defined as

$$d_{12}^2 = (z_1 - z_2)'(z_1 - z_2) \tag{4.15}.$$

Equivalently Mahalanobis distance can be regarded as the Euclidean distance between the rows of Z_1 where

$$Z_1 = ZS^{-1/2} \tag{4.16}$$

and $S^{-1/2}$ is defined in such a way that if post multiplied by its transpose S^{-1} is obtained.

If S is diagonal then the effect of (4.14) is to standardise the variables in Z. In this case, however, PCA is of limited interest since the data are uncorrelated. More generally, Mahalanobis distance gives less weight to variables with larger variances, compared to Euclidean distance, and gives less weight to groups of correlated variables (Jolliffe, 1986, p. 77). This is an important issue in the context of defining similarity between rows in a cluster analysis and is addressed in more detail in Section 8.5.

For illustration the data of Figure 4.3(a) will be examined further; they are the logged iron and aluminium concentrations from a sample of 136 specimens of Roman mortaria found in excavations at Colchester. For these data

$$S = \begin{matrix} .0191 & .0188 \\ .0188 & .0315 \end{matrix} \qquad S^{-1} = \begin{matrix} 126.86 & -75.83 \\ -75.83 & 77.08 \end{matrix}$$

and $$S^{-1/2} = \begin{matrix} 11.26 & 0 \\ -6.73 & 5.64 \end{matrix}$$

Post-multiplying the data matrix Z by $S^{-1/2}$ gives new co-ordinates such that the transformed variables are uncorrelated; the scatter is approximately spherical; and the bulk of the points lie within two standard deviations of the group centroid. This is illustrated in Figure 4.3(b), which shows the transformed data of Figure 4.3(a). The data in Figure 4.3(b) are more suited to some kinds of statistical analysis than those of Figure 4.3(a).

Figure 4.3(a) shows a single group from which S may reasonably be calculated. If more than one group exists in the data, then S should be calculated as a weighted average of the separate covariance matrices for each group rather than for the data as a whole; this requires, ideally, that the groups be of similar shape. This limits the application of Mahalanobis distance in practice since the detection of groups within the data is often the object of the exercise – they are not known in advance to enable S to be calculated. The assumption that groups, even if known, are of similar shape is also often questionable (Chapters 9, 10).

Thus, although one form of biplot discussed in Section 4.4 implicitly uses (4.14), there seems no real reason for preferring it to (4.15) because it treats all the data as a single group and is not really appropriate with groups in the data. If groups are known *a priori*, then other techniques such as discriminant analysis (Chapters 9, 10) are more suitable. It will be seen in the chapters discussing this method that Mahalanobis distance then comes into its own.

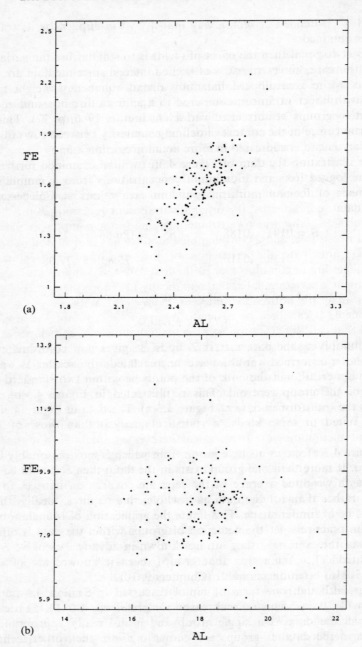

Figure 4.3: Plot to illustrate the effect of transforming to Mahalanobis distance

Note: Diagram (a) shows a plot of the two variables on the original scale; in (b) the same variables are plotted after transforming to Mahalanobis distance.

4.10 Rotation of principal components

Principal component analysis begins with a set of p correlated variables and the methodology transforms these to an uncorrelated set of p new variables. The usual hope is that only a small number of these new variables are 'important' in some sense, so that appreciation of structure in the data can be more readily examined by inspection of these new variables. The new components are mathematical constructs and there is no reason why they should have a physical or archaeological meaning to be useful. Where interesting structure is revealed in the data then the reasons will often be evident either from a biplot, from the use of simple methods such as the calculation of variable means within groups, or from the plotting of component scores against the original variables or functions of them.

Nevertheless, the urge to attribute meaning to the components has often proved irresistible, and a considerable literature has grown around the subject. To anticipate the discussion of this and the next section, two approaches are in common use. In the first case the components, or a subset of them, are manipulated mathematically to reduce the number of coefficients that are noticeably different from zero. With fewer 'non-zero' components it can then be easier to interpret the 'meaning' of a component. I will call this 'rotation of principal components' and clarify what is meant by this in this section. In the second case it is assumed from the outset that the observed variables are determined by a much smaller number of 'fundamental' or 'latent' variables and the objective of an analysis is to identify and interpret these latent variables or 'factors'. This is the aim of 'factor analysis', discussed in the next section.

The two approaches are, in my view, conceptually distinct and call for different methodologies. In practice there is considerable terminological and conceptual confusion between PCA, the subsequent rotation of components, and factor analysis. An attempt to describe the issues involved is made in this and the next section.

For illustrative discussion attention is confined to the case where a two-dimensional representation of the data is satisfactory. Taking the variable or h-plot of Figure 4.1(b) as an example, imagine the two axes on the plot to be separate and rigid lines, attached to the graph by a drawing pin (or similar device) at the origin about which they are allowed to pivot. The main idea in rotating principal components is that the axes are actually rotated about the pivot (origin) with the aim that the coefficients of variables on an axis become either 'large' or 'small'. Thus, the rotated axes will be characterised by variables with high (positive or negative) coefficients and with coefficients on other variables close to zero. If the number of variables with high coefficients is small, so much the better. This is called 'simple structure'. The idea is that the new axes (or rotated components) will be easier to interpret because fewer variables are involved in their definition. Often the effect of rotation will be to cause the new axes to pass through clusters of points evident on an h-plot.

Two main classes of method can be distinguished; either the new rotated axes remain at right angles, in which case the rotation is an 'orthogonal' one, or else the rotation is oblique. Mathematically it is possible to define a rotation in many different ways according to the precise criterion used to define simple structure. These will not be discussed in any further detail except to note that the rotation most commonly applied in practice – Kaiser's (1958) varimax criterion – is orthogonal and tends to force loadings to be large or close to zero.

Opinions on the usefulness of rotation vary. Protagonists claim that it enhances interpretability of an analysis and can assist in the (informal) testing of archaeological hypotheses (Wright, 1989). The idea here is that unrotated components may contain a mix of variables from different variable clusters having different archaeological interpretations, whereas rotation will help identify different, and interpretable, clusters. Antagonists of rotation are worried by the fact that the choice of method of rotation is arbitrary, with different methods giving different results, and interpretation is highly subjective. The fear is that, with sufficient experimentation and ingenuity, anything can be proved. My personal view, for which a supporting example is given in the next section, is that when a two- or three-dimensional representation of the data is reasonable, inspection of h-plots is often sufficient for interpretation without the need for rotation.

Statisticians tend to react to the ideas of component rotation and factor analysis (with which it is confused) with less than enthusiasm. The bad odour that factor analysis has in some quarters is one reason; cynics might say that another reason is because factor analysis has largely been developed and found useful by non-statisticians. Certainly opinions on the subject seem polarised between different disciplines. A constructive, though not uncritical, view is taken by Jolliffe (1989) who notes among other things that (a) rotation destroys the optimality property of principal components that successively account for the maximum variance possible, subject to their zero correlation; (b) there are at least twenty methods of rotation that have been proposed in the literature and results will depend on the one chosen; and (c) that components can be normalised in different ways and the results of rotation will depend on the normalisation used. The differences can be substantial and the normalisations used in some packages such as SPSS-X may often represent the least satisfactory choice (Jolliffe, pers. comm.). Rotation of principal components is common in practice, under the heading of factor analysis, because of its availability in packages such as SPSS-X. Usually the r most important components are selected and rotated. Jolliffe's fourth observation is that (d) the rotated components may well differ according to the value of r that is chosen.

The implication of this discussion is that rotation of principal components leads to very indeterminate results in terms of the exact numerical values obtained, given the dependence on the choice of method of rotation, number of components used and normalisation. Interpretation of a rotated component is also subjective. Statisticians have quailed at the prospect of using the procedures available in the light of such considerations. What has not been studied in detail in an archaeological context (to the best of my knowledge) is the extent to which the theoretically numerous options would lead to similar substantive conclusions in practice. This is worthy of further study.

I suspect – without any proof whatsoever – that not all of Jolliffe's four points, in particular (c) and (d), are widely appreciated among archaeological users. Jolliffe's (1989) proposal to alleviate some of the indeterminacies noted is to rotate only those components associated with eigenvalues that are of similar magnitude. The paper, which is the outcome of research that is continuing, can be consulted for the details.

It must be emphasised, in view of what will be said in the next section, that the rotation of principal components, despite the indeterminacy involved, is a purely mathematical operation that some users find helpful in interpreting their data. No model for the data is assumed. Many archaeological studies have used what is essentially PCA with rotation but have mistakenly called it 'factor analysis', which is a conceptually different technique discussed in the next section. Consequently, conscious uses of PCA with rotation (that do not call it factor analysis) are hard to locate. An exception is Knapp et al. (1988) who analyse the correlation matrix of a set of elemental concentrations of Jordanian Bronze Age pottery and modern clays, and extract four components for subsequent rotation to simple structure. This article, Wright (1989), and texts such as Lawley and Maxwell (1971) can be consulted for comments on and technical aspects of the choice of method of rotation not covered here.

4.11 Principal components and factor analysis

Factor analysis is a subject about which many statisticians have mixed feelings. Chatfield and Collins' (1980, p. 89) conclusion that 'we recommend that factor analysis should not be used in most practical situations' is not untypical. In this section, rather than reviewing archaeological uses of factor analysis in detail, an attempt will be made to explain how it differs from, but is often confused with, PCA, and why it has a bad reputation among statisticians. If used properly, it is distinguished from PCA with rotation by the explicit assumption that the observed variables are determined by a smaller set of unobserved or latent variables – the factors – that the analysis aims to identify.

In writing about factor analysis in archaeology, discussion is made difficult by the considerable terminological variety and confusion that exists. This could be regarded as reflecting widespread conceptual confusion. In this section I will try to distinguish between PCA, PCA with rotation, and proper factor analysis in a consistent fashion. A majority of reported 'factor analyses' in archaeology are really PCA with rotation (Vierra and Carlson, 1981, Section 13) but the term is also sometimes used for unrotated PCA. Some authors use 'factor analysis' as a generic term for all three techniques, and the position is further complicated by it use to describe what is in fact a correspondence analysis (Chapters 5, 6). I have also seen the term 'PCA' used to describe what was a factor analysis.

Some authors, who are clearly aware of the distinctions between the different models, seem to regard 'factor analysis' as a means of enhancing interpretability of principal components. The view taken here, and expounded in the rest of this section, is that factor analysis and PCA should be treated as conceptually and technically distinct methods. Factor analysis requires an explicit model for the data; PCA does not. Factor analysis, correctly executed, should be scale invariant; PCA is not.

In archaeology an influential paper in promoting the use of 'factor analysis' was that of Binford and Binford (1966) who use the technique to identify 'tool-kits' occuring in Mousterian assemblages. Their method is, in fact, PCA with rotation. By 1975 Doran and Hodson (1975, p. 197) describe factor analysis as 'one of the most frequently used multivariate methods in archaeology' and then go on to give an unfavourable assessment of the technique. Cowgill (1977), in a generally favourable review of Doran and Hodson, takes them to task for their attitude to factor analysis. His view is that for practical purposes factor analysis and PCA are very similar, with the former a more interpretable method.

Cowgill's review appeared in *American Antiquity*; several papers from about that period in the same journal looked critically at the misuse of statistics, including multivariate methods, in archaeology (Christenson and Read, 1977; Thomas, 1978; Vierra and Carlson, 1981; Benfer and Benfer, 1981; Braun, 1981; Scheps, 1982). The effect of this onslaught seems to have removed multivariate methods from the journal for a few years; subsequent sightings do not include the use of factor analysis. This backlash against the excesses of a 'quantitative revolution' and attendant dangers has already been discussed in Section 1.4. Factor analysis is one particular baby that has possibly been lost with the bathwater, but the loss is not necessarily one to mourn (see also Orton (1988)). Shennan (1988) devotes about nine pages to the topic, half of which is occupied by detailed discussion of an archaeological application due to Bettinger (1979).

Why should factor analysis arouse strong feelings? It has already been noted that a factor analysis, properly constituted, should involve an explicit model for the data. In practice this model is often left implicit, and there is sometimes the suspicion that users do not realise that a specific model is implied. If a factor analysis model is appropriate, then an appropriate method of estimation (in the eyes of many statisticians), which is invariant to the scale of measurement, is that of maximum likelihood. Once factors are determined, they are usually rotated to enhance interpretation, and an element of arbitrariness in the choice of method of rotation then enters. In practice principal components, which are scale-dependent, are often extracted and rotated, thus combining the worst of both worlds. Either a factor model is appropriate, in which case appropriate methodology should then be used, or it is not, in which case PCA (possibly with rotation) should be used and so called. The widespread confusion between the two techniques betrays a lack of understanding of both and is not merely semantic.

We now examine the methodology in a little more detail to try and make the differences clear. In equations (3.1) and (3.2) the components P_1, P_2 etc. are defined in terms of known variables Y_1, Y_2 etc. and can be written as

$$P_i = \alpha_{i1}Y_1 + \alpha_{i2}Y_2 + \ldots + \alpha_{ip}Y_p \qquad (4.17)$$

for $i = 1, \ldots, p$. Determining the coefficients, α_{ij}, is a simple mathematical problem given the criteria to be satisfied by the components. As a consequence of the mathematics, the variables can be expressed in terms of the components as

$$Y_i = \alpha_{1i}P_1 + \alpha_{2i}P_2 + \ldots + \alpha_{pi}P_p \qquad (4.18).$$

Factor analysis, as usually understood, requires a superficially similar *statistical* model for the data of the form

$$Y_i = \alpha_{1i}F_1 + \alpha_{2i}F_2 + \ldots + \alpha_{ki}F_k + \epsilon_i \qquad (4.19).$$

The F_i represent k (less than p) unknown common factors (or hypothetical or latent variables); the ϵ_i are error terms or specific factors. The fundamental idea is that the observed p variables can all be explained in terms of a (much) smaller number of k unobserved factors or variables in addition to the specific factor; the factors themselves are weighted by the factor loadings, α_{ij}.

An obvious attraction of the factor model is that it holds out the promise of being able to explain complex observed data in terms of a much smaller number of theoretically more fundamental variables or constructs. Unless there are good reasons for believing in the reality of these constructs (in some sense), it is arguable that the factor model should not be used at all.

The differences between equations (4.18) and (4.19) include:

(i) there are fewer factors, F_i, in (4.19) than there are components, P_i, in (4.18) and usually k is expected to be much less than p;

(ii) the factors, as well as the loadings, in (4.19) are unknown and must be estimated;

(iii) equation (4.19) is a statistical model for the data that is dependent, in part, upon the errors which are assumed to be independent but with differing, and unknown variances, σ_i^2;

(iv) estimates are required of k, the factors, the loadings and the error variance that will depend on the assumptions made about the errors, the factors and the relationship between them;

(v) estimates in the factor model are not always immediately interpretable and are usually rotated to get a final solution;

(vi) PCA attempts to account for the variance in the data (e.g.; the diagonal elements of the covariance or correlation matrix); although it will often do a good job of accounting for the covariances as well (Jolliffe, 1986, pp. 11–12), this is not its main purpose. Factor analysis, by contrast, represents an explicit attempt to model the covariances.

Estimation is complex and can be carried out in a variety of ways, as can rotation. Krzanowski (1988, pp. 477–503) can be consulted for an introductory account. Here some of the difficulties with the method that have caused statisticians to be wary of its use are noted.

(a) Historically, factor analysis was developed and used in the social sciences, particularly by psychologists. A great many 'unreliable' methods for determining the α_{ij} and σ_i with 'no mathematical basis' were developed leading to a legacy of 'continuing mistrust among some statisticians' (Krzanowski, 1988, p. 487; Jolliffe, 1986, p. 119; Seber, 1984, p. 222). The statistically 'respectable' method of maximum likelihood estimation (Lawley and Maxwell, 1971) may often be preferable, but this makes strong distributional assumptions about the errors, such as multivariate normality, and has had limited use in archaeological applications.

(b) Some restrictions need to be placed on the values of α_{ij} and σ_i to obtain a solution at all. In general, and unlike PCA, a large number of assumptions need to be made in setting up the factor model. These assumptions include the existence of the factors (Chatfield and Collins, 1980, p. 88).

(c) It is rarely obvious what an appropriate value of k is. Different choices can result in rather different factors being obtained. Similarly, different methods of rotation – of which there are many – may lead to different results. In combination the choice of k and method of rotation lead to many possible end results so that the final choice, for publication, may be highly subjective and difficult to replicate (Chatfield and Collins, 1980, p. 89). Seber (1984, p. 235) suggests that 'unless the true model has a "good" structure ... orthogonal rotation may be worse than useless if the wrong value of k is used' and that 'even if the postulated model is true ... the chance of its recovery by present methods does not seem very great'.

(d) If the maximum likelihood method of estimation is used then the results, unlike PCA, do not depend on whether the data are standardised or not. However, a particularly common method of doing factor analysis in archaeology – 'principal component factor analysis' – is affected by standardisation (see below).

Some of these comments reflect an inherent distrust of the reality of the factor model in most practical situations that is quite open to dispute by subject specialists. Other difficulties arise from the essential indeterminacy of the factor model itself, requiring a set of more or less arbitrary assumptions in order to obtain results. Yet other criticisms might be regarded as related to 'bad practice' rather than the factor model; however, for historical reasons and because of what is available in general purpose statistical packages, such 'bad practice' is common.

In their study of the uses of factor analysis in archaeology Vierra and Carlson (1981) demonstrate that factor analysis (and PCA) can produce patterned and interpretable results even if the data used are randomly generated. This is an illustration of the fact that multivariate methods can impose structure on data. They additionally list details of forty-three examples of factor analysis in anthropology and archaeology up to 1977 from thirty-seven publications. Three of these studies seem to be PCA rather than factor analysis; of the remainder, where the method was identified, 21 out of 37 use 'principal component factor analysis' and 29 out of 36 use 'varimax' rotation including all twenty examples of 'principal component factor analysis' where a method of rotation is given. This combination is the default for 'factor analysis' in the SPSS-X package if rotation is to be attempted (which may help explain its popularity), and it is worth examining this approach in a little more detail.

If the last $(p - k)$ terms in the PCA model (4.18) are lumped together to form an 'error' term, then (4.18) and (4.19) are superficially identical. This motivates the use of PCA to obtain initial estimates of the factors, but ignores the fact that the 'errors' in (4.18) are correlated, unlike (4.19), and in Jolliffe's (1986, p. 122) view has no strong theoretical justification. Since the results of PCA depend on whether or not the data are standardised, factor analysis undertaken in this way, in contrast to the maximum likelihood approach, will be similarly affected.

Two approaches seem to dominate practice. In the first case the specific factor is effectively ignored; a standard PCA is carried out; some value of k is selected; and k components are then rotated to simple structure to assist interpretation. In the second case the contributions of the specific factors are estimated; the diagonal elements of the covariance or correlation matrix of the data are adjusted to account for this; PCA is applied to the adjusted matrix, and this forms the starting point for an iterative procedure from which all the unknown quantities are estimated (see Krzanowski, 1988, pp. 487–88, for details).

Both methods add the scale dependence problem of PCA to factor analysis; the former method ignores an important aspect of the factor model and is just PCA with rotation, not factor analysis. Krzanowski (1988,

p. 488) suggests that 'this method of FA is probably at the root of the considerable confusion between FA and PCA which is exhibited in many disciplines. The one (and probably only) point in favour of the technique is that it does not require any distributional assumptions to be made about the data.'

Methods of rotation were briefly discussed in the previous section. One important point concerning 'principal component factor analysis' and 'varimax' rotation made by Wright (1989) will be reiterated here. In a PCA analysis – unless steps are taken to avoid it – the first component will often have similar loadings on all variables and be readily interpretable as a 'size' component. If this is rotated using the varimax criterion then this pattern may, in the nature of the method, be destroyed, and a pattern of high and low loadings can emerge. In other words a common approach to factor analysis in archaeology is quite capable of destroying an obvious interpretation of the kind that is customarily sought.

In summary, factor analysis is a demanding technique from the point of view of the assumptions that need to be made to get results that are, nevertheless, still open to the accusation that they are arbitrary. For substantive interpretation, either it matters whether factor analysis or PCA is used, or it doesn't. In the former case the model and assumptions used should be made explicit and a good method of estimation chosen. This rarely happens in much of the relevant literature. In the latter case, which some scholars (e.g., Wright, 1989) believe is the rule, PCA is the better understood method and should be used under its own name with or without rotation.

One final point (prejudice?) is that often little is done with a 'factor' other than interpret it. Where a low-dimensional representation of the data is reasonable, such an interpretation may often be possible from a variable (or h-) plot following a PCA. An excellent, if unwitting, illustration of this is provided by Hoffman's (1985, p. 597) analysis of projectile point technology using eight blade-size and four edge-angle variables. It is evident from an h-plot, without rotation, that these form two tightly clustered groups of variables with negative correlation between the clusters. An interpretation in terms of 'factors' is quite possible at this point and does not require the oblique rotation, in which the new axes pass through the clusters, used by Hoffman.

4.12 An example

Data set A4 in Appendix A is a subset of that given in Lukesh and Howe (1978). The measurements relate to the rim, neck and shoulder diameters, height and neck height of sixty cups dated to the Middle Bronze Age from the Apennine culture of the central and southern peninsula of Italy. The cups are dated to an early or late phase of the culture and this is also shown in the table. Lukesh and Howe's concern was to establish whether or not the measurements could be used as a basis for dating assemblages

Table 4.3: Correlation matrix for the data in Table A4

	RD	ND	S	H	NH
RD		.99	.98	.72	.56
ND	.99		.99	.73	.52
S	.98	.99		.77	.56
H	.72	.73	.77		.78
NH	.56	.52	.56	.78	

Key: RD = Rim diameter; ND = Neck diameter; S = Shoulder; H = Height; NH = Neck height.

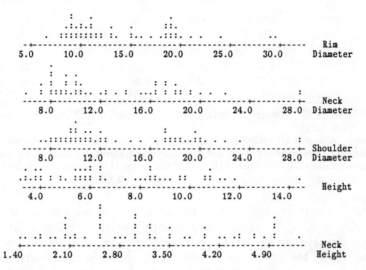

Figure 4.4: Dot-plots of Bronze Age cup dimensions from Table A4

Note: The data are a subset adapted from Lukesh and Howe (1978).

of such cups; here the dating information is ignored and the measurements alone will be used to illustrate various aspects of PCA.

The correlation matrix for the data is shown in Table 4.3; it is obvious that the measurements are highly intercorrelated and, in particular, that two of the first three variables are probably redundant. Dot-plots for the five variables (Figure 4.4) suggest that there are two main groups, which are clearly related to size as measured by the diameters, and two outliers that are particularly large cups. There is no obvious need for data transformation. The standard deviations of the first three variables are in the range 5.1–5.3 cm, for height it is 2.6 cm, and for neck height, 1.1 cm.

With so few, and highly correlated, variables, using a PCA in anger is like using a sledgehammer to crack a walnut, but some useful illustrative points may be made. To begin with an analysis of the covariance matrix, it can be anticipated from the discussion in Section 4.2 and the dot-plots in

Table 4.4: Results of different PCAs of the data in Table A4

Component	Method	Variable				
		RD	ND	S	H	NH
1	A	.57	.55	.56	.22	.07
	B	.47	.47	.48	.44	.36
2	A	−.22	−.22	.01	.89	.35
	B	.31	.34	.28	−.39	−.75
		ND/RD	S/RD	H/RD	NH/RD	
1	C1	.06	.45	.69	.57	
	C2	.08	.61	.93	.76	
	D	−.27	.32	.87	.92	
2	C1	−.73	−.51	.03	.46	
	C2	−.93	−.66	.04	.58	
	D	−.90	−.84	−.31	.26	

Key: RD = Rim diameter; ND = Neck diameter; S = Shoulder; H = Height; NH = Neck height.
Methods: A and B are analyses of untransformed data using the covariance and correlation matrices: other analyses use ratio transformations. C1 and C2 are analyses of the correlation matrix normalised so that the squared coefficients sum to 1 and the corresponding eigenvalue; D is a varimax rotation of C2.

Figure 4.4 that the first three variables will dominate the PCA and that a PCA plot is likely to suggest two groups and two outliers. That this is indeed the case can be seen from the relevant part of Table 4.4 and Figure 4.5(a), where the first three variables have the largest, and similar coefficients. This component could be interpreted as a 'diametric size' component and accounts for 95% of the variance in the data. The second component is dominated by height, so that the resulting component plot in Figure 4.5(a) has a simple interpretation as a height against diametric size plot (and a similar plot could, of course, have been obtained by plotting height against rim diameter). The plot suggests the expected two groups and two outliers.

A similar-looking and even clearer picture is obtained on analysing the correlation matrix of the data in Figure 4.5(b), with group separation on the first axis particularly evident. On looking at the coefficients for the first axis, in Table 4.4, it is clear from their similar size and signs that an interpretation as a general size component is possible. The second component opposes diametric to height measurements, although total height is particularly dominant and can be interpreted as a shape component. In particular it can be shown that scores on the second component correlate highly ($r = 0.72$) with the rim diameter/height ratio. The first component accounts for 81% of the variation and the first two for 96%.

Principal Component Analysis – Specialised Topics

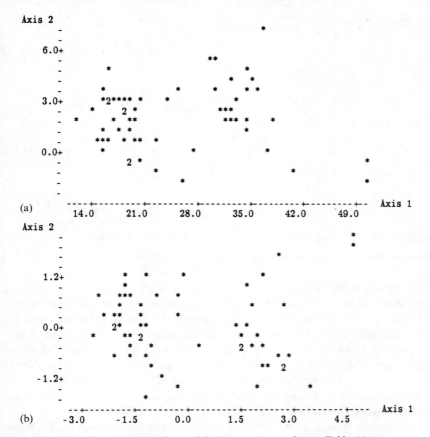

Figure 4.5: Principal component analyses of the Bronze Age cup data in Table A4

Note: (a) is based on an analysis of the covariance matrix of the data, (b) on the correlation matrix.

A common way to eliminate size effects, when they are considered undesirable, is to divide (p-1) of the variables by the p'th chosen to be indicative of general size. Using rim diameter as the divisor thus creates four new variables. Dot-plots of these (not shown) show nothing untoward, and in particular there is no need for further transformation. Undertaking a PCA of the correlation matrix gives the results of Table 4.4 and Figure 4.6. The first two components are of almost equal importance (45% and 41%) and together account for 86% of the variation. The first component tends to be dominated by those ratios involving dimensions of height; the second by ratios involving dimensions of diameters. The interpretation is not, however, clear-cut and will be contrasted with a rotated solution shortly. The figure, in contrast to Figures 4.5(a, b) shows no evidence of grouping. Using the statistic in equation (4.13) to compare plots shows

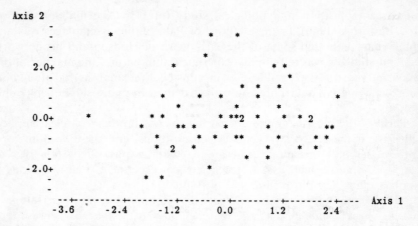

Figure 4.6: Principal component analysis of transformed data from Table A4

Note: The analysis is based on the correlation matrix of the ratios of variables to rim diameter.

that Figures 4.5(a) and 4.5(b) are very similar with $\gamma = 0.09$, and both are very different from Figure 4.6 for which the values of γ are 0.86 and 0.89.

Although not shown on any of the figures, there is no evidence at all for any clear distinction between early and late cups. What is the case is that in Figures 4.5(a, b) most of the early cups were contained in the group associated with small sizes, so that on average they tend to be smaller (a point made, using simpler methods, in Lukesh and Howe (1978)). The general conclusion, taking all analyses into account, is that the main differences between the cups are ones of size rather than shape.

The results reported so far were obtained using the MINITAB package and use the normalisation that the squares of the coefficients of a component sum to 1 (Section 4.4). If the coefficients are normalised such that their squares sum to the associated eigenvalue, then the loadings reported in other packages, such as the SPSS-X FACTOR routine are obtained – these are shown for the analysis of ratios in Table 4.4. They may be contrasted with the results obtained on rotating the first two components (with eigenvalues bigger than 1) using the varimax criterion (the default option in SPSS-X), which are also shown (Sections 4.10, 4.11). The main difference is that the two components are now more clearly associated with ratios involving heights, and ratios involving diameters respectively, so that interpretation is possibly easier (though see Section 4.10 about the dependency of results on the normalisation used).

4.13 Uses of PCA in archaeology

Hodson's (1969) study of the typology of Palaeolithic assemblages is one of the earlier archaeological studies to exploit PCA, while Hodson (1970) is an early application to typological study (of Iron Age fibulae). Binford and Binford's (1966) 'factor analysis' of Palaeolithic assemblages was also influential. Although some of the early work of Hodson and his co-workers and students was admittedly experimental in nature, the manner of display and use of the results (e.g. in constellation analysis) is often both more imaginative and more sophisticated than in many subsequent publications.

Perhaps the most common current use of PCA in archaeological applications is for the analysis of artefact compositions, and these account for a little under half the applications in the bibliography. Shaw (1969) is an early example of such use. A recent critical account of considerations in application is given in Bishop and Neff (1989).

Many of the references to artefact compositional analysis are routine in nature, either using PCA as the main method of analysis or, more commonly, in conjunction with some other multivariate method such as cluster analysis to obtain a more comprehensible picture of groups within the data. Usually analyses based on the leading two components are presented, though three-dimensional displays are attempted in Hughes (1991) and Rauret et al. (1989). Sometimes the higher-order components prove informative; for example, Williams-Thorpe et al. (1984), in their study of archaeological obsidian, discern structure in a plot of the first against fourth components. This last paper provides an example where a knowledge of the 'analytical groups' in the data is probably needed in order to identify them on the plots (i.e., groups are neither disjoint nor obvious), and this is not uncommon.

Analyses of standardised data, possibly after transformation, dominate applications, though the justification for this has been questioned (Bishop and Neff (1989)). Where the data have not been standardised, some form of transformation is usually adopted to avoid the analysis being dominated by a small number of variables. Analyses of this kind, mostly concerned with the effects of tempering on composition, include D'Altroy and Bishop (1990); Leese et al. (1989); Neff, Bishop and Arnold (1988, 1990), and Stopford et al. (1991). Baxter (1989, 1991) and Baxter and Heyworth (1989) compare the results from analyses of both standardised and unstandardised data on glass compositions using a variety of transformations.

D'Altroy and Bishop transform their data logarithmically (to base 10) and, if transformation is used at all, this is the most common approach. Calamioutou et al. (1984) provide an example where only a subset of the variables are so transformed. Usually the decision to transform appears to be based on general principles rather than on data inspection. The studies co-authored by Baxter or Leese, noted in the previous paragraph, use Aitchison's log-centred transformation which has not, otherwise, been much used. The obsidian study of Green and Bird (1989) uses ratios of variables.

Where the presence of outliers is an explicit consideration they are usually handled on a common-sense basis. Thus Cabral et al. (1983) identify five outliers by preliminary data analysis and remove them prior to the PCA. Other papers that explicitly deal with outliers in advance of the PCA include Hughes and Hall (1979). PCA can also be used, subjectively, for outlier detection; a more objective procedure, contrasted with PCA, is given by Leese (1981) but has been little used.

When used in conjunction with cluster analysis, PCA has been employed in a variety of ways. Most usually, 'analytical' groups are obtained using cluster analysis and these are then displayed in two dimensions using a component plot. Of the numerous examples that could be cited, that by Sheridan (1989) is particularly effective. The dendrograms from her cluster analyses are displayed directly above the associated component plots on a scale that allows the implied clusters to be 'read-off' on the plot. This kind of use employs PCA after a cluster analysis. It is also possible to use PCA before a cluster analysis to define new and uncorrelated variables, a subset of which are then used as input to a cluster analysis (see Section 8.5 for a more detailed discussion). Examples include Barrett et al. (1978); Goad and Noakes (1978); Hart et al. (1987) and Storey et al. (1989). An unusual application occurs in Luedtke's (1978) trace element analyses of chert artefacts where principal component scores are used in a regression analysis to try and identify spatial and temporal correlates of sources.

For display purposes it has already been noted that a plot of the first against second component is usual. Biplots or similar visual devices have been relatively neglected (Baxter, 1992a), with exceptions including Berthoud et al. (1979); Mello et al. (1988), and Poirier and Barrandon (1983). Baxter and Heyworth (1991) use h-plots to study differences between chemically distinct groups of glass. For interpretive purposes the use of component rotation or factor analysis is rare. Knapp (1989) and Knapp et al. (1988) provide examples of the former use.

In typological study biplots and h-plots have received wider use, and examples include Hinout (1984, 1985, 1991), Larsen (1988), O'Hare (1990) and Speiser (1989). Hinout's studies of flint artefacts, in addition to metric variables, uses angles of retouch, and such use of mixed variables is quite common in typological study. Angular as well as metric variables occur in Hodson's (1970) study of Iron Age fibulae, for example. Chappell's (1987) study of Neolithic stone-axe morphology uses both metric variables and their ratios, as do Alvey and Laxton's studies (1974, 1977, 1979) of clay pipes. Green's (1980) analyses of flint arrowheads, and O'Hare's (1990) of polished stone artefacts from prehistoric Italy, use weight as well as metric variables. Van Peer (1991) uses presence/absence data, counts of features and metric variables in his analysis of Palaeolithic flakes. Hodder and Lane's (1982) study of Neolithic axe distribution uses, in addition to metric variables, an angular measure and the distance of an axe from its source. A rather different application that uses mixed variables is Fraser's (1983) study of Neolithic chambered cairns which are characterised by variables such as the height above sea level, distance from the coast, and areas of different soil types within a one kilometre radius. Law (1984) uses counts of features (notches) on Maori fish-hooks and transforms these and two of his metric variables logarithmically before analysis.

All the papers cited in the last paragraph, of necessity, base the PCA on the correlation matrix. Although it is no longer essential to do so, studies based solely on metric measures also tend to use the correlation matrix. These include Habgood (1986: aboriginal crania); Impey and Pollard (1985: modern pottery); Larsen (1988: biconical Late Bronze Age urns); Madsen (1988b: Neolithic funnel beakers); Richards (1982, 1987: Anglo-Saxon pottery); Shennan and Wilcock (1975: German bell beakers); and Siromoney et al. (1980: South Indian sculptures).

Most of these studies, and several of those using mixed variables, have to contend with a 'size' effect. Such an effect, associated with the first component, is a main feature of analyses in Chappell (1987); Green (1980); Hodder and Lane (1982); Richards (1982, 1987); Siromoney et al. (1980) and van Peer (1991).

Responses to the discovery or anticipation of a size effect vary. Richards, for example, looks at components other than the first and interprets them in terms of shape factors, whereas Siromoney et al. divide their measures by the facial length of the statue to try and eliminate the effect and then re-do the analysis.

In other cases, where a size effect is anticipated but not of interest, avoiding action is taken prior to analysis. Habgood (1986) undertakes a form of double standardisation, devised by Wright, to remove such effects. (Wright's (pers. comm.) current preference is to avoid such complex pre-treatment of the data and to look for shape factors among components other than the first – assuming this to be a size factor.) Larsen (1988); Madsen (1988b) and Shennan and Wilcock (1975) use ratios of variables.

Turning now to assemblage studies, an early and influential paper using factor analysis (actually PCA with rotation) was that of Binford and Binford (1966). Hodson (1969) and Azoury and Hodson (1973) are early studies that use PCA without rotation. These studies of counts of tool types in Palaeolithic assemblages are characteristic. Hodson (1969) varies the size of plotting symbol to represent scores on the third component. Azoury and Hodson (1973) discuss the choice between the use of data expressed in percentage form or of standardised data, and compare the two. The study is one of the few that attempts comparisons of configurations produced by different analyses, using constellation analysis as the analytic tool.

The influence of Hodson is evident in the work of Pitts (1978a, b; 1979) and Pitts and Jacobi (1979). This last paper and the two 1978 papers report analyses in which flakes are assigned to one of a small number of classes according to their breadth to length ratio (B/L) and assemblages are then characterised by the percentage of flakes in each B/L class. The data in this form are clearly compositional in the sense discussed in Section 4.6. Pitts (1978b) includes analyses of both standardised and unstandardised percentage data; uses plotting devices to represent scores on both the third and fourth components; and uses constellation analysis to compare analyses based on different type-lists of tools. Pitts (1979) also uses compositional data of percentages of tool types within an assemblage, and bases analysis on the covariance matrix. The data used in this paper are given in Table A3 and are discussed in Chapters 2, 3, 5 and 6.

It is common for PCAs of assemblage material to use the data in percentage form that is compositional in the sense of Section 10.4. Other examples include Callow and Webb (1981); Close (1977, 1978) and Gowlett (1986, 1988). Gowlett notes that his analyses were inspired by Isaac's (1977) analysis of Acheulean assemblages which in turn used data given in Binford (1972). Isaac's PCA was undertaken in order to compare it with Binford's factor analysis.

That the compositional nature of the data can pose problems is not always explicitly noted. Callow and Webb (1981) and Close (1977, 1978) use an arc-sin transformation in an attempt to 'de-correlate' the data (Close, 1977, p. 57). Whether or not the percentage data should subsequently be standardised is another issue that does not always receive discussion. Standardisation gives rare and common types equal weight in an analysis; if this is undesirable, then the argument is that the covariance matrix of the (possibly transformed) percentage data should be used.

It is not obvious that PCA as described in this and the last chapter is the ideal method for analysing tool assemblages. Binford (1987, 1989) sees the issue as one of appropriately 'normalising' the data and suggests that this was unsatisfactory in many early applications. In Binford (1987) he states that 'the problem of normalization was essentially solved by using a chi-square calculation. That is chi-square values were calculated for each raw cell count in a matrix. These values were then used as the data for analysis using multivariate techniques.' This approach, which subsequently involves analysis via the singular value decomposition, is described as 'somewhat new and has not been widely discussed in the analytical literature' in Binford (1989). Tantalizingly, the mathematical details are not given, but the description looks very much like that of correspondence analysis given in Section 5.3, and further discussion is provided there.

Binford describes his approach as PCA of normalised data; a form of non-linear PCA would be an alternative description and correspondence analysis (Chapters 5, 6) might also be so described. Gob (1988) compares correspondence analysis favourably to the usual PCA approach and Ringrose's (1988a, b) analyses of faunal assemblages also contrast both approaches.

Callow (1986a, b) and Hutcheson and Callow (1986) characterise flint assemblages by variables such as the mean length and weight of flakes; percentage of types present and ratios of types. Van Peer (1991), in addition to his typological analysis, also treats assemblages in a similar way. Callow's studies are interesting in that they contrast the results of PCA with that of a varimax rotated solution using a three-dimensional (stereoscopic) presentation.

A few of the studies listed in the bibliography are of environmental data. Fall et al. (1981) characterise pollen samples by the percentage of pollen types, subsequently standardised, so that the data is compositional. They claim (p. 300) that their study is the first to conduct such an analysis in an archaeological context. Other palynological applications include Marguerie and Walter (1986) and Prosch-Danielsen and Simonsen (1988), the latter also using charcoal and soil phosphate data.

The two spatial studies listed by Cribb and Minnegal (1989) and Whallon (1984) might equally well have been categorised as assemblage studies. Assemblage data, at a point or within a grid, are used as the basis for a PCA. Scores can then be plotted on the map of a site and used as a basis for contouring to reveal spatial patterns. Whallon presents his analysis primarily to make the point that the results are unsatisfactory in comparison to other techniques.

An extensive analysis of factor analysis applications is not attempted here. Of the studies listed, assemblage comparisons are in the majority. Analyses of artefact compositions occur in Buikstra et al. (1989) and Lambert et. al. (1991); examples of spatial analysis are given in Simek (1987) and Rigaud and Simek (1991). Typological studies include Beneke (1990), Hoffman (1985) and Wynn and Tierson (1990).

Considerations of data treatment in these studies are essentially the same as in PCA. The main point to make here concerns the choices of method of analysis and rotation. It was noted in Section 4.11 that the majority of applications of factor analysis listed in Vierra and Carlson (1981) were really PCA analysis with varimax rotation. Of the studies listed in the bibliography that post-date 1977, the limit of Vierra and Carlson's study, at least two-thirds use this approach (details are not always clear). This suggests that there has been little change in practice in recent years (except that factor analysis is, perhaps, not used as much), and that factor analysis and PCA are still confused by many users.

5

Correspondence Analysis – The Main Ideas

> Correspondence analysis's archaeological sun rose with the paper by Bølviken et al. (1982). Unfortunately their title declares that correspondence analysis is an alternative to principal components – whereas it *is* a principal components analysis, but of a specially transformed matrix.
>
> (Wright, 1989)

5.1 Introduction

Greenacre (1981, p. 119) describes correspondence analysis (CA) as 'primarily a technique for displaying the rows and columns of a two-way contingency table as points in corresponding low-dimensional vector spaces' (e.g., a two-dimensional graph). In the same volume Gower and Digby (1981, p. 93) describe CA as 'appropriate to non-negative data, for example tables of counts and percentages'. Omit the word 'contingency' from the Greenacre definition and the description of CA is very similar to that of a biplot (Section 4.4); and, in fact, CA can be formally applied to any table of non-negative numbers.

Both the biplot and CA are used in much the same spirit. A view of some commentators (e.g., Jolliffe, 1986, p. 88) is that the choice between them depends on the nature of the data being used. Data in the form of counts, such as Table A3, are more appropriately treated using CA; biplots may be more appropriate for continuous data such as those in Tables A1 and A2.

Some treatments of CA emphasise differences from PCA and present it as a superior technique (e.g., Bølviken et al. 1982; Blankholm, 1991, p. 93). A feature of CA that has attracted favourable comment is the possibility of simultaneously representing both the rows and columns of a data matrix as points on the same plot. This allows structure in both rows and columns, as well as their interrelationship, to be examined. Used as a reason for regarding CA as a superior technique to PCA, this neglects both the possibilities of the biplot and the fact that CA can be interpreted as a form of PCA (Jolliffe, 1986, pp. 201–4).

Few of the archaeological applications of CA listed in the bibliography provide technical details, possibly because these can be difficult to follow. Some authors view CA as the PCA appropriate for discrete, as opposed to continuous, data and this standpoint is adopted in this chapter. In particular CA is described in terms of a PCA of appropriately transformed data for both rows and columns. Chapter 6 looks at some features and archaeological applications of CA in more detail.

The method has been (re)discovered in a number of guises but has only become relatively popular over the last twenty years or so. A paper by Hill (1974) describes CA as a 'neglected multivariate method'; thirteen years later Digby and Kempton (1987, p. 70) suggest it is 'probably the most popular ordination technique among ecologists'. Much of the theoretical development has been undertaken in France by Benzécri (1973) and coworkers, and is described in the English language texts of Lebart et al. (1984) and Greenacre (1984). Other, and different, strands of development are reflected in the books by Nishisato (1980) under the title of 'dual scaling', and Gifi (1990). In the archaeological literature it is only in the late 1980s that the method began to be widely used (Section 6.10).

Correspondence analysis has been used far more within some intellectual traditions than others. Scholars in certain parts of Western Europe (excluding Britain) have been far more enthusiastic than statisticians within the Anglo-American statistical tradition. Consideration of the reasons for this follows the published discussion of the paper given to the Royal Statistical Society by van der Heijden et al. (1989).

This aspect of the development of CA is worth noting since it is reflected in archaeological uses. From about the mid-1970s some French scholars have been publishing the results of applications of CA on a fairly routine basis and the *Bulletin de la Société Préhistorique Française* and the *Bulletin du Musée d'Anthropologie Préhistorique de Monaco* are fruitful sources of applications. Section 6.10 discusses the history of the use of CA in the archaeological literature. Bølviken et al. (1982) is often cited as the paper that popularised the technique in the archaeological literature and the best single collection of applications of CA to archaeological material is, similarly, the product of Scandinavian scholars in the collection edited by Madsen (1988a).

Some typical applications of CA are presented and discussed in the next section. This is followed in Section 5.3 by some of the more technical aspects of CA, deliberately mirroring earlier discussions of PCA as far as possible. More detailed consideration of particular points is reserved for the next chapter.

5.2 *Some examples of correspondence analysis in archaeology*

The first example is a reworking of the second of the examples given in Bølviken et al. (1982) (a re-presentation of their first example is given by Shennan, 1988). Table A5, based on their Table 2, shows the absolute frequency of sixteen tool types on forty-three Early Stone Age sites in the Varanger fjord area of Northern Norway. The sites can be classified by their geographical location – nineteen located in the inland fjord area and twenty-four in the outer fjord/coast area. The hypothesis that the data are used to investigate is that sites will 'cluster' according to their geographical location and will be characterised by an emphasis on different tool types which reflect patterns of seasonal transhumance.

The correspondence analysis of the table is given in Figure 5.1 – we shall think of it for the moment (incorrectly, but without much harm) as a PCA in the form of a biplot. I have chosen to represent the analysis as two separate but adjacent graphs; more usually, and in the original presentation, the graphs are superimposed. This is partly a matter of taste; many CA graphs in the literature are, to my mind, too 'busy' in that they contain

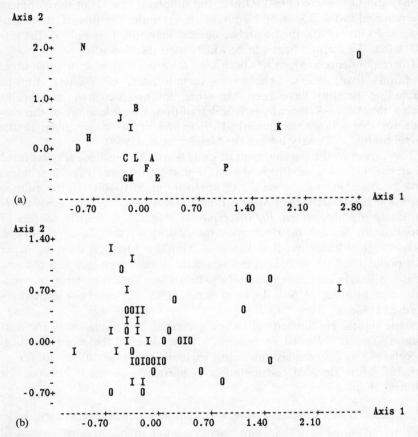

Figure 5.1 Correspondence analysis of the data in Table A5

Note: Diagram (a) is the column plot for tool types, identified by letters as in Table A5; G and M are perturbed very slightly to avoid overlap. Diagram (b) is the row plot for sites with O's identifying outer and I's interior fjord areas, and with overlapping points separated. Explanation on the leading two axes is 19% and 15%.

Correspondence Analysis – Main Ideas

too much detail because of the superimposition and can border on the incomprehensible.

The row plot is interpreted in Bølviken et al. (1982) as suggesting two clusters, the smaller of seven (or five) sites to the right of the graph consisting mainly of sites in the outer fjord area. This is considered to at least partially support the hypothesis of interest, although it must be noted that the larger cluster contains about equal numbers of sites of both geographical locations. In CA terminology the 'proportion of inertia' explained by the axes is 34%; this is identical to the 'variance explained' in a PCA as measured by the eigenvalues (Section 4.4).

Imagining the two graphs to be superimposed with a common origin the sites to the right are particularly associated with tool type K (disc-scrapers) which 'explain 67% of the direction of the first axis' (Bølviken et al., 1982, p. 50; and see Section 6.5). Taking the two graphs into account, a possible subdivision of the large group into two groups is suggested. That consisting of sites in the upper left of Figure 5.1(b) is interpreted as having a different emphasis 'on tools from scrapers and arrows on blade, to burins, knives, tanged and transverse/oblique arrows, microliths, and axes' (p. 50) compared with the rest. This is further interpreted to 'reflect different emphasis on maintenance and procurement in both areas'.

The patterns identified are not obvious from the point scatters alone; they require a knowledge of what the points represent. This is common in applications of CA where clear clustering is not always evident on the graphs produced. It is often the case that output is interpreted on the basis of what is known about the data prior to analysis. That is, CA (as with many applications of PCA), is often used to display or confirm a known or suspected pattern as opposed to discovering unknown grouping within the data.

A second, less detailed, illustration is provided by an analysis of the data in Table A3 that have already been examined in Section 3.3 and Figure 3.10 using PCA. The results of the CA are shown in Figure 5.2. The same story emerges, though in a different way. The earlier PCA was undertaken using the covariance matrix of the percentages and the first axis was largely determined by the numerically dominant tool types A and B. This reflects the fact that in such analyses 'size' effects, such as the high proportion of tool types A and B, tend to dominate the first component. The main feature of the analysis, the unusual nature of site 27 with a relatively high proportion of tool type C, was revealed on the second axis. In the present case it is the first axis which is mainly determined by tool type C (accounting for 72% of that axis – see Section 6.5) and on which site 27 is clearly separated from the others. This arises because in CA 'shape' rather than size effects predominate. This is explained in more detail in Sections 5.3 and 6.1. Here the message of the data is so clear that both analyses reveal it.

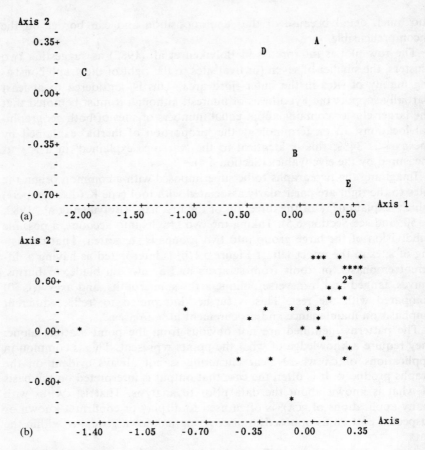

Figure 5.2: Correspondence analysis of the data in Table A3

Note: Diagram (a) is the column plot with tool types labelled as in Table A3; (b) is the row plot with site 27 (Star Carr) to the extreme left. Explanation on the leading two axes is 52% and 37%.

The two analyses so far discussed have aims that are also characteristic of many uses of PCA, namely the identification of groups or unusual values in the data. The third example to follow illustrates the use of CA for seriation, an application now more usual with CA than PCA.

The data set in Table A6 is adapted from Nielsen (1988, p. 48). Rows correspond to seventy-seven female graves from different locations in Denmark and Scania; columns correspond to thirty-nine different ornament types found in the graves, a 1 indicating the presence of the type in a grave. Compared with the original publication, which presented the rows and columns in 'seriation order', several changes have been made for illustrative purposes. Firstly, presence/absence rather than abundance (mainly 0's and 1's but with some 2's and 3's) is shown. Secondly, the matrix has

Correspondence Analysis – Main Ideas

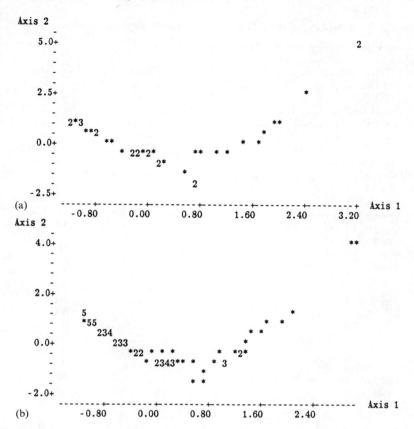

Figure 5.3: *Correspondence analysis of the data in Table A6*

Note: Diagrams (a) and (b) are the column (ornament) and row (grave) plots respectively with numbers showing multiple occurrences.

been scrambled with graves ordered by their original numbering rather than seriation order and ornament types ordered randomly. There is no obvious pattern in Table A6. We shall see shortly how CA can be used to produce a seriation of the data leading to a more informative reordering of the table.

Figure 5.3 shows the results of the CA of Table A6. The curved pattern of the graphs is characteristic (Section 6.6). Interpretation of the pattern shown is an archaeological rather than statistical problem. Ideally such graphs would show an unambiguous ordering of rows/columns reflected by their scores on the first component. External archaeological evidence must then be adduced if the ordering is to be interpreted in terms of relative chronology and if the start and finish of the sequence are to be identified. A pattern such as that in Figure 5.3 does not guarantee a chronological

Table 5.1: Re-ordered version of Table A6 after correspondence analysis

```
1.1.....................................                    ....................1.....1......
11..1...................................                   ...................11...1.111......
..1.111.................................                   ....................1.1...1......
...11.11................................                   ....................1.1...1......
..1..1.1...1............................                   ...................1....1.11......
....1111...1............................                   ....................1.....1......
....1......1............................                   ....................1.....1......
......11...1............................                   ...................1............11..
......111..1............................                   ....................1....111.1......
......1.1..1............................                   ....................1.....1.1......
......1...1.............................                   ....................1.....11......
.....1..1.11............................                   ....................1....1.1......
.........1..1...........................                   ....................1....1....1..
.........1..1...........................                   ....................1.....1.1....1..
.........1..1...........................                   ....................1....11.11..1..
....1....111......1.....................                   ....................1.111..1....
..........111...........................                   ....................11...1...
...........11..1........................                   ....................1.1......
..........111...........................                   ....................1.1......
...........11...1.......................                   ....................111......
..............1111...1..................                   ....................11.1...1..
..............1..1........1.............                   ....................11.1.1.1..
..............1...1...1.................                   ....................11.111.1..
..............1............1............                   ....................1...1.1..
..............1............1............                   ....................1......1.
........1..11........1.111....1.........                   ....................111...11.
..........1..1................1.........                   ....................1.1..
..............1........1.1..............                   ....................1..11..
..............1...11.1..11..............                   ....................11..
.............1...1.11...1...............                   ....................11..
..............1...11.....1.1............                   ....................1....1.1
..............1...11.....1.1............                   ....................1....1.
..............1......1.....1............
..............1.....1...1...1...........
..............1...1.....11.1............
..............1..........111.1..........
...............1.....1.1................
..........1..........1111...1...........
....................1.1.1...............
....................1..11...............
....................1...1...............
...................111111.1..1.........
...................1...1.11...1........
....................11.1....1..........
....................1..1....1...1.1.....
```

Note: This is a reordering of Table A6 after seriation by correspondence analysis and is similar to that in Nielsen (1988). The purpose here is to show the approximate diagonal band achieved in a successful seriation and row and column labels are omitted.

Correspondence Analysis – Main Ideas

interpretation; other influences (e.g., social factors reflected in the relative richness of the graves) might also produce such a result.

In the present case Table 5.1 shows Table A6 reordered according to the scores of rows and columns on the first component. The reordered table is now very similar to that given in the original publication, which can be referred to for a more detailed interpretation. It turns out that the row ordering can be satisfactorily interpreted as a chronological ordering with the columns hence ordered from early to late ornament types.

A number of issues raised by this example, concerning when and why CA works as a method of seriation, interpretation and so on have been omitted here. They are considered at length in Section 6.6.

5.3 Correspondence analysis as principal component analysis

In this section a fairly bald presentation is given of the way in which CA can be viewed as PCA after data transformation. This is followed by a discussion of the ideas that underlie the transformation involved. More detail is provided in Sections 6.2 to 6.5.

The row plot for a correspondence analysis can be obtained by a PCA as follows. Define x_{ij} to be the entry in the i'th row and j'th column of data tables such as A3; $x_i.$ is the sum of the i'th row; $x_{.j}$ is the sum of the j'th column; and $x_{..}$ is the sum of all entries in the table. Thus, from the information given in Table A3, $x_{20.}$, the sum of row 20, is 3899. The column sums $x_{.1}$, $x_{.2}$, etc. are 6936, 5108, etc., and $x_{..}$, the overall sum, is 13770.

The row plot for CA can be obtained by

(a) undertaking a PCA on the transformed (n by p) data matrix with entries given by

$$y_{ij} = [x_{ij} - (x_i.x_{.j}/x_{..})]/(x_i.x_{.j})^{1/2} \qquad (5.1);$$

(b) rescaling the component scores before plotting by dividing the scores for the i'th row by $(x_i.)^{1/2}$.

Thus

$$y_{20,1} = (1458 - (3899 \times 6936)/13770)/(3899 \times 6936)^{1/2}$$
$$= -0.0973.$$

Applying (5.1) to all the entries in rows 20 and 21 of Table A3 gives

Row 20 −.0973   .1077 −.1128 −.0310   .1067
Row 21   .0283 −.0284   .0048 −.0030 −.0129

Note that one effect of this transformation is to reduce the size differences in the original table (e.g., 1458 compared to 28) so that a PCA of the transformed data is much less affected by size differences than a PCA of the original table.

This interpretation of CA is that given (in matrix form) by Jolliffe (1986, pp. 202–3) and presented in this form in Baxter et al. (1990). An even simpler presentation follows if the description of CA by Gower and Digby (1981) is used. This amounts to replacing (5.1) by

$$y_{ij} = x_{ij}/(x_{i.}x_{.j})^{1/2} \tag{5.2}$$

and basing the plot on the second and third (as opposed to first and second) components. The main advantage in using (5.1) rather that (5.2) lies in the way it contributes to an understanding of what CA is about (Sections 6.4, 6.5).

The row plot is, then, obtained as a PCA of the n by p data matrix of the y_{ij} followed by a rescaling of the component scores before plotting. The column plot can be obtained analogously, and in principle, as a PCA of the p by n data matrix obtained on interchanging the roles of rows and columns followed by a rescaling, dividing component scores for the j'th column by $(x_{.j})^{1/2}$. In practice a CA is not carried out like this, but the end result is the same.

Further simplifications result for certain special data structures. In particular, if the data are compositional in the sense that all row sums, $x_{i.}$, are constant then to obtain the row plot (5.1) can be replaced by

$$y_{ij} = (x_{ij} - \bar{x}_{.j})/(\bar{x}_{.j})^{1/2} \tag{5.3}$$

where $\bar{x}_{.j}$ is the mean of column j and there is no need for rescaling after the PCA (Baxter et al., 1990; Baxter, 1991).

The foregoing presentation is simply an interpretation of the end result of the algebra used in CA. Explaining why the algebra takes the form it does is a different, and more complex, matter. The following steps are needed to obtain a row plot.

(a) Row profiles are defined by dividing values by their row sums. This places rows on an 'equal footing' and eliminates 'size' effects (i.e. the number of observations in a row). The resultant analysis thus involves a comparison of row 'shapes'.

(b) A 'distance' between rows is defined for all pairs of rows. In contrast to the Euclidean (or Mahalanobis) distances met with in the discussion in Chapter 4, a so-called chi-squared distance is used (Section 6.2).

(c) Analysis then attempts to reproduce these distances as accurately as possible in two dimensions, but giving more weight to those rows based on larger numbers of observations.

An analogous description could be given for the column plot. The algebraic relationship between the two analyses arises from the use of chi-squared distance and justifies the joint interpretation of the plots. These matters are discussed at greater length in Section 6.2.

A further interpretation of correspondence analysis that may be helpful for those readers with a knowledge of contingency table analysis, and

Correspondence Analysis – Main Ideas

which is elaborated on in Sections 6.3 and 6.4, is as follows. Multiply (5.1) by the square root of the total, x.., then it can be written as

$$y_{ij} = (O_{ij} - E_{ij})/\sqrt{E_{ij}} \qquad (5.4)$$

where O_{ij} and E_{ij} are the observed and expected values for the (i,j) cell in the usual chi-squared test of no association in contingency table analysis. Thus (5.4) is just the square root of terms in this test and is sometimes called a 'residual'. Thus, correspondence analysis can be viewed as a form of PCA of chi-square residuals defined by (5.4). Except for the rescaling below (5.1) Binford's (1987, 1989) description of the PCA analysis of suitably 'normalised' data in tables of assemblages by types seems to be a similar formulation. The resulting 'normalised' data matrix is then analysed via the singular value decomposition. Binford describes the methodology, attributed to his mathematical colleagues, as 'new'. If I have interpreted his formulation correctly it would appear to represent yet another rediscovery of correspondence analysis.

6

Correspondence Analysis – Extensions

6.1 Introduction

Section 6.2 outlines the manner in which CA is often developed in the literature, concentrating in particular on the concept of chi-squared distance. The close connection between CA and the chi-squared statistic is further examined in Section 6.4 which is preceded by a review of the relevant theory in Section 6.3. The quality of a CA plot can be assessed in a variety of ways that are linked to decompositions of the chi-square statistic that measures the 'variation' in a contingency table. This is explained and illustrated in Section 6.5. Remaining sections examine specific archaeological applications of CA. Section 6.6 looks at applications to seriation in more detail; Section 6.7 examines multiple correspondence analysis, the extension to three- and higher-dimensional tables; Section 6.8 examines applications to chemical compositional data. Illustrative applications are given in Section 6.9 and the literature on correspondence analysis in archaeology is reviewed in Section 6.10.

6.2 Correspondence analysis and chi-square distance

In Section 5.3 correspondence analysis was interpreted as a PCA of a data transformation (5.1), in which points are rescaled before plotting. Rows and columns of the data matrix are treated in a symmetrical fashion and often result in plots that are examined together. The effect of the rescaling is that inter-point distances on the plot approximate 'chi-square' distances. Presentations of CA sometimes start from a definition of such a distance as the 'natural' one to use, and CA as presented earlier is the outcome of this and other 'natural' operations on the data. In this section these ideas are sketched in somewhat more detail while, in the following two sections, the close connection between CA and the chi-squared statistic is examined.

A CA begins by converting rows (columns) to profiles by dividing by their sums. The following exposition is confined to rows; columns are treated similarly.

Having obtained profiles and put rows on an equal footing, a 'distance' between rows is defined. A 'natural' definition is the so called chi-squared distance

$$d_{ik}^2 = \sum_j (x_{ij}/x_{i.} - x_{kj}/x_{k.})^2 / (x_{.j}/x_{..}) \quad (6.1).$$

This is not obviously natural to a non-mathematician but has the important invariance property that amalgamating rows with identical profiles will leave distances unaffected (Lebart et al., 1984). The presence of $x_{.j}$ in the denominator of terms in the summation has the further consequence that variables with a large sum are down-weighted in the definition of distance. This removes some of the 'size' effects from an analysis (of the kind discussed in Section 4.5) and leads to the contention that CA involves an analysis of 'shape' rather than 'size'. This effect was noted, from a different perspective, in Section 5.3. Having defined distances between rows, the aim is to approximate them as accurately as possible in a two-dimensional plot in which rows are represented as points (see Appendix B for a discussion of the general idea).

If there are p columns in the original data matrix, then converting rows to row profiles reduces the dimensionality to (p-1). This is because the rows now all sum to 1 and, given any (p-1) values, the p'th is automatically determined. Figure 4.2 provides an illustrative example where similarly constrained three-dimensional data can be seen to lie in two dimensions. The plot involves fitting a two-dimensional plane to the higher (p-1) dimensional data set (Appendix B).

For those familiar with linear regression methods it may be helpful to draw an analogy between this and the least squares method in linear regression. Recall that the simple linear regression model is $y = \alpha + \beta x$ and α and β are determined so that the expression

$$S = \sum_i w_i (y_i - \alpha - \beta x_i)^2 \quad (6.2)$$

is minimised. The w_i are weights reflecting the importance of each observation and are all equal to 1 in ordinary least squares. In this latter case we can view the method as maximising the amount of variation in y 'explained' by x, and the proportion of variation explained is measured by R^2 the coefficient of determination (Shennan, 1988, p. 176).

In obtaining the CA plot it is considered 'natural' to weight profiles in the 'fitting' procedure in proportion to their original row sums (e.g., Greenacre and Hastie, 1987, p. 438; Lebart et al., 1984, p. 36). This has the effect that the analogue of the total variation in the data is proportional to the usual chi-squared statistic for testing the hypothesis of no association in a contingency table – an aspect explored in more detail in Section 6.3. Analogously to the minimisation of (6.2) and maximisation of variation explained, CA attempts to explain as much of the variation, as measured by chi-squared, as possible in two dimensions. The algebra behind all this leads, equivalently, to a PCA of (5.1) followed by a rescaling of the component scores.

In summary a CA involves:

(a) defining row profiles so that rows are on an equal footing;

(b) defining chi-squared distances between rows, which has the effect at the same time of down-weighting the contribution of columns with large sums, thus leading to an analysis of 'shape';

(c) maximising the goodness of fit of the (usual) two-dimensional plot to the full data set in a way which differentially weights profiles in proportion to the original row sum.

Each stage involves some form of scaling/weighting, having a different function in each case, and this can be confusing. The column plot is obtained in an identical manner, with the algebraic duality between row and column analyses justifying their joint presentation.

On the plots obtained distances between rows can, by construction, be directly interpreted as showing the similarity between profiles (assuming the quality of the plot is good). The same is true of the column profiles. The distance on a plot between a row and column point when the plots are superimposed does *not* have this kind of interpretation (e.g., it does not make sense to say site A is closer to type C than to site B). What is the case, as with a biplot, is that away from the origin, row points tend to occupy a similar region of space to the column points identifying the variables on which they have high values relative to the average profile.

6.3 A digression on the chi-squared test

In the next section we shall explore further what is happening in a correspondence analysis and look at useful ways of decomposing the output in Section 6.5. To prepare the ground for this it will help to review the use of the chi-squared statistic for testing the hypothesis of no association between the variables that define the rows and columns of a contingency table (Shennan, 1988, Chapter 6, covers some of the relevant material).

Table A3 is an example of a contingency table with entries in it showing the count of a particular tool type occurring on a site. Suppose, for such a set of data, that the hypothesis that there is no association between site and type is of interest. This is equivalent to the assumption that assemblage compositions (i.e. row profiles) are not significantly distinct. Formally, such hypotheses are often tested using the chi-square statistic of the form

$$X^2 = \sum (O - E)^2 / E \tag{6.3}$$

In (6.3) the O's are the observed values; the E's are the expected values under the hypothesis of no association; and summation is over all cells in the table. The expected values are defined as

$$E = (\text{Row total} \times \text{Column total}) / \text{Overall total} \tag{6.4}$$

or, in the notation of the previous chapter,

$$E_{ij} = x_i.x_{.j}/x_{..} \qquad (6.5).$$

Individual terms of the form $(O - E)^2/E$ identify the importance of an individual cell's contribution to X^2 and sometimes a 'residual' of the form

$$r_{ij} = (O_{ij} - E_{ij})/(E_{ij})^{1/2} \qquad (6.6)$$

is also defined. Large values of r_{ij} identify the observations most responsible for departures from the hypothesis of no association with the sign showing the direction of departure.

As an illustration in Table A3 we have $O_{27,3} = 334$ with an associated row total, $x_{27.} = 920$, and column total, $x_{.3} = 619$, with an overall total of $x_{..} = 13770$. From (6.5), on substituting in these values, $E_{27,3} = 41.357$ which is much smaller than the observed value. This, in itself, suggests there are many more burins on site 27 than is to be expected if there is no association between tool type and site. The contribution to (6.3) is given by $(334 - 41.357)^2/41.357$ or 2071.7; the square root of this, $r_{27,3}$ from (6.6), is 45.5. It is noted in the next section that CA can be viewed as a PCA of such residuals.

Formal significance testing is not of concern here, but recall that the chi-squared test can be invalidated if any of the E's are 'too small'. This is because E occurs in the denominator of terms in the summation (6.3) and small values can artificially inflate the value of X^2 giving a false impression of significance. The value of E can be small if, as (6.4) shows, the row and/or column totals are small relative to the overall total. What is to be understood by the term 'too small' is a matter of some debate but values of E less than 2 or lots of values less than 5 can invalidate a formal chi-squared test. For Table A3, for example, the small number of axes/adzes (type D) would cause problems. Thus $E_{44} = (41 \times 60)/13770 = 0.18$, and several other expected values in the same column are of similar size or smaller. This would invalidate a formal chi-squared test on these data. There is an analogous difficulty with CA that is discussed in more detail in Section 6.5. As with a chi-squared test there would be some justification for omitting type D from a CA analysis, though it will be seen later to have little effect.

6.4 Correspondence analysis and chi-squared

With the foregoing notation in place we shall now reinterpret the analysis given in Section 5.3. There it was simply stated that CA could be obtained from PCAs of the transformation defined by (5.1). On the basis of the notation established in the previous section it is possible to write (5.1) in the form

$$y_{ij} = r_{ij}/(x_{..})^{1/2} \qquad (6.7).$$

Thus, ignoring the constant denominator, CA can be thought of as a PCA of the residuals that arise in a chi-square test of the assumption of no association between the rows and columns of a table.

The row plot in CA thus provides a picture showing which rows most differ from the average row profile, and how. Rows that differ from the average profile in a similar way should, if the CA is a good one, appear in similar positions on the plot. The position on the row plot relative to the positions of the column points on the column plot identifies on which variables, and in which direction, the observed values differ noticeably from those to be expected if rows (e.g., assemblages) did not differ significantly.

Recall, from Chapter 4, that the eigenvalue sum in a PCA of a data matrix **Y** has an interpretation as the total variation in that data set (on the scale of measurement used). In presentations of CA this quantity is usually called the 'total inertia' and will be denoted by I. The total variation can be defined as the sum of the squared transformed values (Appendix B) so that in the present case, from (5.1),

$$I = \sum_{ij} y_{ij}^2 = X^2/x_{..} \qquad (6.8).$$

The individual eigenvalues that are obtained in a CA or PCA are interpreted as the proportion of inertia or variation explained by the associated component (or axis).

Thus a CA can be seen as an attempt to define new variables that explain as much as possible of the departure of a table from the form it would have if there were no association between rows and columns. The resulting plots show which rows and columns most contribute to the departure from this 'null hypothesis'.

To summarise, the chi-squared statistic is a statistic used to measure the departure of a contingency table from the hypothesis of no row and column association, and CA is an attempt to 'explain' as much of this statistic as possible in a low number of dimensions (usually two). In the sense that (5.1) is the transformation that leads naturally to X^2 as the measure of variation in the data, it is the appropriate transformation to use for a CA/PCA of a contingency table.

6.5 Decompositions of the inertia

It has already been observed that the total inertia, I in (6.8), can be decomposed into the contributions of the eigenvalues associated with the different components extracted. Other decompositions of the inertia are both possible and useful and are described and illustrated here. The presentation is modelled on that of Underhill and Peisach (1985).

Squared values of (5.1) define the contribution of an individual value to the inertia; the sum of such values across a row or column defines the row or column inertia. These may be expressed as a proportion of the total inertia.

Considered as PCAS based on (5.1), the squared values of the coefficients

Correspondence Analysis – Extensions

of the k'th component are interpreted as the contribution of the associated row (column) to inertia along the component. These may be expressed as a proportion of the inertia explained by the component (i.e. the associated eigenvalue) and are sometimes termed 'absolute contributions to inertia'. They measure the importance of a row or column in defining a component.

Conversely, the importance of a component in defining a row or column can also be measured and is termed the 'relative contribution to inertia'. For the k'th component and i'th row the relative contribution to inertia is defined as the square of the i'th coefficient on component k divided by the sum of the squared i'th coefficients across all components. These may best be appreciated in the example to follow.

Finally, as noted in Section 6.4, the use of the chi-squared statistic can be invalidated if any of the row or column sums are too small and lead to small values of E. Similar considerations arise in CA which can be unduly influenced by rows or columns of a table with low totals. Underhill and Peisach (1985) define the row (column) mass as the row (column) total divided by the overall total and counsel against using rows (columns) for which the mass is small.

As an example the data of Table A3, the correspondence analysis for which was given in Figure 5.2, will be examined in more detail. Table 6.1 shows the eigenvalues (or squared singular values) expressed as a proportion of their sum (the total inertia) and as cumulative proportions. The first two axes account for 89% of the total inertia so that the representation is a good one.

Table 6.1: Inertia 'explained' by the correspondence analysis of Table A3

Axis	Eigvals	Prop.	Cum. Prop
1	.22	.52	.52
2	.16	.37	.89
3	.03	.07	.95
4	.02	.05	1.00

Note: The third and fourth columns show the proportion and cumulative proportion of the inertia 'explained', based on the eigenvalues or squared singular values of the second column.

Table 6.2: Statistics summarising the quality of fit of the correspondence analysis of Table A3 to the columns

Column	Mass	Quality	Inertia
A	.50	1.00	.19
B	.37	.88	.15
C	.04	.99	.46
D	.00	.04	.05
E	.08	.72	.15

Table 6.3: Contributions to inertia of the columns of Table A3

	First axis			Second axis		
Col.	Coord.	Rel.	Abs.	Coord.	Rel.	Abs.
A	.16	.16	.06	.37	.84	.43
B	−.05	.02	.00	−.39	.87	.36
C	−2.06	.98	.87	.16	.01	.01
D	−.35	.02	.00	.31	.02	.00
E	.43	.22	.07	−.64	.49	.20

Note: The coordinates and relative and absolute contributions to inertia of the columns associated with the first and second axes are shown.

Table 6.4: Statistics summarising the quality of fit of the correspondence analysis of Table A3 to the rows

Row	Mass	Quality	Inertia
1	.009	.96	.007
2	.006	.97	.006
3	.005	.55	.013
4	.003	.99	.005
5	.004	.02	.003
6	.004	.99	.006
7	.008	.79	.002
8	.021	.15	.016
9	.003	.96	.006
10	.066	.90	.050
11	.019	.50	.016
12	.009	.99	.013
13	.003	.99	.006
14	.110	.98	.124
15	.006	.22	.001
16	.005	.98	.008
17	.003	.87	.001
18	.061	.73	.069
19	.202	.77	.069
20	.283	.99	.109
21	.002	.96	.004
22	.013	.04	.023
23	.004	.02	.001
24	.006	.96	.006
25	.002	.98	.004
26	.013	.95	.006
27	.067	.98	.384
28	.036	.78	.019
29	.005	.96	.010
30	.007	.75	.001
31	.002	.52	.001
32	.006	.99	.009
33	.005	.89	.002

Correspondence Analysis – Extensions

Looking at the variable plot (Figure 5.2(a)) it can be seen that the first axis contrasts tool type C with the rest, while the second axis contrasts A and D with B and E. The 'quality' column in Table 6.2 shows that A and C are almost perfectly represented in the plot with values of 1.00 and 0.99, and D hardly at all with a value of 0.04. Type D has a very small mass of 0.00 (see the previous section) and might have been omitted from the analysis but also has little influence on the analysis. It can be seen that C accounts for just under half (46%) of the total inertia in the data, with A, B and E more or less equally responsible for the remainder.

Table 6.3 shows that the first axis is almost entirely determined by type

Table 6.5: Contributions to inertia of the rows of Table A3

Row	Coord.	Rel.	Abs.	Coord.	Rel.	Abs.
1	.16	.09	.001	.51	.87	.015
2	.21	.11	.001	.60	.86	.015
3	−.14	.02	.000	−.81	.53	.019
4	.30	.14	.001	.74	.85	.010
5	−.09	.02	.000	−.03	.00	.000
6	.22	.08	.001	.75	.91	.016
7	−.13	.18	.001	.25	.61	.003
8	.19	.12	.004	−.10	.03	.001
9	.02	.00	.000	.89	.96	.016
10	.10	.03	.003	.53	.87	.118
11	−.29	.24	.007	−.30	.26	.011
12	.26	.10	.003	.76	.88	.032
13	.31	.13	.001	.80	.86	.013
14	.22	.10	.024	.65	.88	.295
15	.04	.01	.000	.14	.21	.001
16	−.83	.93	.015	.20	.05	.001
17	.11	.10	.000	.31	.77	.002
18	−.50	.53	.071	−.31	.20	.037
19	.26	.47	.063	−.20	.29	.054
20	.19	.23	.049	−.35	.76	.225
21	.02	.00	.000	.89	.96	.011
22	.08	.01	.000	−.16	.03	.002
23	.05	.02	.000	.01	.00	.000
24	.26	.16	.002	.57	.79	.012
25	.17	.04	.000	.79	.94	.010
26	.23	.28	.003	.35	.67	.010
27	−1.54	.98	.723	.05	.00	.001
28	−.32	.44	.016	.27	.34	.018
29	.06	.00	.000	.89	.96	.025
30	−.03	.01	.000	.23	.73	.002
31	.04	.02	.000	.24	.50	.001
32	.30	.13	.002	.78	.86	.022
33	−.41	.78	.004	.16	.12	.001

Note: The coordinates and relative and absolute contributions to inertia of the rows associated with the first and second axes are shown.

C which 'explains' 87% of the inertia of the axis. The axis itself accounts for 98% of the inertia of C. The second axis is mainly determined by A, B and E in that order of importance with most of the inertia of A and B and about half of that of E 'explained' by the axis. Overall the plot provides a very good picture of the relationship between A, B, and C; a good (72%) picture of E; and almost no information on D. As this latter contributes little to the inertia it could, however, have been omitted from the analysis at the start.

An equivalent analysis of the row information can be undertaken, though the larger number of rows makes a succinct summary less straightforward. From Table 6.4 the quality of representation of most rows on the leading two axes can be seen to be good. The main exceptions are rows 5, 8, 15, 22 and 23. Observation 27 stands out as making by far the greatest individual contribution to the inertia (38%), with observations 14 (12%) and 20 (11%) also noticeable. From Table 6.5 it can be seen that 27 explains 72% of the inertia on the first axis and is almost entirely accounted for by this axis, while 14 and 20 explain 52% of the second axis and are themselves largely accounted for by this axis.

6.6 Correspondence analysis and seriation

A common use of CA is for the purpose of seriation (see Section 6.10). In this section we examine why and also look at some potential limitations of the procedure for this purpose.

An incidence matrix such as Table A6 is defined to be a two-way Petrie matrix if in each row there is a single block of 1's with 0's to either side (except at the edges) and the rows can be so arranged such that this pattern is similarly obtained for the columns. A simple example is illustrated in Table 6.6. The definition may be extended to a general (two-way) Petrie matrix in which each row contains a peak value on either side of which (except at the edges) values steadily decrease, and the row arrangement is such that a similar property holds for columns.

Kendall (1971) used the term 'Petrie matrix' in a problem where rows correspond to tombs; columns to artefacts; and the aim was to seriate rows such that the ordering corresponds to the chronology of the graves.

Table 6.6: An incidence matrix in the form of a two-way Petrie matrix

```
1 1 1 0 0 0 0 0
1 1 1 1 0 0 0 0
0 1 1 1 0 0 0 0
0 0 1 1 1 0 0 0
0 0 0 1 1 1 0 0
0 0 0 1 1 1 1 0
0 0 0 0 1 1 1 1
0 0 0 0 0 0 1 1
```

Correspondence Analysis – Extensions

The terminology recognises the achievements of the archaeologist Flinders Petrie (1899) in formulating and treating this kind of problem.

The two-way Petrie matrix is an ideal form for which both rows and columns can be seriated perfectly. Subsequent chronological or other interpretation requires an appeal to external archaeological evidence. In the example used, the hope is that the row order reflects the relative chronology of the burials and the column order a chronological development of types. The appeal of CA for treating such problems is that, in the ideal case described, it will recover the correct ordering. In practice the raw data matrix will be jumbled up (e.g., Table A6) and CA attempts to recover a useful ordering of the data, as in the third example of Section 5.3.

Hill (1974) gave an early example of the use of CA for seriation, applying the method to the same data as Kendall (1971). This had little immediate impact on archaeological practice (Bølviken et al., 1982) but such use is now common (Section 6.10).

One reason for this popularity is that if a matrix can be put into (general) Petrie form in the manner described, then CA will recover the ordering on the first axis, allowing for the subsequent possibility of a chronological interpretation. A proof of this fact is sketched for incidence matrices by Hill (1974) while Greenacre (1984, p. 229) notes the extension to the general case. It has been observed (e.g., Djindjian, 1985a) that CA is similar to other seriation methods, either in being formally equivalent to them or in leading to similar results in the ideal case.

The theoretical merits of CA as a method of seriation have persuaded many authors of its merits for handling real data, and the example of the previous chapter is typical of published applications. The curved or 'horseshoe' pattern of the plot (also known as the 'arch' or 'Guttman' effect) is characteristic (Figures 5.3, 6.1). It arises for reasons discussed by

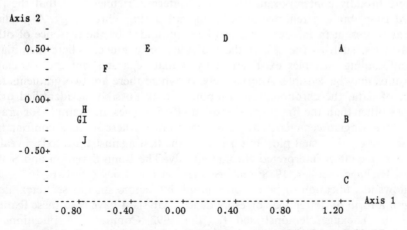

Figure 6.1: Correspondence analysis showing the seriation of the data in Table 6.7

Hill (1974) and Greenacre (1984, Section 8.3). Given a chronologically-ordered data matrix, of sites by type for instance, sites at either end of the spectrum will have nothing in common and will be 'forced' to be approximately equidistant on the usual correspondence analysis plot. It can be shown that in idealised cases that are approached by some real data, the second axis is approximately a quadratic function of the first, the third axis approximately cubic, and so on.

The horseshoe effect has occasioned concern in areas of application such as ecology, partly because distances on the CA plot do not reflect real ecological distances (Hill and Gauch, 1980). In archaeology this would correspond to a concern that only relative, rather than absolute, chronologies are recovered. Given that the usual archaeological concern, where CA is used, is with relative chronology, the horseshoe effect does not seem to be a serious problem. A more practical concern is that with large data sets points are 'squeezed together' at either end of the plot so that details can be difficult to discern. There have been some attempts to develop methods that avoid the horseshoe effect, such as 'detrended correspondence analysis' (Hill and Gauch, 1980). There is some suspicion that this last method, as available in the package DECORANA, is rather 'arbitrary' (Digby and Kempton, 1987) and may introduce spurious features into an analysis (Greenacre, 1984, p. 232). Since the horseshoe effect is natural to CA in the presence of seriation structure, and since in many archaeological uses an unambiguous ordering is all that is wanted, there may be little need to worry about it.

The diagonal band in Table 6.6, an approximation to which occurs in Table 5.1, can be thought of as reflecting a gradient. Its identification as a chronological gradient, together with the late and early ends, needs to be justified by appeal to external archaeological evidence. Other kinds of gradient are possible. A simple hypothetical example would arise if artefact types were broadly contemporaneous but of differing 'richness', so that the row ordering obtained reflected social rank rather than chronology. A gradient that is present in the data can also be confounded by the presence of other variables, such as the sex of the individual in a grave. Where identifiable confounding variables exist, an analysis that separates out the associated groups may be sensible. Alternatively, or where there are two gradients in a set of data, the chronological component may possibly be identified using axes other than the first two. Djindjian (1985a) gives an example for graves in Merovingian cemeteries of north-east France where sex is a confounding variable on the usual plot, but a plot of the first against third axis reveals a horseshoe effect interpreted chronologically. The Bonn Seriation and Statistics Package (Scollar, 1988; and see Fletcher and Lock (1991, p. 167)) now allows identification of points on a graph by factors such as sex, etc., along with a form of detrended correspondence analysis. I have seen these features of the package demonstrated (at the 1992 Computer Applications in Archaeology Conference) but have not used them.

Table 6.7: An artificial abundance matrix

Row					
A	4	0	95	0	1
B	6	0	75	0	19
C	8	0	56	0	36
D	10	40	50	0	0
E	26	54	20	0	0
F	43	51	5	1	0
G	44	31	0	25	0
H	65	30	0	5	0
I	66	29	0	5	0
J	88	12	0	0	0

Source: Inspired by Laxton (1990). The rows represent sites and the columns types with entries the percentage of a type within an assemblage. Rows are ordered so that the data is in Q matrix form.

The 'nice' properties of CA for seriation have generated some enthusiasm for its use. A warning note is sounded by Laxton (1990) who has identified a potential problem when abundance matrices are used.

Suppose an abundance matrix is scaled so that its entries are the percentage of a type within an assemblage, as in Table 2.2. It is in Q-matrix form if 'down each column the quantities never strictly decrease and then increase again' (Laxton, 1990). An artificial example of such a matrix is given in Table 6.7. It is less restrictive than the two-way Petrie matrices discussed earlier in that no column structure is imposed. Nevertheless, the data satisfies a seriation model, and any competing method of seriation should be able to recapture the correct row order. (The data can be viewed as reflecting a case where types come into and go out of fashion, some being long-lived and others short-lived. The relative abundance reflects the popularity within a given context and the row order plausibly reflects chronological change.)

Figure 6.1 shows the row plot from the CA of these data. Reading off the row order from the horseshoe or from the first axis which gives (CBADEFHIJG) does not give rise to a Q-matrix even though one is possible. The example uses percentage abundance (of the row total), but the figures could equally well be raw abundances. The example demonstrates that CA is not guaranteed to seriate an abundance matrix correctly even when a 'correct' ordering is possible. To understand why the method does not work in this case, we pursue it in a little more detail using earlier results.

Since we are dealing with compositional data (all rows sum to 100%) the CA for rows is equivalent to a PCA of y_{ij} as defined in (5.3) with the 'distance' between rows defined as

$$d_{ik}^2 = \sum_j (x_{ij} - x_{kj})^2 / \bar{x}_j \qquad (6.9).$$

For any three rows with A, B and C in correct seriation order, so that A is 'nearer' to B than C in the Q-matrix, we would ideally like

$$d_{AB}^2 - d_{AC}^2 < 0 \tag{6.10}$$

and for this to be reflected in the CA plot. Now, in fact, it is quite possible for this condition to be violated if A and C are at either end of the seriation with B in the middle. This helps account for the horseshoe effect since the plot may need to 'bend round' for A to be nearer to C than B. That A and C are visually separate arises because of their relationship to intervening points. Problems potentially arise when (6.10) is violated locally for triplets that are close to each other in seriation order. Laxton's (1990) work and the example given here show that this is not inevitable. To see why rewrite the left hand side of (6.10) as

$$\sum (x_{Bj} - x_{Cj})(x_{Bj} + x_{Cj} - 2x_{Aj})/\bar{x}_j \tag{6.11}.$$

For (6.10) to be violated, some terms in the summation must be positive and this requires

$$x_{Bj} > x_{Aj}, x_{Cj} \quad \text{and} \quad x_{Bj} + x_{Cj} > 2x_{Aj} \tag{6.12}$$

for some j. This is a necessary but not sufficient condition.

For Table 6.7 equate A, B and C with rows 1, 3 and 4. Then $d_{13}^2 = 270$ and $d_{14}^2 = 133$ violating (6.10). This means that row 1 is closer to row 4 than to row 3 in terms of the distance implicit in CA so that rows 1, 2 and 3 are 'flipped over' on the plot relative to their correct seriation order. The individual terms in the summation (6.11) are $-.6$, -64.8, -16.7, 0 and 218.6 so that the 5th type is that responsible for the reversal of order.

To maximise the chance of violating (6.10), positive terms in (6.11) should be 'large' and negative terms 'small'. The former is most likely if one or more types either reach a sharp peak, of relative abundance, in context B, or increase dramatically between A and B declining thereafter; and if $\bar{x}_j$ is relatively small. Conversely, negative terms will be small for large $\bar{x}_j$ associated with strictly increasing or decreasing relative abundance in which there is little change from A to C.

In archaeological terms this implies that CA is most likely to invert the true seriation order for abundance matrices consisting of relatively common types showing little change in use over longish periods of time (as reflected in the contexts); and relatively uncommon types, possibly short-lived, that peak sharply in popularity in one period. The example used here is an artificial one, but illustrates that CA is not obviously the ideal tool for seriation that has sometimes been implied (Djindjian, 1985a; Madsen, 1988b), at least for abundance matrices.

This discussion of seriation has treated correspondence analysis as a mechanical means of seriation and it is often used as such. Hodson (1990,

Correspondence Analysis – Extensions

p. 35) has observed that 'it would be naive to suppose that a mechanical seriation can by itself act as a substitute for a well conceived archaeological sequence'. He lists several considerations that should be taken into account in constructing a sequence that are not readily incorporated into a mechanical sorting procedure. If the relative abundance of a common type changes little over time, for example, it could well be omitted from any attempt at seriation.

Mechanical procedures can be viewed as producing starting-points that the archaeologist is at liberty to manipulate, in any way that seems appropriate, to get an archaeologically relevant sequence. Djindjian (1985a), for example, has noted the possibility of omitting rows and columns of the data matrix in order to improve the horseshoe effect. This process, of omitting variables that give rise to problems, is something I have seen demonstrated in conference presentations and occasionally generates unease. There is a danger of circularity of argument, in that variables are then only used if they give rise to a chronological ordering that conforms to the investigator's beliefs about the data. Ideally, omission or inclusion of variables should be justified by archaeological reasoning, of the kind indicated in Hodson (1990), and not simply on the basis of 'nice' results as judged by the aesthetics of the horseshoe.

6.7 Multiple correspondence analysis

Many archaeological applications of CA have been to two-way tables of the kind used in Section 5.3. It is possible to generalise the method to three- and higher-dimensional tables, and this topic – multiple correspondence analysis (MCA) – is now discussed. The method as usually applied is first considered, followed by a discussion of its limitations. Some practical examples are given in Section 6.9.

Multiple correspondence analysis as usually understood is ordinary correspondence analysis applied to data coded in a special way. If we have a set of contexts, graves for example, and a single type of object, then either (a) each grave can be coded 1 or 2 according to whether the object is present or absent, giving rise to a single column of information for the type, or (b) we can create two columns, one of which indicates if the type is present (coded 1 if it is and 0 otherwise) and one of which indicates if it is absent (0 if it is and 1 otherwise). A hypothetical example is given below, where the left hand side is coding of type (a) and the right hand side coding of type (b):

$$
\begin{array}{ccc}
1 & & 1\ 0 \\
1 & & 1\ 0 \\
2 & \rightarrow & 0\ 1 \\
1 & & 1\ 0 \\
2 & & 0\ 1 \quad \text{etc.}
\end{array}
$$

Table 6.8: Three way cross-tabulation of variables 2, 3, 4 in Table A7

		Variable 4					
		1			2		
		Variable 3					
Variable 2		1	2	3	1	2	3
1		1	3	0	17	5	1
2		0	3	0	3	10	2

The two approaches contain equivalent information. Multiple correspondence analysis as often used is simply correspondence analysis as already discussed applied to data coded in the second form.

For a realistic application, data published by Engelstad (1988, p. 82) in the collection of papers by Madsen (1988a) will be used. This is summarised, in modified form, in Table A7. It defines a six-way crosstabulation. There are six variables, each of two to four categories. The categories relate to morphological and other attributes of Stone Age pit houses in Arctic Norway. Table A7 is a convenient way of summarising the six-way tabulation. These data, with other sets, are used in an investigation of house typology in the original article.

If variables 2, 3 and 4 (depth by size by form) are cross-tabulated, for example, then Table 6.8 results. Multiple correspondence analysis (MCA) provides a means of representing such data graphically. It is possible to generalize CA to three- and higher-dimensional tables in several ways (Greenacre, 1988). The approach used here is simply to apply the usual CA algorithm to the two-way indicator matrix representing the data, and this is obtained in a manner similar to the artificial example outlined above. For each variable, j, having p_j categories, define p_j new variables that take the values 1 or 0 according to whether or not a particular row falls into that category. This results in a new n by p indicator matrix of 1's and 0's where p is just the total number of categories across all variables.

To illustrate, the first four rows of the data for variables 2, 3 and 4 in Table A7, having 2, 3 and 2 categories respectively, give rise to the following indicator matrix (on the right hand side):

```
1 2 2          1 0    0 1 0    0 1
1 1 2    →     1 0    1 0 0    0 1
1 1 2          1 0    1 0 0    0 1
2 1 2          0 1    1 0 0    0 1
```

The full indicator matrix for Table A7 is shown in Table A8. An ordinary CA of this gives rise to the results shown in Figure 6.2. The first two axes account for 34% of the inertia in the data; there is no obvious clustering into house types on the row plot – the effect being looked for – and no simple interpretation of the variable plot. Engelstad (1988, p. 83) notes

Correspondence Analysis – Extensions

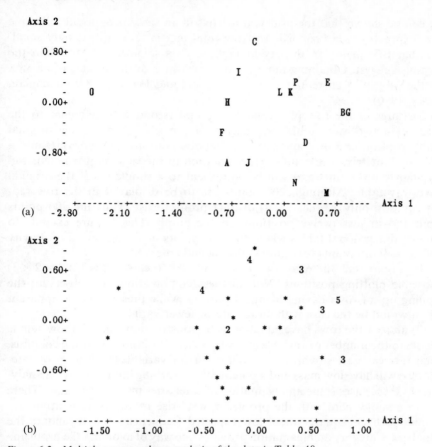

Figure 6.2: Multiple correspondence analysis of the data in Table A8

Note: Diagram (a) is the variable plot with labels as in Table A8; (b) is the row (house) plot with numbers showing multiple occurrences arising for reasons discussed in the text.

that the variables do not group as might be expected had they 'followed the normative house type descriptions in use today', and observes that the houses analysed form a very heterogeneous group.

Several features of this analysis are characteristic of MCA and these must now be examined. It can first be noted that it would be quite possible to analyse a two-way cross-tabulation in this fashion, provided one had access to information on the characteristics of the individuals grouped in the cells of the table. In an MCA of such data, the column representation involves the same points used in the joint representation of simple CA, and the two analyses are essentially equivalent (Greenacre and Hastie, 1987, p. 443). Despite this equivalence, the inertia 'explained' by the first two axes of an MCA is typically much less than that explained by the simple CA.

It can be shown that the principal inertias in an MCA, as opposed to simple CA, expressed as a proportion of the total inertia, are misleadingly small, so that the quality of display in Figure 6.2 is somewhat better than the figures suggest. Obtaining more realistic measures of the success of an MCA is the subject of current research (Greenacre and Hastie, 1987; Greenacre, 1988, 1990).

Features of the row plot can also be understood with reference to the case of a two-way tabulation analysed as an MCA. Suppose the original table is of order 3 by 4 and contains 200 observations. This converts into a 200 by 7 matrix. The column representation in the MCA, as just noted, will contain seven points and will be equivalent to a simple CA of the original two-way table. Although 200 points are to be displayed in the row representation only 12 (i.e. 3×4) distinct profiles exist. Thus the 200 points 'pile up' in just twelve positions on the graph. This feature extends to higher-dimensional tables where the row points lie at the centre of gravity of the column points (categories) that define observed profiles.

In Figure 6.2 there are 45 rows and 288 (i.e. $4 \times 2 \times 3 \times 2 \times 2 \times 3$) possible plotting positions. Not all these combinations are observed; the 'piling up' of rows having identical profiles, while present, is less apparent than would be the case with more data or fewer variables.

In an MCA the rows have equal mass by construction since the row sum is equal to the number of variables in every case. This is not so for the columns, and rare categories (which may well occur for variables with a lot of categories) will have low mass and a potentially disturbing influence on the analysis. In these cases some amalgamation of categories may be desirable. There is an analogy here with the problems that arise in analysing contingency tables with chi-square where low expected frequencies can invalidate the analysis. Category amalgamation provides one way of avoiding the problem.

Some archaeological uses of MCA avoid this problem by constructing variables with equal numbers of categories having equal numbers of observations in each category. Thus, in a study of the morphology of Iron Age funerary urns in Aquitaine, Mohen (1980) characterises urns by a set of measurements, such as height, and ratios related to shape. Each variable is then divided into eight classes, the class boundaries being chosen to have equal numbers in a class. This ensures that categories, as well as urns, have equal weight in the subsequent MCAs.

Similar studies include Verjux and Rousseau's (1986) analysis of Mousterian side-scrapers using variables such as length, breadth, angle of retouch and so on. Each variable is subdivided into between three and five categories defined, in most cases, to have equal numbers. Some illustrative examples are given in Section 6.9 and references to other applications discussed in Section 6.10.

Archaeological applications of multiple correspondence analysis as understood here are mostly to be found in the French literature – not necessarily named as such and with technical detail often omitted. There are some limitations of multiple correspondence analysis to which Greenacre (1988, 1990) and Greenacre and Hastie (1987) have drawn attention. One problem, already mentioned, is that the usual measures of the 'success' of an analysis based on the inertia explained can badly underestimate the quality of an analysis. This is the subject of research (Greenacre, 1988) that is yet to permeate the archaeological literature. A second problem is that the interpretation of distances between variable markers on the plot as approximating to chi-squared distance, that is applicable in simple correspondence analysis, does not carry over to multiple correspondence analysis.

6.8 Correspondence analysis of chemical compositions

Correspondence analysis can be formally applied to any table of non-negative numbers. This includes tables of artefact compositions such as Table A1 and A2 and, although CA is less obviously suited to these kind of applications, its use has attracted some attention.

In the geo-statistical literature, Teil (1975) and Valenchon (1982) have advocated the use of CA for such compositional data. The studies by Peisach et al. (1982) and Gihwala et al. (1984) are archaeological applications to south-western African pottery and ores (in the first paper) and eighteenth- and nineteenth-century pottery and glass (in the second). Co-authors of these papers, Underhill and Peisach (1985), in a useful expository article, apply CA to trace element data, and their methodology is applied in an archaeological context by Jacobson et al. (1991). Baxter and Heyworth (1989), Baxter et al. (1990) and Baxter (1991) have explored the use of CA with data where the full composition, summing to 100%, is used. The motivation in these last three papers was to compare alternative PCA and CA-based approaches to analysis in the presence of the statistical difficulties arising with compositional data. Gratuze and Barrandon (1990) report a successful CA (i.e. 90% explanation on the first two axes) of a 57 by 12 data matrix on the compositions of Islamic glass weights and stamps, but provide no technical detail.

Some points concerning the use of CA arising from these articles are now summarised. Firstly, if the full composition is used, then the row analysis is equivalent to a PCA of (5.3). This differs from a PCA of the correlation matrix in the denominator of the transformation used (e.g., compare with (3.4)). If the mean of each variable is equal to its variance, then the analyses are the same. Baxter (1991) has noted that, empirically, analyses using the major and minor elements only (up to about twelve) often seem to approximate to this condition so that PCA of the correlation matrix and CA of the raw data give similar results. Unpublished work suggests that this empirical similarity does not hold if trace element data is also used.

Secondly, correspondence analysis may be applied after transforming the data in some way. For example, Peisach et al. (1982) use the square roots of trace element data, while Underhill and Peisach (1985), using similar data, advocate 'pre-processing' it by dividing each number by the element mean. This gives each element equal mass in the analysis. If this treatment is then applied to the full composition, Baxter et al. (1990) show that an approximation (to first order) to a PCA of the transformation proposed by Aitchison (1986) results (see Section 4.6). Empirical evidence of the similarity between the two approaches is given in Baxter (1991). The CA approach has the advantage over PCA in that it is unaffected by the presence of zeroes in the data. Both approaches can be dominated by a small number of the numerically least important elements, for reasons analysed in Section 4.6. The results may or may not make archaeological sense; even when they do, however, this involves ignoring much of the information in the data.

The use of chi-squared distance is not obviously a natural one for data of the kind under discussion. For this reason, and those just outlined, CA cannot be advocated as the main tool of analysis for compositional data. It can produce informative results, emphasising a subset of the variables in the data set, but should be used in conjunction with other approaches, such as PCA, that give more equal emphasis to the variables. An alternative approach using multiple correspondence analysis is noted and discussed in the next section.

6.9 Examples

The first example illustrates the use of multiple correspondence analysis for typological study, comparing it with PCA in the process. The data used are adapted from Madsen's (1988b) analysis of Danish Neolithic pottery. This consisted of the co-ordinates of eight landmarks from the profiles of 135 pots; horizontal co-ordinates were divided by rim diameter and vertical co-ordinates by height to eliminate size effects. This results in a set of fourteen ratios that are used for analysis. A subset of seventy-two pots comprising six typological groups, taken from Madsen's Figure 8 (excluding untyped specimens) are used. A PCA of the standardised data, similar to Madsen's own analysis, was undertaken, and is shown in Figure 6.3 where the larger groups A, B and C are reasonably separated apart from odd specimens. The smaller groups D, E, F are intermediate between B and C without being obviously distinct. Madsen (1988b, p. 20) regards this attempt to make a 'morphological division using basic measurements' as 'quite successful'. The first two eigenvalues account for 71% of the variation in the data.

The same data were analysed by multiple correspondence analysis in two different ways. Firstly, each variable was divided into four equal-sized groups of eighteen pots according to the value of the variable (i.e. the eighteen smallest would be coded 1; the eighteen largest 4 etc.). As

Correspondence Analysis – Extensions

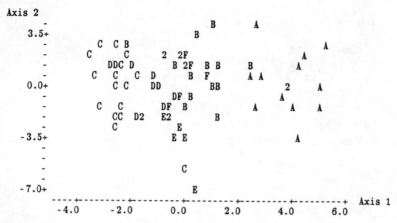

Figure 6.3: *Principal component analysis of Neolithic pottery data*

Note: Data are adapted from Madsen (1988b) with labels identifying one of six pottery groups and numbers identifying multiple occurrences. Analysis is of the correlation matrix of the data.

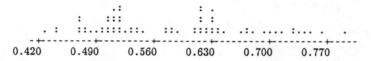

Figure 6.4: *An example to show 'natural' grouping in the dimensions of Neolithic pottery data*

Note: Figure 6.3 is based on a PCA of ratios of landmark data of neolithic pot data. For one of the variables used Figure 6.4 show 'natural' groups that can be used as the basis for the definition of size categories in a multiple correspondence analysis.

described in section 6.7, each variable can be converted to four indicator variables giving rise to an indicator matrix of fifty-six columns. An advantage of this approach is that all size categories within variables receive equal weight; however, 'natural' structure within the data is ignored. Thus with variable 8, for example, in Figure 6.4, there appear to be two fairly clear modes in the data. Dividing the data at just less than 0.49, where there is the hint of an anti-mode, and 0.56 and 0.66 where there are 'gaps', gives four groups of size 10, 24, 23 and 15 which better respects the structure of the data. This, admittedly subjective, procedure was applied to other variables where appropriate – using between three and five categories with a minimum group size of nine – leading to an indicator matrix with fifty-two columns. This was used for the second multiple correspondence analysis.

As it happened, both analyses gave virtually identical results as measured by the correlation between axes for the two analyses or by Sibson's coefficient (equation 4.13), so only the latter is reported. Figure 6.5

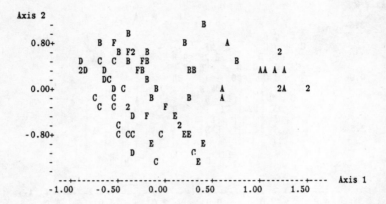

Figure 6.5: Multiple correspondence analysis of the data analysed by PCA in Figure 6.3 using the first and second axes

Note: Letters label pottery groups; numbers show multiple occurrences.

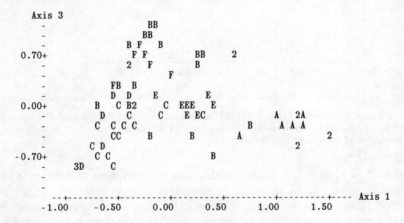

Figure 6.6: Multiple correspondence analysis of the data analysed by PCA in Figure 6.3 using the first and third axes

Note: Conventions are as in Figure 6.5.

shows the object plot and may be compared with Figure 6.3 for the PCA. The two plots are in fact broadly similar, with the correlation between the corresponding pairs of axes in excess of 0.9. Two of the larger groups, B and C on the plots, are perhaps more coherent in the PCA plot. The fifth group clustered together quite well on plots of the third against first axes for both analyses, and that for the multiple correspondence analysis, which was particularly clear, is shown in Figure 6.6.

In the light of these results, the two analyses were compared using three-dimensional plots in Figure 6.7. This is easier said than done. To enable

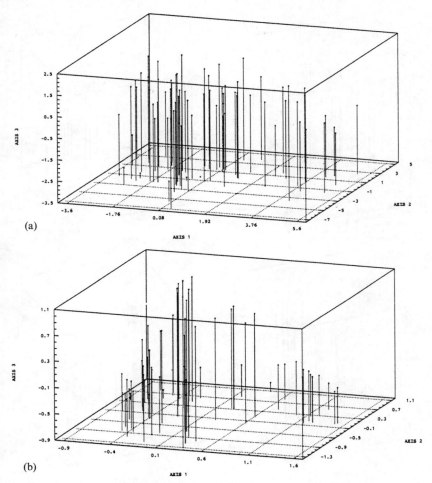

Figure 6.7: Three-dimensional plots of principal component and correspondence analyses of Neolithic pottery data
(a) The principal component analysis
(b) The correspondence analysis

plots to be labelled (in STATGRAPHICS), and for detail to be seen they were split in two. Identifying three-dimensional patterns on the two-dimensional page is also not straightforward. The main advantage of the correspondence analysis over the PCA is in the separation out of most of the fifth group.

In the figures, points are labelled according to a predetermined archaeological typology. Were this unknown, both analyses would lead to the identification of a group consisting mainly of pots of the first type; correspondence analysis would identify a small group based on specimens

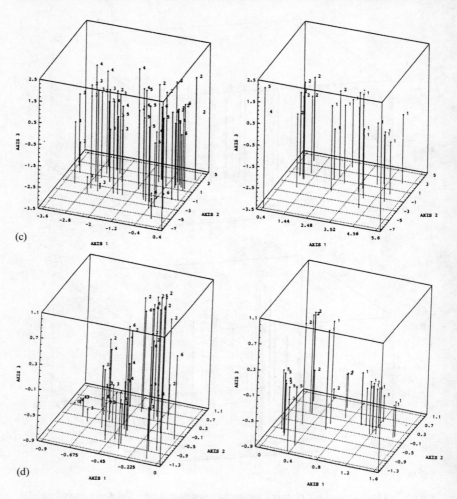

(c) The PCA in (a), labelled and divided into two parts
(d) The correspondence analysis in (b), labelled and divided into two parts

of the fifth type; and both methods would identify a large group containing most members of the third group – possibly more clearly in the case of PCA. Other identifications are of a more subjective nature and partially dependent on the angle of view adopted. This example illustrates the use of multiple correspondence analysis for typological study. Neither it, nor PCA, is clearly superior in this instance.

It is possible, in an exactly analogous way, to apply multiple correspondence analysis to compositional data of the kind in Table A1. I have not seen any examples of such use, though it would not surprise me if they

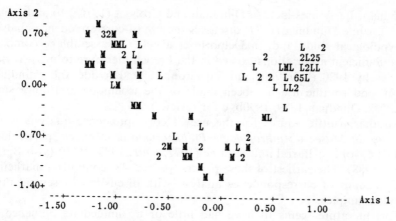

Figure 6.8: Multiple correspondence analysis of the data in Table A1

Note: The letters L and M label the sites Leicester and Mancetter with numbers showing multiple occurrences.

exist. In a spirit of experiment, and for comparison with the PCA of Figure 3.2, such an analysis is briefly reported.

The eleven variables were each divided into three size groups, taking account of natural modes where these were evident. After a correspondence analysis on the thirty-three-column indicator matrix derived from this, Figure 6.8 was obtained. It suggests even more clearly than Figure 3.2 that there are three groups in the data, two largely composed of specimens from Mancetter and one of specimens from Leicester. There is a very strong correlation of 0.96 between the first axis and that for the PCA. Correlation on the second axis is weaker at 0.53, and Sibson's coefficient of 0.32 does not suggest very close agreement. The group structure appears similar primarily because it is the first axis that is responsible for it in the analyses.

This correspondence analysis approach to the analysis of compositional data removes the need to worry about data transformation and the effect of outliers. It can clearly produce sensible descriptive output. Whether or not the output is sensitive to the possibly subjective choice of class boundaries and number of classes, and whether or not it can consistently produce useful results, awaits further research.

6.10 Correspondence analysis in the archaeological literature

Diffusion of the use of correspondence analysis through the archaeological literature has been an odd and rather prolonged process. Hill (1974) is an early paper, written for a statistical audience, that includes an application to a seriation problem, but applications really begin in the French literature in the mid-1970s and are particularly associated with the work of

Djindjian. Early uses include Djindjian and Croisset (1976a, b) and Hours (1976), while Djindjian (1977) discusses the use of correspondence analysis in typological study and, in Delporte et al. (1977), assemblage comparisons. Sufficient applications existed in the French literature to merit review articles by 1980 (Djindjian and Vigneron, 1980; Leredde and Djindjian, 1980), and routine use has been made of the technique in France since then (see Djindjian 1989; 1990b, c for reviews).

Another fruitful source for the use of correspondence analysis is the *Bulletin du Musée d'Anthropologie Préhistorique de Monaco*, particularly with the work of Barral and Simone (1977, 1981, 1989, 1990) (also Barral et al., 1988). The earliest of these papers (pp. 38–48) contains a mathematical account of correspondence analysis with discussion of several illustrative examples.

This literature seems to have had little or no immediate influence on archaeological practice outside France and Monaco, and it was left to Scandinavian scholars writing in English to introduce correspondence analysis to a wider audience. Although Holm-Olsen and Solheim (1981) is earlier, it is usually the *World Archaeology* article by Bølviken et al. (1982) that is cited as one of the first applications of correspondence analysis in archaeology. Although the seminal work of Benzécri and his colleagues (including Greenacre) is acknowledged, the French archaeological applications are not. Indeed, the paper notes that 'some examples in Hill's paper are the only published applications [in archaeology] as far as we know' (p. 41). Since the early 1980s Scandinavian scholars have used correspondence analysis on a routine basis, and edited collections of applications are to be found in Madsen (1988a) and Jørgensen (1992a).

The South African contribution, through the influence of Greenacre, can be traced to the French connection. This includes the papers by Peisach et al. (1982); Gihwala et al. (1984); Avery and Underhill (1986), and Jacobson et al. (1991).

French and Scandinavian scholars are responsible for about two-thirds of the articles in the bibliography, roughly in the ratio 2 : 1. Excluding these and papers already referenced, most of the remainder date from 1987 and usually later. In total, about 10% of the references predate 1980 and about 60% (30%) are 1987 (1990) or later.

One reason for this piecemeal and tardy development has presumably been linguistic barriers. Another, perhaps more important, influence has undoubtedly been the absence until recently of correspondence analysis algorithms in the better-known general-purpose statistical packages. Potential users have either had to have the computing and mathematical knowledge to write their own routines, or else have had to have access to locally developed packages of possibly limited circulation.

Gorecki et al. (1984) is an early Australian contribution and Wright (1985), a co-author of the paper, published an algorithm for correspondence analysis subsequently incorporated into the MV-ARCH package. Wright is thanked by Annable (1987) for his assistance with her correspondence analyses. Disqualifying this latter paper on the grounds of antipodean assistance would appear to leave Ringrose (1988a, b; also 1992) as the first detailed (and late) 100% British application to appear in the archaeological literature (Djindjian, 1989). Subsequent British use has grown, if not into a flood then a stream, and accounts for about a sixth of the references. Most of the remaining references are by scholars based on the European mainland.

The single American application in the references that refers to correspondence analysis by name is Ollendorf (1987). This (or my ignorance of the American literature) is surprising, though perhaps less so since (in terms of published papers) it took twelve years for correspondence analysis in archaeology to cross the English Channel, and the Atlantic is somewhat wider. Two papers by Binford (1987, 1989) listed in the bibliography for PCA and discussed in Sections 4.13 and 5.3 do, however, describe an approach to PCA that appears very similar to correspondence analysis. The implication of Binford's discussion is that published applications of his approach are few and post-date 1985.

To turn now to the content of the papers, few include technical expositions of the methodology. Exceptions, in chronological order and varying in detail, include Barral and Simone (1977), Holm-Olsen and Solheim (1981), Bølviken et al. (1982), Peisach et al. (1982); Baxter et al. (1990), Ihm (1990) and Ringrose (1992). The general absence of detail often leaves the impression that correspondence analysis is something of a 'black-box' technique and may have contributed to its slow acceptance.

Assemblage comparisons (about 60%) dominate the use that has been made of correspondence analysis. Seriation is a common aim, and consideration of the reasons why correspondence analysis is useful for this is given in Djindjian (1985a, 1990a), Hill (1984), Ihm (1990) and Madsen (1988b; 1989), with a warning note concerning the seriation of abundance matrices sounded by Laxton (1990).

In seriation, assemblages may be characterised by the abundances of artefact types or by their presence/absence. Nielsen (1986, 1988), for example, characterises Late Germanic Iron Age female graves on the basis of abundance of types of ornament, and Bech's (1988) analysis of burials and pottery inclusions is in similar vein. Højlund (1988) seriates pottery lots characterised by abundance of morphological types; and Holm-Olsen (1988) seriates regions characterised by the abundance of different types of sites. This last reference is interesting in that a geographical rather than chronological seriation is obtained. Except for Nielsen (1986), the analyses described in this paragraph so far all occur in the edited collection of Madsen (1988a), whose own seriation of Early Bronze age hoards (Madsen 1988b) is based on material seriated and discussed in Vandkilde (1988).

Other seriation studies listed in the references include Gruel et al. (1981); four studies in Jørgensen L. (1992a) of grave assemblages in different parts of Europe dating to between the fifth and eighth centuries AD (by Jørgensen, A.; Jørgensen, L.; Palm and Pind, and Stilborg); and Schlachmuylder (1985: mesolithic assemblages). Bayliss and Orton (1988) seriate Medieval church bells – considered here as an assemblage of stamps – that are characterised by the presence or absence of stamps. The majority of studies noted limit presentation to correspondence analysis plots based on the first two axes and the associated inertias, and use none of the interpretive machinery discussed in Section 6.5. Djindjian (1985a) presents an example, using graves from Merovingian cemeteries, where, because of the confounding effect of sex differences, seriation structure is present on the first and third, rather than second, axes.

Where the result of a seriation study is not clear-cut, an interpretation as a classification of sites may be possible. Holm-Olsen (1988), for example, looks at her results from both points of view. Often classification, or simply a display of relationships between assemblages, is what is aimed at. Several studies on lithic material from various periods of (mostly French) prehistory include analyses of this kind (Bosselin and Djindjian, 1988; Delporte et al., 1977; Djindjian, 1985b, 1986, 1988a; Mohen and Bergougnan, 1984; Voruz, 1984). Characteristically, plots based on pairs from up to the first five axes have been interpreted, and analyses have been complemented with cluster analyses based on the axis scores (Chapter 8). Gob (1988) presents summary details of several similar analyses, mainly to compare correspondence analysis favourably with PCA.

A number of applications apply correspondence analysis to the study of prehistoric rock art (Barral and Simone, 1989; Sauvet and Sauvet, 1979; Soggnes, 1987). In the first paper an 'assemblage' is a carved slab characterised by the abundance of each of five categories of sign; the second paper similarly characterises ten periods/regions by the abundance of fourteen types of animal drawing; the last paper characterises sites by the presence/absence of different types of boat picture.

Ceramic assemblages have been studied by Gebauer (1988) and Orton and Tyers (1990, 1991); faunal assemblages by Denys (1985) and Ringrose (1988a, b; 1992); mixed assemblages of faunal and material remains by Bertelsen (1988a, b), and ceramic and lithic remains within different zones of a site by Hamard (1984). Barclay et al. (1990) present numerous analyses from excavations of Medieval Winchester, England, relating finds types or material to context and date. Some of these studies have interesting technical aspects that will shortly be noted in more detail. Beck and Shennan (1991) examine grave assemblages in order to try and detect associations between the presence of amber and other materials; this paper is a rare example of the use of detrended correspondence analysis in the bibliography.

Other uses of detrended correspondence analysis listed are the study of pollen assemblages within strata by Grönlund et al. (1990); and of plant remains by Van der Veen (1987) and Lange (1990). Jones (1991) reprints an analysis from this last paper. These studies are at the interface between archaeology and ecology – a subject where the use of correspondence analysis borders on the routine. In this general area phytolith studies by Ollendorf (1987) of leaves, and Powers (1988) and Powers et al. (1989) of peat and faecal remains, may be noted. These papers study modern material as an aid to the interpretation of remains within the archaeological record. A similar strategy is adopted by Avery and Underhill (1986), who study seasonal patterns of beached seabirds as an aid to understanding bone remains.

Miscellaneous applications include Stopford et al. (1991), who use correspondence analysis to display the relationship between groups derived from a cluster analysis of chemical compositions of Medieval tiles, and groups of the same material derived on archaeological grounds. In much the same vein Baxter (1992b) uses multiple correspondence analysis to compare several groupings derived by multivariate methods applied to the chemical compositions of archaeological glass.

Muñiz (1988) studies bone fractures arising in butchering, characterising assemblages by morphotypes treated as present or absent. Jensen's (1986) study of the use of unretouched late Mesolithic blades tabulates polish categories against edge angle intervals and compares scores on the first axis for each of four sites to demonstrate consistency between sites.

Typological studies (about 20%) are the next most important category of application, with the earliest applications referenced being Djindjian and Croisset's (1976a, b) work on Mousterian bifaces. They appear to apply the method directly to metric measurements. Barral and Simone (1981) and Barral et al. (1988) operate similarly, their respective analyses of handaxes and knapping products leading to the identification of a small number of ratios suitable for descriptive purposes. Tuffreau and Bouchet (1985), in another lithic study, use ratios of metric variables directly, but also employ a multiple correspondence analysis (Section 6.7) based on qualitative variables.

Most of the other studies listed use some form of multiple correspondence analysis. Djindjian (1977) and Dive (1986) give details of the coding procedures needed and illustrate the procedure using lithic material; Boutin et al. (1977), Decormeille and Hinout (1982) and Verjux and Rousseau (1986) are similar kinds of study. Mohen (1980) uses the same methodology to study the morphology of Iron Age funerary vessels from Aquitaine.

The typology of structures has also been investigated using multiple correspondence analysis. Englestad (1988) studies pit houses in Arctic Norway, and her data has been used as the basis for an illustrative example in Section 6.7. Holm-Olsen's (1985) analysis of farm mounds in North

Norway characterises them by variables such as acreage (classified into three categories), the quality of the fishing, etc.

Spatial analyses are still uncommon in accessible published applications. Studies that do exist often provide little technical detail (Chataigner and Plateaux, 1986). The general idea is to apply correspondence analysis to 'assemblages' defined at points, or within cells, of an area, and to use scores derived from the analysis as a basis for 'contouring' the map, possibly using cluster analysis (Djindjian, 1988b). Defining assemblages at a point is not straightforward (Section 8.6), nor is the choice of unit size if areal units are used. The methodology awaits further empirical application and critical evaluation. In a paper primarily concerned with another technique of spatial analysis – local density analysis – Johnson (1984, p. 88) uses correspondence analyses of matrices of association indices between artefact types to select a suitable scale for spatial analysis. Djindjian (1990c) lists a number of other spatial studies within the French literature. A comprehensive review of different approaches to intra-site spatial analysis, including correspondence analysis with applications, is provided by Blankholm (1991).

The analysis of artefact compositions has also attracted relatively little attention. The series of studies by Baxter and co-workers, using the full compositions of a range of data sets, suggests either that results will be similar to those obtained using PCA, or that they will be determined by a small subset of elements if trace, as well as major and minor elements are used (Baxter 1991, 1992b; Baxter and Heyworth, 1989; Baxter et al., 1990; Heyworth et al., 1990). Another approach, using trace-element data only and transforming the data in ways noted in Section 6.8, is exemplified in Peisach et al. (1982); Ghiwala et al. (1984), and Jacobson et al. (1991). Gratuze and Barrandon (1990) report a successful application to trace-element data for Islamic glass weights and stamps, but give no technical details of the analysis.

The majority of the analyses listed in the bibliography do little more than interpret the analysis based on graphical presentation along with the inertias of the leading axes. Decompositions of the inertia, usually to identify which variables most contribute to the definition of an axis, occur in Bertelsen (1988a), Holm-Olsen and Solheim (1981), Holm-Olsen (1988) and Orton and Tyers (1990), for example. Ringrose (1988a, b, 1992) has introduced the technique of 'bootstrapping' an analysis to assess its stability into archaeology, though at present the method is not readily applied.

Finally, the work of Barclay et al. (1990) and Orton and Tyers (1990) which allies the exploratory technique of correspondence analysis to the modelling of data using log-linear models should be noted. Previously the two approaches have been opposed to each other and to some extent associated with 'French' and 'Anglo-American' approaches to data analysis. Their reconciliation is a topic of current research in the statistical literature, and the papers cited suggest that their complementary use is about to permeate the archaeological literature.

To summarise, at the time of writing (mid-1992), it is evident that after a long period of gestation correspondence analysis has – with the exception of the USA – come to be widely accepted as a useful tool for the exploratory analysis of archaeological data. The breadth of applications listed in the bibliography testifies to its versatility. With increased availability in computer packages it is likely that routine use of the technique will expand rapidly in the near future, and there is already strong evidence for this. Useful developments of the methodology already exist in the wider literature (e.g., improved approaches to multiple correspondence analysis and relationships to log-linear modelling) and it is to be hoped that these will diffuse into archaeological practice in a shorter time than the basic technique has taken to do so.

7

Cluster Analysis – The Main Ideas

7.1 Introduction

Accessible general accounts of methods of cluster analysis are provided by Everitt (1980), Aldenderfer and Blashfield (1984) and Gordon (1981). Cluster analyses were among the earlier multivariate approaches to be systematically deployed by archaeologists (e.g., Hodson et al., 1966) and accounts by and for archaeologists are given in Doran and Hodson (1975) and Shennan (1988). Pollard (1986) gives a brief discussion of applications to the provenancing of pottery using chemical compositions. It is clear from the review in Section 1.5 that cluster analysis is the most widely applied multivariate method in archaeology. This and the next chapter concentrate mainly on the technical aspects of cluster analysis. The general area of typology and classification has been widely debated in the archaeological literature, and for contributions to this debate Adams (1988); Adams and Adams (1991); Aldenderfer (1987); Carr (1985); Christenson and Read (1977); Cowgill (1990b); Hodson (1980); and Whallon and Brown (1982) may be consulted.

Cluster analysis is not a single method; it is a generic term for a wide range of techniques. In archaeology the most common use is to classify a set of 'individuals' (e.g., artefacts, assemblages, graves, etc.) into subgroups such that individuals within a group are similar to each other in some sense and different from individuals in other groups. Commonly, 'different' implies that there should be a clear 'gap' between the groups rather than merely a partition or dissection of a continuum, although this is perhaps an over-optimistic expectation in many cases.

After choosing variables and deciding how, if at all, to transform them, three main steps are involved in a typical cluster analysis:

(a) the (dis-)similarity between all pairs of individuals is measured;

(b) a rule, or algorithm, for grouping individuals on the basis of their (dis-)similarity is selected and implemented;

(c) a decision concerning the number of clusters present is made.

The first two steps can be carried out in many different ways and this gives rise to the many different approaches that exist. Implementation can be carried out within several statistical packages. The most extensive and

Cluster Analysis – Main Ideas 141

widely used is the CLUSTAN package, and Version 3.2 (Wishart 1987) is that used for the analyses undertaken here.

No attempt is made at an exhaustive discussion of cluster analysis, and some previous familiarity with the method – at the level of Shennan's (1988) discussion – would be helpful. This chapter concentrates on those methods that have found most favour in practice, and examples to illustrate some of the more common measures of (dis-)similarity and clustering will be given. Many archaeologists (and statisticians) have reservations about cluster analysis methods (Cormack, 1971), and a more critical look at the problems involved is undertaken in Chapter 8.

7.2 Some applications of cluster analysis in archaeology

7.2.1 Chemical compositions and hierarchical agglomerative methods

Cluster analysis is the multivariate workhorse in the analysis of the chemical composition of artefacts. This is attested to by the analysis of the contents of the journal *Archaeometry* and of proceedings of conferences on *Archaeometry* undertaken in Section 1.5. (see also Slater and Tate, 1988; Budd et al., 1991). Of those applications listed in the bibliography, over 60% post-date 1985, so that the methods described here are still current. The methods used have also been widely applied to other archaeological problems.

A caricature of a typical application is as follows:

(a) raw or log-transformed (to base 10) data are standardised;

(b) dis-similarity between objects is measured by squared Euclidean distance, d_{ik}^2 defined in (4.7), by d_{ik}, or by some variant of these (e.g., divided by the number of elements used in the comparison);

(c) objects are clustered by either average linkage cluster analysis or Ward's method.

The nature of these last two procedures will be explained and illustrated shortly. Two other procedures that are sometimes used and will also be illustrated are single linkage and complete linkage cluster analysis.

The four procedures are all examples of hierarchical agglomerative clustering procedures. In such methods all individuals are initially considered as separate clusters, so that we begin with n clusters. These are successively amalgamated to form (n-1), (n-2) clusters, etc., until in the end all individuals form a single cluster. This type of procedure is in contrast to hierarchical divisive methods and k-means methods discussed in later sub-sections. A problem with all methods is to decide how many 'real' groups are in the data; this is not easy and is discussed in Section 8.3.

The four clustering algorithms mentioned are now considered in order of their complexity.

(a) Single linkage (or nearest neighbour) – at any stage of clustering those two clusters are amalgamated that contain the two individuals having the greatest similarity.

(b) Complete linkage (or furthest neighbour) – the similarity between two clusters is defined by the pair of individuals (one from each cluster) that are least similar. At any stage the most similar clusters according to this criterion are joined.

(c) Average linkage (or unweighted pair groups method) – label any two clusters A and B with n_A and n_B members respectively. The similarity between two clusters is defined as the average (dis-)similarity between the $n_A n_B$ distinct pairs of individuals, one from each cluster. The most similar clusters are amalgamated at any given stage.

(d) Ward's method (or error sum of squares) – a measure of the amount of variability within a cluster is defined as

$$S = \sum_i \sum_k (y_{ik} - \bar{y}_k)^2 \tag{7.1}$$

where $\bar{y}_k$ is the mean of the k'th variable in the cluster. The overall variability is the sum of these terms across all clusters. Any amalgamation of clusters will increase the overall variability, and the two clusters that are merged at any stage are chosen to produce the least increase.

Simple numerical illustrations of these methods are given in some of the references cited (e.g., Everitt, 1980; Shennan, 1988). The last two methods, unlike the first two, take account of cluster structure in the clustering

Table 7.1: Chemical compositions of Roman glass from Norway

Specimen	Oxide (%)							
	Ti	Al	Fe	Mn	Mg	Ca	Na	K
A	.10	2.0	.80	1.50	1.18	6.28	18.0	.58
B	.10	2.0	.50	1.40	1.16	6.41	18.4	.43
C	.10	2.0	1.00	1.20	.77	7.00	19.0	.61
D	.20	2.0	.70	1.20	.90	6.12	19.3	.36
E	.09	1.8	.95	1.00	.70	6.20	16.2	.45
F	.09	1.8	1.10	.90	.68	6.00	16.1	.44
G	.08	1.7	.60	1.40	.71	6.35	17.6	.37
H	.08	1.7	.60	1.30	.70	6.20	17.2	.32
I	.05	1.5	.20	.02	.53	6.20	18.9	.45
J	.30	1.8	1.00	1.40	1.01	8.80	18.1	.53
K	.30	2.2	1.00	1.90	1.06	6.20	18.6	.34
L	.35	2.8	1.20	2.00	.96	5.90	18.5	.37
M	.30	2.5	1.00	2.00	.96	6.70	18.5	.41
N	.07	1.5	.45	.95	.58	6.85	17.5	.35
O	.07	1.5	.45	1.00	.78	6.25	19.4	.27
P	.08	1.6	.50	1.10	.65	6.20	17.5	.37
Q	.06	1.3	.33	.85	.50	5.90	16.8	.29
R	.35	2.2	1.00	1.50	1.20	6.50	18.0	.40
S	.07	2.0	.40	1.20	.80	6.00	18.0	.30

Source: Christie et al. (1979)

Cluster Analysis – Main Ideas

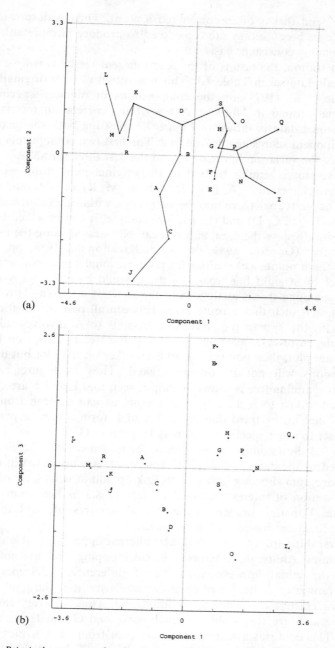

Figure 7.1: *Principal component plots for the analysis of the Roman glass found in Norway shown in Table 7.1*

Note: The upper diagram shows the minimum spanning tree for the data.

procedure, and this is often considered a merit. Their widespread use in practice arises because they are considered to produce interpretable results on a reasonably consistent basis.

For illustration, the results of the four different methods will be applied to the small data set in Table 7.1. This is a subset of data originally given in Christie et al. (1979) on the compositions of nineteen specimens of Roman glass found in Norway. A PCA of the correlation matrix of the untransformed data is shown in Figure 7.1, plotting both the second and third component scores against the first. The first two components account for 68% of the variation in the data, and the first three for 82%.

The first plot is similar to that in the original publication where two groups were suggested – (A, B, C, D, J, K, L, M, R) and the rest. The second plot suggests that there may be subgroups within these such as (K, L, M, R) and (A, B, C, D) and (E, F), for example. It can be asked how well the first plot displays the data, and this can be assessed using the minimum spanning tree (Gordon, 1981, pp. 37–8; Krzanowski, 1988, pp. 102–4). Think of the n points as locations (in p-dimensional space) connected with each other by straight line 'roads' – there would be $n(n-1)/2$ roads. The minimal spanning tree is analogous to the shortest set of $(n-1)$ roads that can be chosen such that a route exists between all pairs of locations with no loops in the system (i.e. travelling from B to A involves taking the same route, in reverse, as from A to B). If the tree is shown on the two-dimensional plot, then points close to each other on the plot but distant in p dimensions will not be directly linked. Thus in Figure 7.1(a) the minimum spanning tree is shown; it can be seen that E and F are closest to G but not linked to it directly. The reason, as can be seen from Figure 7.1(b), is that in the third dimension E and F form a remote pair, somewhat closer to other specimens, such as H, than to G.

Figure 7.2 shows output from each of the four methods of cluster analysis applied to the data of Table 7.1. The results are presented in the form of a dendrogram showing how objects link up and at what level of similarity. A question of interest, referred to Section 8.3, is how many clusters there are. For the moment some typical features of, and differences between, these dendrograms will be noted.

The first thing to notice is the rather different appearance of some of the dendrograms. Although, in terms of broad grouping, they are not too dissimilar, the initial impression is one of difference. Adjacency on the dendrogram is no indicator of chemical similarity; it is the level at which objects or clusters combine that is important. Thus, in every case, specimens E and F are very similar to each other and G and H are also very similar. This confirms a feature that is evident from the PCA plot. By contrast, specimens D and L are adjacent on Figures 7.2(b–d) but belong to clusters that only amalgamate at a high level of dis-similarity. They are not especially similar and this is also confirmed by reference to Figure 7.1.

Cluster Analysis – Main Ideas

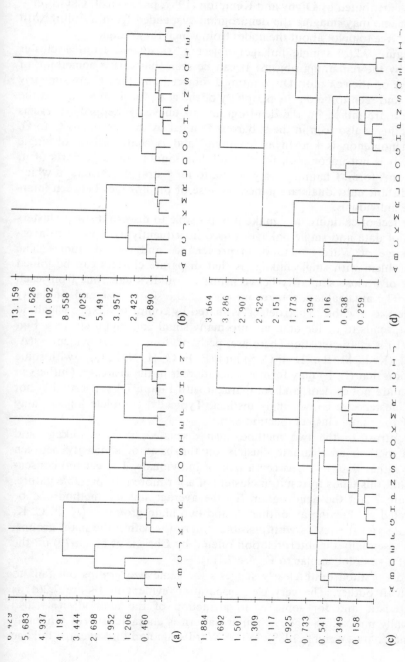

Figure 7.2: Dendrograms for four different cluster analyses of the data in Table 7.1

Note: The analyses (a), (b), (c) and (d) are for single linkage, complete linkage, average linkage and Ward's method.

Several authors have written about the problems of interpreting dendrograms (e.g., Digby, 1985; Pollard, 1986) which have been likened to mobiles – an idea attributed by Digby and Kempton (1987, p. 139) to J. C. Gower – in which one may imagine the dendrogram suspended from a ceiling with branches free to pivot about the nodes from which they dangle.

In Figure 7.2(a) (single linkage) object J is isolated from a cluster formed by the remaining objects. It can be seen that this is something of an outlier on the PCA plot. Discounting J, object I and then C are similarly isolated and seen to be on the periphery of the PCA plot. If one looks at the way these are joined on the dendrogram, an unevenly stepped pattern is evident; this is also seen in the subgroup defined by the specimens G to O. This phenomenon is known as 'chaining' and is characteristic of single linkage. It is not uncommon for the full dendrogram, or large parts of it, to have this form. Chaining can give rise to long straggly clusters in which rather distant individuals are joined because of the linkages between intermediate individuals.

The effect of chaining may make it impossible to discern distinct clusters in a set of data and single linkage is used infrequently, compared to average linkage or Ward's method, in provenance and related studies. The main problem with single linkage is that disparate clusters can be joined because of a close similarity between just two individuals, one from each cluster. The amalgamation takes no note of the shape, size and similarity of clusters as a whole. This 'defect' is shared by complete linkage which, in contrast, tends to divide data up into many small and 'tight' clusters. Figure 7.2(b) is characteristic. There are, perhaps, four clusters evident – (A, B, C, J), (K, L, M, R), (D, I, O, S) and (E, F, G, H, N, P, Q) – with some suggestion that (E, F) may form a separate pair in the last case. Outliers in the data are not evident, and comparison with Figure 7.1 suggests that not all the clusters are convincingly distinct. Typically, complete linkage may identify too many clusters (Section 8.2).

In contrast to the two methods just discussed, average linkage and Ward's method amalgamate clusters on the basis of similarity between groups rather than just between a pair of individuals. The general consensus is that this tends to result in clusters of a far more interpretable nature. In Figure 7.2(c) the dendrogram for the average linkage method can be interpreted as showing an outlier, J, and two main groups – (A, B, C, K, L, M, R) and the rest – with possible subgroups within the main groups. This is essentially the interpretation offered by Christie et al. (1979) on the basis of a PCA plot similar to Figure 7.1(a).

Ward's method quite clearly suggests the same two groups but fails to suggest an outlier. The very clear separation evident in Figure 7.2(d) is characteristic and for some is an attraction of the method; it is also potentially misleading since a similar pattern is characteristic of randomly patterned data (Wright, 1989). While Ward's method and average linkage produce similar results here this is not inevitable, and Pollard (1986) produces examples where they differ in a substantive way. There are a number of potential problems with these methods, particularly with artefact compositional data, and these are discussed in Chapter 8.

7.2.2 Relocation methods and k-means – general ideas and an application

Hierarchical cluster analyses can impose unwarranted structure on a set of data. An individual, once within a cluster, remains there. This facilitates the presentation of results in the form of a dendrogram but is not necessarily optimal, even according to the criterion used for clustering.

A hierarchical structure may be expected for the data dealt with in disciplines such as taxonomy (within which many ideas of cluster analysis originated – Sneath and Sokal, 1973) but is not necessarily appropriate for many archaeological applications (e.g., 'Amr, 1987; Doran and Hodson, 1975; Hodson, 1970; Knapp et al., 1988). This has led some archaeologists to abandon hierarchical cluster analysis as a useful research tool; others have proposed alternative methodologies.

One proposal is the use of 'k-means' analysis. The basic principles are simple:

(a) decide on an appropriate number of clusters – k;
(b) allocate individuals to k clusters, randomly or otherwise;
(c) define some measure, Q say, of the quality of clustering;
(d) move individuals between clusters to improve Q and stop when Q is optimised.

It is not usually obvious what k should be, and basing analyses on several (consecutive) values of k is common. Random initial allocation of individuals to clusters is possible (e.g., Wishart, 1987; Wright, 1989) or a hierarchical clustering procedure could be used to find a starting-point. The algorithms used to improve the initial clustering are not guaranteed to find a global optimum of Q so that several analyses starting from different allocations are usually recommended. Needless to say, Q can be defined in several ways so that 'k-means' is a generic rather than specific technique.

Where one of the methods of the previous section is used to provide a starting allocation to k groups, individuals can be relocated to improve the chosen clustering criterion Q, and k-means in this guise often goes under the heading of 'relocation' methods. Ward's method, for instance, lends itself well to relocation since it is naturally associated with a method of cluster construction that defines a criterion to be optimised, Q_w say – the sum across clusters of terms defined by (7.1) – that relocation can only improve.

This may be illustrated by continuing the example of Section 7.2.1. For Ward's method at the two- or three-cluster level, no relocations are needed. At the four-cluster level the clusters are (A, B, C, J), (D, G, H, I, N, O, P, Q, S), (E, F) and (K, L, M, R). It is possible to improve Q_w by moving individual D from the second to the fourth cluster. This is not evident from the dendrogram (Figure 7.2(d)) where individuals D and O are joined at an early stage; from the PCA plots, D is not obviously closest to either group but is initially in the second group because of its association with O.

Further discussion of k-means methods occurs elsewhere in this and the following chapter, but three additional points can be made here. The first is that refinements can be made such that not all individuals need be clustered, but can be placed in a 'residue' of individuals that are not really similar to other clusters. The second point is that archaeological use of relocation is patchy even though, for some commonly-used methods, it can only improve the final classification – one reason may be the difficulty of graphical representation if many individuals are reallocated (though see Wolff et al. (1986) for an attempt). The third point is that if some clustering algorithm is used to generate a starting-point for relocation, it seems sensible to use a method consistent with the measure to be optimised. Some published applications begin with single linkage, for example, and then optimise Q_w; the initial use of Ward's method seems more appropriate in this case.

7.2.3 Spatial analysis and k-means clustering

Multivariate methods used in archaeology have often been perceived, by archaeologists, to be 'borrowed' from other disciplines and not necessarily appropriate to, or 'congruent' with, the structure of archaeological data and the questions that the data are to help answer (e.g., Whallon and Brown, 1982; Whallon, 1984; Carr, 1985; Aldenderfer, 1987). Cluster analysis has attracted particular criticism because the 'philosophy' or 'logic' that underpins many specific techniques has been seen as embedded within a discipline – taxonomy – whose concerns differ from and are inappropriate to archaeological data. More prosaically, cluster analysis techniques may impose inappropriate (hierarchical) or non-existent structure on classifications of archaeological data.

This has led to a perceived need for methodology that is 'congruent' or 'concordant' with archaeological data and problems, and this was a particular concern of quantitatively-minded archaeologists in the 1980s (Orton, 1988, 1992). One candidate for such a methodology is the k-means clustering of spatially distributed data introduced by Kintigh and Ammerman (1982) and subsequently developed and applied by them and others (Ammerman et al., 1987; Simek, 1984a, 1984b, 1987, 1989; Simek and Larick, 1983; Simek et al., 1985; van Waarden, 1989; Gregg et al., 1991).

I have not had cause to apply this methodology in anger, and in the course of writing this book three good critical and comparative studies of k-means spatial clustering and related methods became available (Blankholm, 1991; Gregg et al., 1991; Kintigh, 1990). These, which may be commended

Cluster Analysis – Main Ideas

to readers, allow me the luxury of doing little more than sketch the outline of the methodology here with further brief discussion in Section 8.6.

The basic idea in spatial k-means clustering (or pure locational clustering in the terminology now preferred by Kintigh (1990)) is simple. Given the spatial co-ordinates (x_i, y_i) of $i = 1, 2, \ldots, n$ artefacts, the aim is to identify, in an objective fashion, clusters of artefacts that may, for example, be associated with work areas or structures (that may or may not be otherwise visible in the archaeological record). Initially all artefacts are treated as a single cluster; the spatial centroid of the cluster is located; and the squared distance of each object from its centroid is determined. The artefact most distant from the centroid is then used as the 'seed' for a new cluster; artefacts closer to this than the original cluster centroid are reallocated to the new cluster whose centroid is then calculated. This gives two clusters. The sum of squared distances of each object from its centroid is the measure, Q, that is to be optimised (similar to Q_w). This procedure continues in a step-wise manner – at any stage, with k clusters, the two nearest are amalgamated to give (k-1) clusters; artefacts are reallocated and Q recalculated to see if the previous allocation to (k-1) clusters can be improved on.

The approach was devised as an 'heuristic' one which can provide a sensible way of identifying compact clusters of spatially distributed material that may subsequently be inspected, at different levels of clustering, for archaeological meaning. Numerous examples occur in the references cited and further discussion is given in Section 8.6.

7.2.4 Methods for binary data

The methods discussed so far have been applied to continuous data. The clustering algorithms may equally be applied to binary data (e.g., the presence or absence within a grave of a particular kind of artefact), provided a suitable similarity coefficient can be defined. A concrete example is provided by Table A6 where the presence or absence of thirty-nine types of ornament in each of seventy-seven female graves is recorded. Each type of ornament may be regarded as an attribute that is, or is not, possessed by the individual (grave). For any pair of individuals a table of the form Table 7.2 may be constructed where a is the number of attributes two individuals have in common; b is the number of attributes possessed by the first but not the second individual, etc. Thus, for the first two rows and the first and last rows of Table A6 the tables are

	+	−			+	−
+	1	1	and	+	0	2
−	2	35		−	2	35

Such tables form the basis of a large number of possible (dis-)similarity coefficients – Wishart (1987) lists twenty-five – few of which are regularly used in archaeological practice.

Table 7.2: General form of table showing attributes possessed by two individuals

		Individual j	
		+	−
	+	a	b
Individual i	−	c	d

The main issue that arises is whether or not mutual absence of an attribute should be regarded as indicative of similarity between individuals or not. For many purposes, particularly with large numbers of uncommon attributes, the view is that it should not. This has led to the Jaccard coefficient, defined as

$$S = a/(a + b + c) \qquad (7.2)$$

being commonly used. Mutual absence of an attribute is disregarded in the definition of similarity. Thus, for the two examples above, S is 0.25 and 0, showing that the first row is more similar to the second than to the last row.

In mortuary studies that produce data similar to Table A6, Jaccard's coefficient has been used by Pearson (1989) and Rothschild (1979), with Ward's method and average linkage respectively; Pyszczyk (1989) groups sites on the basis of presence or absence of artefacts using complete linkage. Sognnes (1987) study of boat picture types in Bronze Age rock art is in a similar spirit as is Croes' (1989) study of style in basketry. Perry et al. (1985) cluster insect assemblages using Ward's method on the basis of species presence or absence.

If mutual absence of an attribute is indicative of similarity then the simple matching coefficient

$$S = (a + d)/(a + b + c + d) \qquad (7.3)$$

might be used. For the two examples above this gives S = 0.92 and 0.90. There is little differentiation between individuals, all of which show a high degree of similarity, and this is one reason why (7.3) is used infrequently. Early uses include the study by Hodson et al. (1966) of La Tène fibulae and Smith's (1974) study of settlement morphology. Use of any of the other possible coefficients listed in Wishart (1987) is rare, though see Tainter's (1975) mortuary study and Hynes and Chase (1982) on aboriginal influence on plant communities.

The foregoing studies all use hierarchical agglomerative methods; they are instances of polythetic techniques in which cluster formation at any stage is determined by all the attributes used, and individuals in a cluster are generally similar but not necessarily identical. An alternative approach − not widely used in general but which has found particular favour in mortuary studies − is hierarchical monothetic divisive clustering. In this case individuals are initially treated as a single cluster; this is subdivided

Cluster Analysis – Main Ideas

into two, three, etc. clusters until the desired level of clustering is achieved. A cluster is split at any stage on the basis of whether or not individuals possess a single attribute so that all individuals in a cluster are characterised by the presence or absence of a set of diagnostic attributes. Methods differ in the way an attribute is selected for the basis of cluster division at any stage.

The popularity of this approach seems to have arisen because patterning in burial remains is recognised as a useful tool for the study of past social organisation (O'Shea, 1985), and the particular structure that can be obtained in a monothetic divisive clustering is potentially congruent with theories about the way in which social distinctions are reflected in death (Brown, 1987). At its simplest – and to use an artificial example of O'Shea's (1985) – graves of the top stratum of society will contain artefacts from one set of types; lower strata will be characterised by different types that are exclusive to a particular stratum. If 'diagnostic' types can be identified, the graves can then readily be grouped into different social strata. This is clearly an ideal and highly simplified structure, but it does provide motivation for the use of monothetic divisive procedures.

One approach, association analysis, is based on a table similar to Table 7.2 calculated for each pair of attributes rather than individuals; thus, i and j are attributes, and a is the number of individuals that possess both attributes, etc. A chi-square statistic can be calculated for each pair of attributes, i and j, as

$$\chi_{ij}^2 = n(ad - bc)^2/(a + b)(a + c)(b + d)(c + d) \tag{7.4}$$

where n is the cluster size. The general idea is that the bigger this is, the more closely two attributes are associated. Equation (7.4) is summed across all values involving attribute i for each value of i in turn. The largest value obtained will, it is hoped, identify the attribute that best predicts the presence or absence of other attributes and the cluster is then subdivided according to the presence or absence of the attribute. The method was proposed by Whallon (1972) for pottery typology and is used in O'Shea's (1984) mortuary studies. A detailed illustration is given in Shennan (1988, pp. 220–5). There are technical problems with this method that are discussed in Section 8.7.

Currently more favoured is the use of the information statistic. For a single cluster of size n and p attributes that are either present or absent, let f_i be the number of individuals for which attribute i is present and $(n-f_i)$ the number for which it is absent. The information associated with the attribute is then defined as

$$n\log(n) - f_i\log(f_i) - (n - f_i)\log(n - f_i) \tag{7.5}.$$

Note that this is zero when all individuals possess the attribute and will be a maximum if half have the attribute. The total information in the cluster is just the sum of this over all attributes which is

$$\text{pnlog}(n) - \sum_{i=1}^{p} [f_i \log(f_i) - (n - f_i)\log(n - f_i)] \qquad (7.6).$$

This may be summed over all classifications to get the total information in the classification. The ideas involved here come from the field of information theory, and 'information' has a technical meaning as a measure of the amount of disorder in a cluster. A little thought will show that if a cluster is defined for a subset of attributes such that all individuals are identical and possess all attributes (e.g., $f_i = n$ for all i) then the information content measured by (7.5) will be zero. Thus at any stage of division of a cluster a split is made on the basis of the attribute that leads to the greatest reduction in information content.

Peebles (1972) is one of the earliest uses of monothetic divisive clustering using the information statistic in mortuary studies. Other studies include Jones (1980); King (1978); O'Shea (1984, 1985) and Tainter (1975). Critiques of these methodologies are considered in Section 8.7.

7.2.5 Mixed data and Gower's coefficient

All the methods discussed so far have been based either solely on continuous data or on binary data. It will sometimes be the case that one wishes to undertake a cluster analysis using both kinds of data and also multi-attribute data. This can be done using the clustering algorithms previously described provided a suitable similarity coefficient can be defined. Gower's (1971a) coefficient of similarity may be used in such cases.

For the k'th variable we need to define the similarity between two individuals, i and j, and a weight; call these s_{ijk} and w_{ijk}. If the variable is continuous with range r_k then

$$s_{ijk} = 1 - |z_{ik} - z_{jk}|/r_k \qquad (7.7)$$

with a weight of 1. For binary variables the value of s_{ijk} may be taken as 1 if the individuals match; the weight is 1 for a positive match where both individuals possess the attribute and 0 or 1 for a match where neither individual possesses the attribute. The decision in this latter case depends on whether or not negative matches are deemed to be indicative of similarity (see the earlier discussion of the Jaccard coefficient). For multi-attribute data s_{ijk} is 1 if both individuals possess the same attribute, with weight 1, and 0 otherwise. Gower's coefficient is then defined as

$$S_{ij} = \sum_k w_{ijk} s_{ijk} / \sum_k w_{ijk} \qquad (7.8).$$

Doran and Hodson (1975) describe this coefficient and note its potential promise for archaeological use without giving an example. Philip and Ottoway (1983), noting the reference in Doran and Hodson (1975), state that Gower's coefficient 'was not used on archaeological problems until it was tested by Ottoway (1981)'. The 1983 paper uses metric variables such as length, and multi-attribute variables such as type of terminal and shoulder profile, in a study of Cypriot hooked-tang weapons in an attempt to define chronologically coherent groups to compare with previous typologies. Some success is claimed although this is achieved after a degree of experimentation with different weighting schemes to account for distinctions between primary and secondary attributes. That such manipulation 'by the user to satisfy his own prejudices' is a possible objection is acknowledged in the paper and defended (p. 132).

Rice and Saffer (1982), in a pottery provenance study, use continuous trace element data and multi-state and binary attributes descriptive of various physical and stylistic features of highland Guatemalan whiteware pottery. Again, success in producing 'meaningful interpretable groupings' (p. 407) is claimed. Palumbo (1987), in a study of mortuary practices and social structure at Jericho, uses binary and multi-state data; an advantage of using Gower's coefficient that is exploited is that with binary data co-absence of an attribute can be treated on its merits and one is not confined to using one of (7.2) or (7.3) exclusively for all attributes.

Shennan (1988, p. 207) observes that in recent years Gower's coefficient has 'found fairly extensive archaeological use'. In fact I have been able to find few published applications, and several of these are of an admittedly exploratory nature so that the opposite may be the case. One possible reason for this is that while attractive in theory, the coefficient is not much use in practice and gives too much weight to nominal variables (Orton, pers. comm). This assessment is fairly general amongst statisticians with experience of the cluster analysis of archaeological data with whom I have discussed the matter.

8

Cluster Analysis – Some Problems

8.1 Introduction

In the previous chapter some of the more common clustering methods used in archaeology were presented. In this chapter, issues deferred so far concerning the choice of numbers of clusters, cluster validation (Section 8.3) and cluster comparison (Section 8.4) are addressed. This is preceded in Section 8.2 by a discussion of issues related to the choice of similarity coefficient and the properties of clustering algorithms.

Many scholars, both statisticians and archaeologists, are sceptical about the value of the methodology generally and as applied to archaeological data. Some specific problems that are particularly pertinent to archaeological applications are discussed in the remainder of the chapter. Section 8.5 looks at the problems posed by correlated data, and possible answers to these problems, including cluster analysis of principal components, are also discussed. Limitations of k-means spatial clustering, and alternatives, are noted in Section 8.6. Section 8.7 examines monothetic divisive methods with particular reference to mortuary studies. Section 8.8 briefly reviews some statistically 'respectable' alternatives to the techniques most used in practice, and Section 8.9 includes a detailed example to illustrate some of the matters discussed in the chapter. Uses of cluster analysis in the archaeological literature are reviewed in Section 8.10.

8.2 Characteristics of clustering algorithms

Among the attractions of clustering techniques is their apparent 'objectivity' and ability to reveal multivariate structure in data without prior assumptions about that structure. In fact, as Gordon (1981, p. 122) notes, the problem is that 'each clustering criterion is predisposed to finding particular "types" of clusters, and may well considerably distort the data towards this ideal. Clustering criteria have not escaped dependence on an underlying model for the data; it is simply more deeply buried.' In this section some properties of the more commonly used methods are discussed.

8.2.1 What types of cluster are to be expected?

In the first instance it may be noted that what constitutes a 'cluster' is by no means evident. Everitt (1980, pp. 59–64) gives several two-dimensional examples where cluster structure is clearly visible to the eye but not necessarily detectable by some, or even most, of the available algorithms. Cormack (1971, p. 329), in a much quoted phrase, observes that two basic ideas are involved in what constitutes a cluster: 'internal cohesion and external isolation', but these are rarely defined. Cluster shape is also important; many of the algorithms used in archaeological applications are predisposed to find spherical clusters, but there is no reason to believe that real clusters will have this shape (Section 8.5). What follows presents a view of the kind of clusters that seem to be expected in archaeological applications.

In provenance and related studies based on chemical compositions of artefacts, the general expectation, or hope, seems to be that artefacts of different origin will form externally isolated clusters in multivariate space. These clusters may be of different size and shape and, for reasons discussed in Section 8.5, are more likely to be elliptical than spherical. This raises questions about the choice of definition of similarity and clustering method that are considered further in Section 8.5. Although the use of hierarchical methods is common, it is the final classification selected, rather than the tree which links them, that is usually of interest. Knapp et al. (1988) express the view that the main purpose is the definition of 'boxes' within which to locate specimens, and a similar view is implicit in 'Amr's (1987) advocacy of k-means methods.

In mortuary studies where different groups are characterised by the possession of attributes (i.e. grave goods) exclusive to their stratum, clear cluster separation seems to be expected. The possession of attributes that cross-cut strata, or reflect aspects of society other than status, clearly complicates matters and may militate against the use of popular clustering techniques. This is discussed further in Section 8.7.

In studies of artefact typology, even supposing that an artefact may be adequately characterised by measurements of the kind suitable for a cluster analysis, several possibilities exist. At one extreme it might be hoped that clearly distinct types will result in internally coherent and externally isolated clusters; however, division into internally cohesive but not necessarily isolated groups may also be useful for descriptive purposes. A recent paper by Cowgill (1990b) discusses the issues involved in some detail.

High correlations between measurements that define an artefact can cause problems similar to those arising in the analysis of artefact compositions. This can arise in the use of metric measurements of pots or flint artefacts, for example, where all measurements may be highly correlated with height or length. This issue of variable 'redundancy' is discussed, for example, in Christenson and Read (1977).

In k-means clustering of artefact locations, a cluster analysis is hardly needed at all if the different clusters are visually distinct. The method thus seems most useful when the density of artefact coverage over a surface is variable and may include adjacent clusters. What the algorithm actually achieves is a dissection into internally cohesive non-overlapping clusters that need not be externally isolated. The restriction to non-overlapping clusters is a possible limitation of the method that may be partially overcome in the manner of representing and interpreting results (Section 8.6).

8.2.2 Choice of similarity measure

With continuous data, the use of some form of Euclidean distance as a measure of dissimilarity is extremely common. For similar reasons to those discussed in Section 4.2, some form of standardisation is also common. Seber (1984, p. 355), in a discussion of standardisation, notes that 'the question of scaling is clearly a difficult one and there is a general lack of guidance in the literature'. One problem is that scaling can actually dilute differences with respect to those variables that best discriminate between clusters; the ideal transformation to avoid this problem cannot, however, be defined without a knowledge of the clusters (Seber, 1984, p. 352). Cormack (1971, p. 325) condemns the 'unthinking use of scaling' and provides a concise discussion of some of the problems with Euclidean distance. The issues involved are considered further in Section 8.5.

While standardisation to zero mean and unit variance is common, division by the range of a variable is also possible and is used in the definition of Gower's coefficient. The range is sensitive to marked skewness in the data and to outliers, and Seber (1984, p. 358) counsels against the use of Gower's coefficient in these circumstances. Possible remedies include data transformation and/or the omission of clear outliers in the data. As a general point Everitt (1980, p. 103) notes that many clustering algorithms are sensitive to outliers and it may be best to try and identify and remove these before analysis.

In defining distance between clusters Seber (1984, pp. 364–5) suggests that it is appropriate that the definition applies to clusters of single objects. This implies that squared Euclidean distance is appropriate for Ward's method, while Euclidean distance is appropriate for single, complete and average link although the squared version may also be used. In practice squared Euclidean distance is often used with average linkage without discussion; Wright (1989) makes this particular point and recommends the 'less distorting' use of true (rather than squared) Euclidean distance with average linkage. Euclidean distance is a special case of the Minkowski metric

$$d_{(ik)}^{(\lambda)} = \left[\sum_{j=1}^{p} |y_{ij} - y_{kj}|^{\lambda} \right]^{1/\lambda} \tag{8.1}$$

for $\lambda = 2$; the case of $\lambda = 1$, the so-called Manhattan or city-block metric, is the other one commonly mentioned. Although rarely used in published archaeological applications, Bieber et al. (1976) observed that its use led to satisfactory results in provenance studies of ceramic compositions. Compared to the city-block metric, Euclidean distance will give greater emphasis to larger differences between variables, and squared Euclidean distance even more so.

For binary data and hierarchical agglomerative clustering, the main issue in defining a similarity coefficient – that of deciding on the import of co-absence of an attribute – has already been discussed in Section 7.2. Often co-absence is not considered to be indicative of similarity and Jaccard's coefficient is preferred. Here it is assumed that the data are genuinely binary. Abundance data on the numbers of each type within a context, for example, can be recorded as presence or absence. This, however, involves a loss of information as well as 'biasing' the coefficient in favour of contexts containing large assemblages. This is because in these cases many types are likely to be present so that in comparing two such contexts 'a' in Table 7.2 and hence S in (7.3) is likely to be large in comparison to contexts with smaller assemblages (Doran and Hodson, 1975, p. 142). Given a set of attributes for which, in some cases, co-absence is thought to be indicative of similarity, Gower's coefficient may be used (e.g., Palumbo, 1987).

In monothetic divisive clustering the main choice, in practice, has been between the use of association analysis and the information statistic. The method is often used in mortuary studies and a fuller discussion of the merits of the two choices, and of the clustering technique, is given in Section 8.7.

8.2.3 Properties of clustering algorithms

Initial discussion will focus on hierarchical agglomerative methods that result in a dendrogram. In Seber's (1984, p. 372) view, such a representation is 'likely to be useful only if the data are strongly clustered *and* have a hierarchical structure' (my emphasis). This has long been recognised within archaeology, and the notion that hierarchical structures are appropriate for many typical applications has been rejected by some scholars (e.g., Hodson, 1970) in favour of k-means methods. It was suggested at the start of this section that in some applications externally isolated clusters are expected to exist, and there is often no reason to believe that these should form a hierarchy. Despite this, hierarchical agglomeration remains the method of choice in many studies (Section 7.3).

That dendrograms may be thought of as analogous to mobiles with adjacency on the diagram implying nothing in terms of similarity has already been noted. This allows some reordering of individuals for the purposes of display, and some possibilities are discussed in Digby (1985).

In discussing methods, it can be useful to think of them as 'space-contracting' or 'space-dilating' (e.g., Seber, 1984, p. 373), even though use of such terms is hampered by lack of a formal definition (Cormack, 1971). Different clustering methods tend to impose different structure on data. If we imagine the individuals, in their original state, to be leading an unfettered existence in some multivariate space then, roughly speaking, space-contracting methods may force an unwonted intimacy upon them, perhaps blurring cluster distinctions; space-dilating methods, by contrast, may impose unwelcome separation and identify too many or non-existent clusters.

Single linkage is a classic case of a space-contracting method, often resulting in the well-known phenomenon of chaining, in which real distinctions between clusters may be lost because of a few intermediate individuals that link them. It has been widely accepted since the early days of experimentation with archaeological data that single linkage cannot be recommended for general use (Hodson et al., 1966; Hodson, 1970; Celoria and Wilcock, 1975), though Orton (pers. comm.) has observed that it can be useful in limited but well-defined circumstances with strong and simple structure and possibly a large number of objects. Bayliss and Orton's (1988) analysis of Medieval church bells characterised by stamps on them is one such example.

Complete linkage is at the opposite extreme and is very much a space-dilating method. It tends to produce small, compact and spherical clusters (e.g., Digby and Kempton, 1987, p. 127) that may have no real existence and is, additionally, sensitive to minor changes in the rank order of the proximities (Seber, 1984, p. 375). On general grounds there is no strong case for its regular use, though it occasionally occurs in archaeological publications (e.g., Cavalloro and Shimada, 1988; Lintzis and McKerrell, 1979; McManamon, 1982; Tyldesley et al., 1985).

Average linkage lies between these two extremes and avoids their main problems. Seber (1984, p. 375) notes that it tends to be space-contracting, but is clearly much less so than single linkage. Together with Ward's method, it is the most commonly used agglomerative method in archaeological practice. Apart from its 'nicer' properties compared to single and complete linkage, it is also considered to produce interpretable results on a consistent basis.

Ward's method is known to be a space-dilating method with a tendency to produce spherical clusters of similar size. It can do this even if the data are completely random (see Section 8.3), and examples are given in the literature to show that it can completely miss very simple structure involving small numbers of non-spherical clusters (e.g., Everitt, 1980, p. 92).

The choice between Ward's method, average linkage and any other method is not a simple one. If clusters do exist that are reasonably spherical then Ward's method will be good at finding them and will display them in a clear fashion. If no structure exists Ward's method may be misleading.

That it is a recommended method in the CLUSTAN package (Wishart, 1987, p. 19) presumably reflects a preference for it over other methods. Pollard (1986, p. 73), by contrast, expresses a preference for average linkage based on numerous pottery provenance studies. The preference is not clear-cut, however, and he gives examples where Ward's method, but not average linkage, produces interpretable results and vice versa.

Krzanowski (1988, pp. 94–5) warns of the dangers of analysing data using several methods and picking out that which gives the best results. He suggests that careful consideration should be given to the objective of an analysis and to 'select from the outset the clustering method that best meets this objective'. This is a counsel of perfection, not easily followed in many archaeological studies where no one method is clearly best. My own preference – in the context of provenance and similar studies, where hierarchical methods are to be used – would be to employ both Ward's method and average linkage, reporting the results of both, and assessing their similarity (the subject of Section 8.4). With fairly clear cluster structure unaffected by the problems of correlated data (Section 8.5) the two methods are likely to give substantively similar results, with average linkage more likely to reveal outliers in the data. Section 8.9 gives a detailed example.

If cluster analysis is used then, in general (and excluding monothetic divisive procedures which are discussed in Section 8.7), k-means procedures seem more appropriate than hierarchical methods for the kinds of clusters that often seem to be expected in archaeological studies. For many archaeological problems of a non-archaeometric nature this conclusion is unexceptionable and has long been advocated (e.g., Hodson, 1970; Doran and Hodson, 1975; Kintigh and Ammerman, 1982). For archaeometric studies it also holds – there is no basis in general to expect a hierarchical structure to obtain, and relocation of the final clusters derived hierarchically (a form of k-means) can only improve the classification according to the clustering criterion (typically Ward's method or average linkage) used.

Seber (1984, p. 381) notes a tendency for k-means algorithms to produce clusters of a similar size. Procedures based on sums of squares – which includes k-means clustering of spatial data – will tend to find spherical clusters even though there is no particular reason to expect artefact scatters, for example, to be spherical (Wright, 1989).

8.2.4 Conclusions regarding clustering algorithms

Cormack's (1971, p. 321) review of classification began by stating that computer packages of classification techniques had 'led to the waste of more valuable scientific time than any other "statistical" innovation'. The techniques he reviewed included most of those discussed in Chapter 7 that have been widely used in archaeological applications. On technical grounds alone (ignoring archaeological considerations), many difficulties exist.

At one extreme, in studies of artefact compositions and particularly ceramics, it might be argued that the hierarchical methods typically used are not congruent with the structure to be expected; the dissimilarity coefficient usually employed is not especially appropriate for the data; and the two main methods of choice, Ward's and average linkage, can produce different results and either create artificial structure or obscure real structure. This rather pessimistic view is discussed at greater length in Section 8.5.

Similar problems arise with such methods applied to other archaeological data. Monothetic divisive procedures are well known to lack robustness; k-means procedures are capable of imposing inappropriate structure on the data.

Faced with these difficulties, the widespread use of cluster analysis in archaeology has to be justified by pragmatic and empirical rather than theoretical considerations. Do the methods produce interpretable results that cannot be produced by other and simpler means, or do they significantly speed up the process of data analysis and interpretation? Archaeologists and archaeological scientists are the best judges of this. For some kinds of problems some have answered 'No'; Adams (1988), for example, doubts whether typological studies that depend on automatic methods of classification have improved on more traditional approaches. One hopes, however, that the widespread and continuing use of the methodology represents an implicit 'Yes' rather than confirmation of Cormack's view.

The methods can be treated as quick exploratory means of imposing some order on data that can subsequently be analysed for substantive content. Different possibilities may be listed.

(a) Structure in the data is reasonably clear and captured by all the competing methodologies, so that one is reasonably confident that the revealed structure is 'real'.

(b) In practice theoretical difficulties either do not apply or are of little practical consequence. Some demonstration of this would usually be desirable.

(c) Notwithstanding inherent problems in the methodology, the results obtained may be satisfactory in that they can be given a sensible archaeological interpretation. For example, in published provenance studies it is often the case that a clear number of clusters are apparent in the final dendrogram and that these may be associated with material of different origin. In these cases it is likely that different methods or relocation would have little effect on the substantive conclusions. In any case it is desirable that the outcome or interpretation of an analysis is validated in some way. That is, any apparent structure revealed by a cluster analysis should be confirmed as genuine, rather than purely an accidental product of the methodology. These matters are considered in the next sections.

8.3 Assessing cluster validity

It can be difficult to determine the number of 'real' clusters from typical output and any decision should therefore be validated in some way. Aldenderfer (1982) and Whallon (1990) discuss aspects of this problem, but otherwise there are few explicit discussions of the topic in the archaeological literature. It is not adequate to judge the numbers of clusters by inspection of the dendrogram; as Seber (1984, p. 388) notes, 'a large change in the fusion level of a dendrogram is a necessary but not a sufficient condition for clear-cut clusters'. This can be graphically illustrated by reference to Figure 8.1(a) which shows a random scatter of points based on two

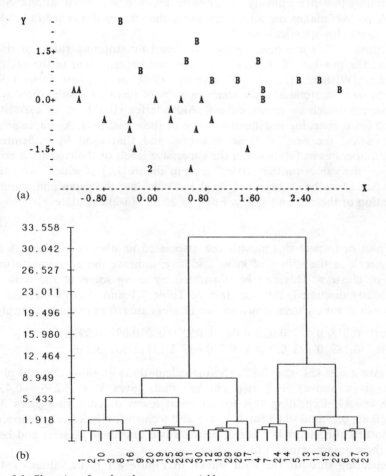

Figure 8.1: Clustering of random data on two variables

Note: Diagram (a) is a plot of two randomly-generated variables labelled according to the clusters suggested by Ward's method in diagram (b).

randomly and independently-generated normally-distributed variables. Figure 8.1(b) shows the dendrogram obtained by using Ward's method; this would often be interpreted, incorrectly, as showing two quite distinct, externally isolated clusters. Wright (1989) notes that this phenomemon is typical.

As Everitt (1980, pp. 64–6) observes, the problem of determining the 'correct' number of clusters is a difficult one for which no completely satisfactory solution exists. Solutions based on statistical hypothesis-testing procedures described in Everitt (1980) and Seber (1984), for example, make strong distributional assumptions about the data (i.e. multivariate normality) that are unlikely to apply in many practical situations. Seber (1984, p. 389) makes the additional point that the methods are only likely to be useful for spherical clusters.

Mojena (1977) has investigated two heuristic stopping rules for determining the number of clusters, and these are implemented in the CLUSTAN package (Wishart, 1987). Both Everitt (1980, p. 67) and Seber (1984, p. 388) state that one of these seems 'worthy of further consideration' without saying which or giving details. Aldenderfer (1982), in an expository article for archaeologists, illustrates one of these methods. As implemented in CLUSTAN, the first of these methods, and that used by Aldenderfer (1982), operates as follows. Let the successive levels of the clustering criterion (i.e. the values on the vertical scale in Figure 7.2) at which individuals or clusters fuse be $(z_1, z_2, \ldots, z_{n-1})$ and calculate the mean and standard deviation of these, $\bar{z}$ and s_z say. For $j = 2, \ldots, (n-2)$ calculate

$$k_j = (z_{j+1} - \bar{z})/s_z \qquad (8.2)$$

and plot or inspect this against the associated number of clusters. A natural break in the values of k_j will, ideally, indicate the appropriate numbers of clusters. This can be illustrated by using some of the data sets previously discussed. For the data of Table 7.1 and Ward's method, the values of z_j for eighteen down to two clusters are (from CLUSTAN output)

0.010 0.018 0.023 0.028 0.047 0.052 0.060 0.092 0.097
0.107 0.189 0.193 0.353 0.387 0.639 1.141 1.501 3.236

and give $\bar{z} = 0.454$, $s_z = 0.808$ (basing calculations on more decimal places than given above). For two clusters (8.2) gives $k_j = (3.236 - 0.454)/0.808 = 3.44$. Repeating this for different levels of clustering gives 3.44, 1.29, 0.85, 0.23 and -0.08 for 2, 3, 4, 5 and 6 clusters respectively. Inspecting or plotting this suggests a break between two and three clusters and hence that there are two clusters in the data.

Repeating this for the data shown in Figure 8.1 gives values of 4.79, 1.39, 1.03, 0.51 and −0.01 for 2, 3, 4, 5 and 6 clusters, suggesting clearly that there are two clusters. Unfortunately, in this case the data are known to be random, so that the rule supports the misleading impression given by the dendrogram. Aldenderfer (1982), citing other studies, notes that the rule seems to favour clustering procedures biased towards the discovery of compact spherical clusters.

The second of Mojena's rules is described as 'statistically superior' by Wishart (1987); one reason is that calculations of $\bar{z}$ and s_z in (8.2) are influenced by the criterion value, z_{j+1}, that they are supposed to test. The second rule is somewhat more complicated but has the form

$$k_j = [z_{j+1} - E(z_{j+1})]/s_j \qquad (8.3).$$

Here $E(z_{j+1})$ is the expected value of z_{j+1}, an expression for which is given by Wishart (1987, p. 157); it and s_j are based on the r values of z_i preceding z_{j+1}. The value of r can be chosen by the user; if no choice is made CLUSTAN will produce output for a succession of choices. Results are more difficult to interpret than for the first rule but are intended to identify candidates for significant partitions. For Ward's method applied to the data of Table 7.1 the largest values of k_j with $r = 8$ are 3.61, 2.94 and 2.59 with 2, 4 and 7 clusters. The 3.61, for example, shows that the criterion value for the two-cluster partition is about 3.5 standard deviates higher than the mean of the preceding eight criterion values. The result supports that of the first rule in that the two-cluster partition is most favoured. For the randomly generated data of Figure 8.1 with $r = 12$ the values of k_j are 6.53, 5.13 and 3.59 for 2, 5 and 4 clusters leading to similar (erroneous) conclusions to the first rule.

There is little evidence from the archaeological literature that Mojena's rules, or other formal testing procedures, have been widely used. Whallon (1990), whose paper is a rare exception to this generalisation, makes a similar point. The rules are in a similar spirit to that in which the value of the fusion criterion is plotted against the number of clusters – a procedure that has been advocated for k-means methods. Seber (1984, p. 388) states that such plots are often unclear and, as their interpretation is subjective, may help little in reaching a definitive answer. Rather than attempting to 'cut' the dendrogram at a single fusion criterion, Digby and Kempton (1987, p. 138) suggest that 'cutting certain joins on the dendrogram rather than choosing a single similarity threshold' may lead to more interpretable results. This idea has been explored with archaeological data by Whallon (1990).

In practice, if assessment of the number of clusters goes beyond the mere inspection of a dendrogram (which may mislead), then informal and subjective methods tend to be used. Since the term 'subjective' is sometimes used in a pejorative fashion I should emphasise that I do not regard it as such. For example, in Ottaway's (1974) study of impurity patterns in Armorico-British daggers her statement that the fusion was broken 'more or less arbitrarily' on the grounds of metallurgical sense is an admirably honest and clear statement of what was done. This could usefully be emulated in other studies where the grounds for choosing a certain level of clustering are sometimes unclear.

Having said this, the logic involved in subjectively determining the numbers of clusters in this way needs to be monitored carefully to avoid 'circular' reasoning. The success of archaeological cluster analyses is often judged by the ability to reproduce known archaeological groups. Thus in provenance studies, where the material studied is known to form groups from distinct locations, the success of a cluster analysis is sometimes judged by its ability to reproduce these known groups. It is not always clear what can be achieved in such cases, apart from confirming that archaeologically-defined groups are chemically distinct. Negative results are, I suspect, not reported or not believed (constraining the ability to learn from application of the methodology and constituting a form of publication bias); where negative results are reported this is usually deemed to reflect badly on the methodology rather than being substantively informative.

One approach to this problem is to include individuals known to group together as a subset of the total number studied. It is expected *a priori* that such material will cluster together and it can be used to monitor a clustering. Thus if, in a provenance study, specimens of known origin were included with unprovenanced material, a clustering could be terminated at the point where known groups were broken up. Examples in the literature include Djingova and Kuleff (1992) who include a reference sample in their study of Bulgarian Medieval glass, and Gunneweg et al. (1991) who include ten samples of coal fly ash that should group together in a study of Middle Eastern pottery.

In summary, no generally applicable solution to determining the appropriate number of clusters exists. Reliance on the form of the dendrogram is unsafe; monitoring the value of the fusion criterion as clusters change is often unclear; and formal and objective methods are rarely used (and will often be inapplicable). This means that informal and subjective criteria, based on subject expertise, are likely to remain as the most common approach. In published studies practice could be improved by making such criteria more explicit than is sometimes the case.

Determining the number of clusters is an important aspect of assessing cluster validity. Other graphical procedures can also be useful and some of these have been discussed in Aldenderfer (1982). Principal component or correspondence analysis plots can be used to supplement a cluster analysis and the use of the former is quite common (Sections 1.5; 4.13). Clear structure on the component plot (or plots if more than the first two components are used) will serve to confirm the reality of structure and will sometimes suggest that the cluster analysis is partitioning the data rather than identifying distinct clusters. Absence of structure on the plots does not, however, imply that there are no clusters, since these may well exist in

Cluster Analysis – Some Problems

a higher number of dimensions than are capable of representation on the plots. In addition, as Alvey and Laxton (1974) and Aldenderfer (1982) note, PCA can blur distinctions between clusters.

Discriminant analysis, in contrast to PCA, will emphasise distinctions between known groups and is also a common way of graphically complementing a cluster analysis. It is discussed at length in Chapters 9 and 10. Its main drawback, in the context of cluster analysis, is that it can produce too optimistic a portrayal of cluster separation and may mislead as to the genuineness of clusters. This problem is discussed and illustrated in Section 10.8.

8.4 Comparing clusterings

Rather than relying on a single cluster analysis method, for which there is no overwhelming theoretical basis, it will often be sensible to examine competing methods. If these produce similar results this will increase faith in the reality of the clusters. In reporting results it is desirable that all the analyses undertaken be commented on and that any differences or negative results be explained. At all costs the temptation to report the 'best' result only (and as if it were the only analysis) should be avoided.

Clusterings can be compared in a variety of ways. For output in the form of dendrograms careful inspection will often reveal similarities and differences. If, say, a Ward's method output suggests a smallish number of clear clusters, the individual labels on the output can be highlighted using different coloured pens, according to their cluster. Similar highlighting on dendrograms for other methods will quickly reveal how similar or dissimilar clustering is, despite the often different appearance of the dendrograms.

This approach, though difficult to report formally, is quicker than it sounds and I have used it with success on data sets of over 150 individuals. In a comparison of Ward's method against average linkage, for example, the latter method typically suggests outliers not evident on the dendrogram for the former. Clustering is usually clearer for Ward's method but, overall, the method suggested above will often show that the results are essentially the same. The larger Ward's method clusters may be subdivided by average linkage and the subgroups will sometimes belong to different higher-level clusters. It is rare for the results of one method to completely scramble those of another. (These observations are based on my own research and unpublished student project work on artefact compositional data sets; an example is given in Section 8.9.)

More formally, with two methods, results may be compared in the form of a contingency table. If equal numbers of clusters are obtained in each case then, ideally, most or all of the individuals will fall into cells on the leading diagonal of the table (assuming clusters are appropriately ordered). In practice, outliers or clusters with few members may cause some problems and may need to be 'discounted' before forming clusters. This is illustrated in Section 8.9.

Digby and Kempton (1987, pp. 147–9) suggest that the contingency table might also be run through a correspondence analysis, if its message is not obvious, to get a visual display of the clusterings. This kind of approach is illustrated in Stopford et al. (1991). For comparing three or more clusterings, Digby and Kempton suggest that multiple correspondence analysis of a three- or higher-dimensional table relating the clusterings may be useful. An example of this approach is given in Baxter (1992b).

For output in the form of a dendrogram the co-phenetic correlation coefficient is sometimes used either to compare different analyses or to assess the extent to which an analysis distorts the original similarity matrix (e.g., Magne and Klassen, 1991). Shennan (1988, p. 231) gives a simple example and the measure is available in the CLUSTAN package. The (dis-)similarity between two individuals in a clustering is measured by the value of the clustering criterion at which they fuse on the dendrogram. The co-phenetic correlation coefficient is just the correlation between these similarities for two dendrograms, or between these and the original similarities.

Table 8.1 shows the value of the co-phenetic correlation between each of the methods used in Figure 7.2 and the original similarity matrix. The comparison of methods with the data is typical for this type of data (Harbottle, 1976; Pollard, 1986) – good but by no means perfect. The similarity between the methods varies and is good between Ward's method and complete linkage, for example, both of which differ from single linkage.

The appropriateness of the co-phenetic correlation for many archaeological problems may be questioned. Essentially it is a measure of the similarity between two complete trees rather than of the final clusters inferred from those trees. It has earlier been argued that for many archaeological applications a hierarchical structure is inappropriate (even if hierarchical methods aid in cluster identification). Put another way, two techniques may well lead to identical clusters as judged by the user, but the dendrograms may well be different and measured as such by the

Table 8.1: The co-phenetic correlation coefficient for different cluster analyses of the data in Table 7.1

	SL	CL	AL	W	Data
SL	1				
CL	.33	1			
AL	.75	.82	1		
W	.28	.97	.80	1	
Data	.60	.63	.70	.60	1

Note: SL = Single link; CL = Complete link; AL = Average link; W = Ward's method.

8.5 Problems with correlated variables

co-phenetic correlation because of what happens higher up the tree than the approved clustering.

It was noted in the previous chapter that the very large number of applications of cluster analysis to artefact compositions, with few exceptions, use Euclidean distance (4.1) or some variant of it as a measure of dissimilarity. Additionally, many applications use Ward's method of analysis which works best with, and will tend to produce, spherical clusters. Where variables are highly correlated, clusters will be non-spherical, and the use of both Euclidean distance as a measure of similarity and Ward's method as a clustering technique is open to question.

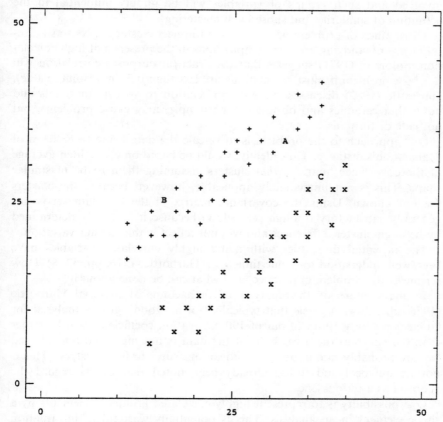

Figure 8.2: An artificial example to illustrate problems that can arise when clustering highly correlated variables

Note: Points A and B are clearly in the same cluster while C is in a distinct cluster. As measured by Euclidean distance A is, however, closer to C than B and is likely to cluster with C when using a similarity measure based on Euclidean distance.

There are several potential problems with Euclidean distance that are concisely summarised in Cormack (1971, p. 326). One problem can be seen by reference to Figure 8.2 where two clear clusters based on a pair of highly correlated variables are shown. Individual A, although clearly in the same cluster as B, is closer to C as measured by Euclidean distance. Several examples in the literature exist, or are cited, that demonstrate that Ward's and other methods will cluster A with C and fail to recapture the true structure (e.g., Everitt, 1980; Everitt and Dunn, 1991, p. 112; Bishop and Neff, 1989).

A related problem arises with several variables that are highly correlated. If one of these makes a high contribution to d_{ik} in (4.1) then it is highly likely that the others will as well; in a sense 'double counting' is involved and such groups of variables will be unduly influential in the definition of similarity and subsequent clustering.

These kinds of problem have led some to reject cluster analysis as a useful means of studying artefact compositions in the presence of high correlation. Solomon (1971) suggests that for practical purposes correlation will not be a problem if most correlations are less than 0.5 (in absolute value). Harbottle (1982) discusses the issue in relation to specific materials and notes that ceramics and obsidian, for example, may cause problems, but not jade or turquoise.

One approach to the problem is to rescale the data using some form of Mahalanobis distance. This, ideally, would be based on a weighted average of the covariance matrix *within* clusters, assuming them to be of similar shape. This is not immediately applicable, however, because the clusters are not known. Using the covariance matrix of the full data set is not generally satisfactory (though see below) because it is largely determined by between-cluster differences and will not reflect within-cluster variation.

The potential difficulties with many highly correlated variables have been well understood for some time (e.g., Harbottle, 1976, pp. 52–8). How serious is the problem in practice, and what can be done about it?

In the context of the analysis of archaeological material Harbottle (1976, pp. 53–4) suggests that typically 'for a "good" group analysed for 20 elements only 10 or 20 out of 190 correlation coefficients will be larger than 0.8' so that 'the great bulk of the data is not highly correlated' and 'we are probably safe in using Euclidean measures to find clusters'. This is not unequivocal and it has already been noted that others regard the problem as a serious one.

One possibility is to represent highly correlated groups of elements by a single element in an analysis. This is potentially wasteful of information ('Amr, 1987) and also raises the problem of identifying variables that are highly correlated within all of the unknown groups.

No satisfactory formal solution to the problem seems to exist or to have been widely used in archaeology. Some that have been proposed are based

– implicitly or explicitly – on transformation of the original data using some form of Mahalanobis distance. The general idea is that a suitable transformation of the data will convert elongated (or, in general, hyper-ellipsoidal) clusters, such as those in Figure 8.2, to spherical clusters well suited to standard clustering algorithms.

As already noted, practical difficulties arise in obtaining a suitable estimate of the variance-covariance matrix, S in (4.14), that will enable the transformation to take place. One approach that has been occasionally used or suggested is to undertake a principal component analysis first and to carry out the cluster analysis on the principal components or a subset of them (Barrett et al., 1978; Chappell, 1987; Goad and Noakes, 1978; Hart et al., 1987; Hutcheson and Callow, 1986; Law, 1984; Storey et al., 1989; Bishop and Neff, 1989; Wright, 1989). Since PCA simply involves the definition of new axes in a space and does not affect distances, a cluster analysis on unstandardised components will be the same as on the original data. Using a subset of the components will simply approximate to the original analysis. There may be some merit in conducting a cluster analysis on components *standardised* to have equal variance. The idea here is that groups of highly correlated variables – that lead to 'double counting' with Euclidean distance – may be accounted for by a single component that is then weighted equally with other components corresponding to other variables, or variable subsets, in the analysis (Jolliffe, 1986, pp. 163–4). Jolliffe also notes that using such standardised components is equivalent to using Mahalanobis distance based on the original variables; it has already been suggested that this will not, in general, be satisfactory because it will mainly reflect between-cluster, rather than within-cluster variation. Nonetheless Bishop and Neff (1989) give an example similar to Figure 8.2 and state that cluster analysis on principal components will successfully recover the structure. This is probably because, in the instance they cite, clusters are of similar shape, highly elongated and fairly close to each other, so that the covariance matrix for the data as a whole reflects within-cluster covariance reasonably well. This may not be true in general.

Where an explicit justification for clustering principal components is given, the hoped removal of correlation effects seems to be the main reason. Whether or not components are standardised is not always made clear. In non-archaeometric applications cluster analysis on factors from a factor analysis, rotated principal components (Sections 4.10, 4.11) or correspondence analysis has sometimes been used (Ciolek-Torello, 1984, 1985; Diaz-Andreu and Fernandez-Miranda, 1991; Gaillard et al., 1986; O'Shea, 1984; Perry and Davidson, 1987). If factor analysis proper is used (Section 4.11), then the idea is that the factors represent latent variables that underlie the observed data, having discounted 'noise' effects, and clustering is based on these (hopefully interpretable) variables.

Other procedures, designed in part to cope with the correlation problem, assume a statistical model for the data. These often suppose that variables are normally distributed within clusters and that clusters are of a similar geometrical size and shape. Given an initial random allocation to clusters, the variation and covariation within clusters can be measured. Iterative procedures are then used to relocate specimens to minimise internal variability in some sense while maintaining cluster separation. These ideas are discussed with a little more technical detail in Section 8.8.

8.6 Spatial clustering and related topics

In k-means spatial clustering a pattern of clusters is imposed on a two-dimensional scatter of points in an attempt to define areas of high density. The need for this, given that the data are readily inspected by eye, may be queried. Ammerman et al. (1987, p. 211) suggest that 'we can search informally by means of visual inspection for meaningful patterns in our distribution maps' but if 'the maps we are working on have many points we may find . . . that our time is endlessly consumed by the business of pattern matching'. Spatial clustering is viewed 'as a means of conducting the search for patterns in a distribution map' that may not lead directly to the solution of an archaeological problem but forms a basic step in the resolution of such problems (see also Kintigh and Ammerman, 1982).

The problem of determining the appropriate number of clusters is addressed in Kintigh and Ammerman (1982). They randomise the ordering of the x and y co-ordinates separately and recombine them to get a random set of locations that is also subjected to cluster analysis. In published applications two such randomisations are often used. For both these and the original data the logged value of Q (to base 10), the clustering criterion optimised, is plotted against k as k varies. Divergence of the plot for the real data from the plots for the random data indicates the existence of spatial clustering. Levels of clustering worthy of further study, of which there may be several, are indicated by inflections in the graph.

Apart from the papers already cited, examples are given in papers by van Waarden (1989), Gregg et al. (1991), many of the references to Simek's work in the bibliography, and the book by Blankholm (1991). Solutions with few clusters will sometimes quite clearly amalgamate spatially distinct concentrations of points, whereas solutions with lots of clusters may arbitrarily subdivide large concentrations into smaller spherical clusters. Both features are apparent in the paper by van Waarden (1989) and, in part, may be attributable to the tendency of the method to impose spherical structure.

In the original paper by Kintigh and Ammerman (1982), Ammerman et al. (1987) and many of Simek's papers, the clusters, once defined, are associated with a radius, and circles with these radii are drawn round cluster centroids to aid interpretation. This reflects the tendency noted in the last paragraph. Wright (1989) observes that this is rather restrictive since clusters are often self-evidently not circular. His MV-ARCH package allows ellipses rather than circles to be placed about cluster centroids as a truer reflection of structure.

Whallon (1984) laid down a number of desiderata for spatial clustering procedures if these are to be congruent with archaeological needs. The imposition of contiguity in Kintigh and Ammerman's (1982) algorithm; the handling of only one type of artefact/structure at a time; and (possibly) the imposition of spherical structure all appear to violate these (Whallon, 1984, pp. 243–4). Whallon's own approach, termed 'unconstrained spatial clustering', is to:

(i) define smoothed density maps across a site for each of p artefact types;

(ii) use these to obtain the set of absolute artefact densities at each of n points, $(x_{i1}, x_{i2}, \ldots, x_{ip})$ say, for $i = 1 \ldots n$ where the n points may be artefact locations or grid points or squares;

(iii) convert each set to relative frequencies by dividing by their sum;

(iv) cluster locations, without regard to contiguity, using Ward's method for example, and measuring inter-location similarity on the basis of the relative frequencies defined at (iii);

(v) plot cluster membership on a map to see if spatial patterning of areas with a similar relative artefact composition is revealed.

It will be seen that step (iii) reduces the data to compositional form and introduces potential technical problems of the kind discussed in Section 4.6. Ridings and Sampson (1990) use Whallon's approach in a spatial analysis of Bushman pottery decorations and note the compositional problem. They reference Aitchison's (1986) work on the topic, but conclude there are too many zeroes and small values in the data for it to be applied. The compositional constraint introduces negative biases into the correlations between artefact types – this is evident in Whallon's (1984) Table 1 based on five variables and his Figures 7–13. This, in turn, means that variables are not independent, and raises questions about the suitability of the clustering procedures used for the reasons discussed in Section 8.5. Other applications of the unconstrained clustering approach are given in Blankholm (1991), Cribb and Minnegal (1989) and Gregg et al. (1991).

One problem associated with the smoothing procedure noted in step (i) above is that the composition of the derived assemblage, over a grid for example, can include artefacts that quite clearly do not occur in the grid (Kintigh, 1990). Whether this and other theoretical difficulties are of practical importance is perhaps an empirical matter, and more studies may be needed for this to be decided. The recent studies by Blankholm (1991) and Gregg et al. (1991) that contrast the k-means and unconstrained spatial clustering approaches are useful in this respect. The two approaches are judged to be complementary and, in the latter paper, prove capable of recovering structure from 'complete' anthropological data where results may be judged against what is known of occupation at the site. The data are successively degraded to simulate material of the kind surviving in the archaeological record and, with sufficient degradation, the methods fail to recapture the original structure. This, however, is a fundamental problem related to the inferences that it is possible to draw from the archaeological record, and not a reflection on the methodologies used. Blankholm's study, including other methods, also judges them to work well in practice.

Koetje (1991) has extended the k-means approach to the analysis of data recorded three-dimensionally, but few published examples of this approach yet exist. It was noted in the previous chapter that I have no experience of using the methods discussed in this subsection. The paper by Gregg et al. (1991) and book by Blankholm (1991) provide a useful source of examples for those interested, and may be complemented by the critical review of Kintigh (1990) for a fuller discussion of technical aspects and problems with the two methodologies.

8.7 Mortuary studies and monothetic methods

Discussion of the limitations of monothetic divisive procedures has been deferred to this section because its main use seems to have been in the context of mortuary studies. Critical accounts of the statistical methodology from an archaeological standpoint occur in Doran and Hodson (1975), Braun (1981) and Brown (1987) and comparative studies of methodology include Jones (1980), O'Shea (1984, 1985) and Tainter (1975); the following discussion draws on these.

Most such studies listed in the bibliography that were discussed in Section 1.5 use binary data. In terms of relevant data that may be to hand this may be wasteful of information. If abundance data are converted to presence/absence data, then large assemblages will typically contain more attributes that are present than smaller assemblages. That this will 'bias' Jaccard's coefficient of similarity (7.3) in favour of the comparison of large assemblages has already been noted. If continuous or multi-state variables are converted to binary variables, or not used, then information is obviously being lost. The seriousness of this for the results obtained is a matter for dispute; it is an aspect of the more general problem of deciding what variables to include in an analysis.

Another general issue is the choice between monothetic and polythetic clustering procedures, which is partly a reflection of the theoretical stance that is adopted towards mortuary practices (Doran and Hodson, 1975; Brown, 1987). Examples of the polythetic approach include Pearson et al. (1989) and Rothschild (1979), who use the Jaccard coefficient with Ward's method and average linkage respectively. Brown (1987, p. 301) suggests that 'polythetic and monothetic procedures have generally provided confirmatory support of each other'.

Although monothetic divisive procedures can be viewed as congruent with theories of how social stratification is reflected in death (Brown, 1987), there are serious doubts about their robustness (Digby and Kempton, 1987, p. 130). This can be because outliers or rare attributes lead to progression down the wrong branch of a hierarchy (Seber, 1984, p. 376) so that early partitions may easily cut through natural groups (Gower, 1967). There is far less scope for relocation to improve the final classification than in polythetic procedures (Everitt, 1980, p. 68).

If monothetic divisive clustering is used, then the choice is usually perceived to be between association analysis and the information statistic (Section 7.2) with published studies since Peebles (1972) tending to favour the latter (e.g., King, 1978; O'Shea and Zvelebil, 1984). There are good reasons for this since association analysis has been very strongly criticised for technical reasons to do with the treatment of co-absence of an attribute as indicative of similarity; sensitivity to skewness in the data; and poor practical performance (e.g., Cormack, 1971; Doran and Hodson, 1975; Seber, 1984).

An alternative approach to the analysis of grave assemblages, the 'status table' approach of Hodson (1977), that could be thought of as a form of cluster analysis may briefly be noted. Functional types that occur within graves are assigned a status index and graves can then be ordered or grouped on the basis of the indices of types within them. Ordering according to the highest status type within a grave or on the basis of the sum of indices of types within a grave is possible (Duncan et al., 1988). This reference describes the Institute of Archaeology (London) statistics packages (IASTATS) within which the approach is available, and a published example is given in Hodson (1990).

8.8 Other approaches

The methods discussed so far have overwhelmingly dominated archaeological practice (Section 8.10). Here, other approaches with links to more formal statistical methodology, that encompass the k-means approach, are noted. Everitt (1980, pp. 40–6), Gordon (1981, pp. 49–53) or Seber (1984, pp. 379–87) can be consulted for further information.

The total scatter or dispersion of a set of data may be defined similarly to the covariance matrix, except that the denominator is ignored. Call this matrix $\mathbf{T}$. It can be shown that, for a given partition into k clusters, $\mathbf{T}$ can be decomposed into two parts, $\mathbf{T} = \mathbf{W} + \mathbf{B}$ say, where $\mathbf{W}$ is a measure of the scatter within clusters and $\mathbf{B}$ is a measure of the scatter between clusters (these ideas are discussed more formally in Chapters 9 and 10). A 'good' clustering is associated with tightly-defined and well separated clusters, which corresponds to $\mathbf{W}$ being 'small' and $\mathbf{B}$ being 'large' in some sense. Since 'small' and 'large' admit of several definitions, this has led to several proposed methods of analysis that involve minimising some function of $\mathbf{W}$ and/or $\mathbf{B}$.

The trace of a matrix is the sum of its diagonal elements and in the case of **W** can be shown to be equivalent to the sum of a set of terms of the kind defined in (7.1). This sum is just the criterion, Q_w, minimised in k-means (spatial) clustering with the error sum of squares as the clustering criterion. Thus, minimising the trace of **W** defines the k-means procedures and raises the same problems about determining k that have already been described. If the variables within clusters are assumed to be uncorrelated, normally distributed and of equal variance, with the variance the same from cluster to cluster, then minimisation of the trace of **W** is equivalent to maximum likelihood estimation (Gordon, 1981, p. 51).

This makes the method a statistically attractive one if the assumptions hold, and helps explain why methods based on the sums of squares criterion will tend to produce similar-sized spherical clusters. Conversely, if the assumptions, which are very strong ones, do not hold, then minimising the trace of **W** is not optimal in the statistical sense. For many archaeological applications the assumptions are unlikely to be valid.

If the assumption of no correlation is dropped but it is assumed that clusters are of the same shape and geometrical size, then maximum likelihood is equivalent to minimising the determinant of **W**. This also tends to produce clusters of roughly equal size (Gordon, 1981, p. 52). Other criteria that have been used include maximisation of the trace of $\mathbf{BW}^{-1}$. Some of these criteria are available in CLUSTAN using the INVARIANT procedure and are illustrated in Section 8.9.

An attraction, to statisticians, of the procedures outlined is that a model of the data is made explicit which allows the cluster analyses to be linked to standard statistical methodology and enables the performance of the techniques and their limitations to be better understood. For archaeologists the practical drawback is that much real data may not even approximately satisfy the assumptions needed for the methods to be optimal ones. Similar considerations concerning assumptions about the normality of the data and form of **W** are likely to limit the practical application of other statistical modelling approaches, such as the Bayesian methodology of Buck and Litton (1991).

8.9 An example

To illustrate some of the ideas of this chapter on a realistic example, the forty-six Mancetter specimens from Tables A1 and A2 will be used. Dotplots for each element (not illustrated) show that specimen 12 has unusually high values of Mg, K and P; 19 is very high on Cr; 30 is very high on K; and 38 is very low on Ba. The elements Sb, in particular, and Pb are skewed with high peaks at low values. The PCA of untransformed standardised data suggests that specimens 9, 12 and 42 form a small outlying group on the first component, while 19 is outlying on the second component (Figure 8.3(a)). Specimens 12 and 19 are also very clear outliers on

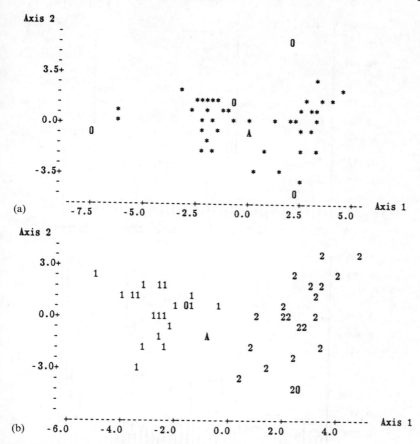

Figure 8.3: PCAS of the Mancetter data from Tables A1 and A2

Note: Both diagrams show PCAS of the Mancetter glass waste analyses extracted from Tables A1 and A2 using all variables. The analyses are of the correlation matrix and the lower diagram omits four outliers identified by the upper analysis. Numbers in the lower diagram label specimens according to the groups suggested by PCA and Ward's method of cluster analysis. An 0 indicates a point suggested as an outlier by initial data analysis; the A identifies a point relocated after the Ward's method cluster analysis.

the fourth component (not illustrated). This apart, there is the suggestion of two groups which becomes clearer on omitting the four outliers on the first PCA plot (Figure 8.3(b)). This analysis, and the conclusions drawn from it, can be done in less than five minutes at a terminal using MINITAB.

For illustrative purposes the outliers will be retained in the cluster analyses to see if they are detected. Figures 8.4(a) and 8.4(b) show the results of a Ward's method and average linkage cluster analysis on the data after standardisation and using Euclidean distance as the similarity measure. These look rather different, though on closer inspection turn out

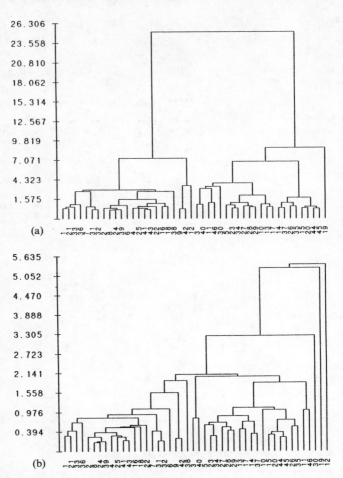

Figure 8.4: Cluster analyses of the Mancetter data from Tables A1 and A2

Note: (a) shows a Ward's method analysis of the data and (b) an average link analysis.

to be more similar than appears at first sight. The dendrogram for Ward's method suggests, as is typical, two clear clusters. The outliers evident on the PCA plot (9, 12, 42) and 19 are to the right of the left and right clusters respectively and would be detached from them if a four-cluster level were selected. Apart from outliers, the division suggested by Ward's method is identical to that on the PCA plot in Figure 8.3(b) – dividing groups at 0 on the first component – except for specimen 28 (A on the figure). If relocation, using Q_w, is used, this specimen is the only one relocated, so that the grouping suggested by Ward's method and the PCA are the same.

If the dendrogram for the average link method is examined, three outliers (12, 19, 30) are suggested. Appearances notwithstanding, cutting the dendrogram to give five clusters gives two large clusters (running from 1 to 38 and 3 to 46 on the plot) that, apart from the outliers, are identical to the Ward's clustering. This is all summarised in Table 8.2.

Are these clusters real? They have been determined by inspection using the dendrogram for Ward's method as a reference point, and decisions have been informed by a knowledge of the PCA and prior data inspection. Mojena's first rule for Ward's method quite clearly suggests a two-cluster solution with a value of 7.87 compared to the next highest of 1.67. For Mojena's second rule the decision is the same, since the two-cluster solution is clearly favoured (with r = 28 the value is 8.95) but a five-cluster solution (with a value of 3.37) is also a possibility (subdividing the cluster to the right). With average linkage the first rule suggests two, three or four clusters, with values between 3.85 and 2, and the second rule leads to

Table 8.2: Results of different classifications obtained by cluster analyses of the Mancetter data from Tables A1 and A2

			Method							Method			
No.	1	2	3	4	5	6	No.	1	2	3	4	5	6
1	1	1	1	1	1	1	24	1	1	1	1	2	3
2	1	1	1	1	1	1	25	1	1	1	1	2	1
3	2	2	3	2	3	2	26	2	2	2	2	1	3
4	1	1	1	1	1	1	27	2	2	2	2	2	2
5	2	2	3	2	2	1	28	1	2	3	2	2	1
6	1	1	1	1	2	2	29	2	2	3	2	2	1
7	1	1	1	1	3	2	30	2	2	2	0	1	2
8	1	1	1	1	1	3	31	1	1	1	1	2	2
9	0	1	1	1	3	3	32	1	1	1	1	2	1
10	2	2	2	2	2	3	33	1	1	1	1	1	1
11	2	2	2	2	3	3	34	2	2	3	2	3	3
12	0	1	1	0	3	2	35	2	2	2	2	3	2
13	2	2	2	2	1	3	36	1	1	1	1	1	3
14	2	2	2	2	2	3	37	2	2	2	2	2	3
15	2	2	2	2	1	3	38	1	1	2	1	2	1
16	1	1	1	1	1	3	39	1	1	1	1	2	1
17	2	2	2	2	2	1	40	2	2	3	2	3	2
18	1	1	1	1	3	2	41	1	1	1	1	2	3
19	0	2	0	0	1	2	42	0	1	1	1	3	1
20	2	2	2	2	2	3	43	1	1	1	1	1	3
21	1	1	1	1	2	1	44	2	2	2	2	2	3
22	1	1	2	1	3	2	45	2	2	2	2	2	3
23	2	2	3	2	3	2	46	2	2	1	2	3	3

Note: Mancetter specimens are numbered sequentially; entries in the table identify cluster membership with a 0 indicating an outlier. The six methods used were (1) PCA; (2) Ward's method; (3) Ward's method on the first six standardised principal components; (4) Average linkage; (5), (6) The INVARIANT procedure in CLUSTAN to minimise trace $\mathbf{W}$ and maximise trace $\mathbf{BW}^{-1}$ respectively.

similar conclusions. In this case between one and three outliers are separated from the bulk of the data leaving the rest undifferentiated.

So far the potential problems caused by correlations have been ignored. Of the 231 correlations 4% are greater than 0.7 and a further 15% are greater than 0.5. According to the criterion noted earlier that was suggested by Harbottle (1976), very few correlations are greater than 0.8, so using Euclidean distance may not be a problem.

In an admittedly experimental manner analyses were rerun using the INVARIANT routine from the CLUSTAN package to minimise the determinant of W and maximize the trace of BW^{-1} (Section 8.8), requesting a three-cluster solution and using ten different random starting positions. Results, which are summarised in Tables 8.2 and 8.3 (b, c), were worse than useless, being neither similar to each other nor to the Ward's method analysis. Plotting cluster membership on the component plot obtained by PCA (suggesting two groups similarly to Figure 8.3(b)) showed members mixed between the groups. Regarding the PCA as defining two reasonably clear groups, it is evident that variables do not have the same distribution within groups. Several variables have variances at least ten times greater in one group than the other. The clusters defined by the two methods, by contrast, have much more similar variances across variables. The methods assume equal variances and covariances within groups and appear to try and 'force' this on the solution; where 'natural' groups do not possess this property, the output may be of little value. Analysis was also undertaken on a logarithmic scale, omitting Sb which had a large number of zeroes, but conclusions remain the same.

Cluster analysis on principal components was also examined. The first six components were extracted and scores standardised before being clustered using Ward's method; as Table 8.2 and 8.3(a) show, results at the

Table 8.3: Cross-tabulations of selected clusterings from Table 8.2 to show the relationships between methods

		Method (3)					Method (5)		
(a)		1	2	3	(b)		1	2	3
Method (2)	1	22	2	0	Method (2)	1	8	10	6
	2	1	13	7		2	5	11	6
(c)		Method (6)							
		1	2	3					
Method (5)	1	4	2	7					
	2	9	3	9					
	3	1	8	3					

Note: Methods (2) and (3) are Ward's method using standardised data and standardised principal components respectively; methods (5) and (6) are based on the INVARIANT procedure in CLUSTAN to minimise trace W and maximise trace BW^{-1} respectively. An outlier suggested by method (3) is omitted in Table (a).

four-cluster level largely reproduce those of the earlier analysis. Specimen 19 is isolated as an outlier; cluster 1 corresponds closely to cluster 1 from the previous analysis, and 2 and 3 to the original cluster 2.

This example, which in many ways is typical, is not a particularly good advertisement for cluster analysis. Ward's method, in this instance, produces sensible and interpretable results that could be deduced more quickly and with more detail from the PCA. This will not always be the case. Average linkage is more successful at identifying some of those outliers identified by prior data inspection, but interpretation of the dendrogram is not easy. The methods of Section 8.8 proved of little use, though more extensive evaluation of their merits with archaeological data than is possible here would be of interest.

Having defined two groups, and a number of outliers, a substantive analysis would need to identify the chemical reasons for the difference and assess the significance of these. The second group, for example, contains specimens that are generally low in Sb (including all those where Sb was below the level of detection) and Pb compared to the first group, and averages a higher value of Ca (7.6 with a standard deviation of 0.4) than the first group (6.9 with a standard deviation of 0.3).

8.10 Cluster analysis in archaeology

The structure of this review differs in certain respects from those in Chapters 4, 6 and 10. First of all, the literature containing applications of cluster analysis is vast, to the extent that I gave up trying to record everything that I came across (in contrast to correspondence analysis, for example, where a special effort was made to locate applications). Secondly, much of what has been published tends to either repeat what has been done elsewhere; gives limited technical detail of a statistical nature; and/or results in little more than a single dendrogram with minimal comment. In short, a lot of the literature is, from a purely statistical standpoint, boring. This problem of quantity and (statistical) quality meant that I was unwilling to re-read, in addition to my original notes, all the papers in the bibliography. Some of what follows is, accordingly, more impressionistic than in the other review sections and I hope I will be forgiven this dereliction of scholastic duty. A further point is that specific applications have been noted in earlier sections of this and the previous chapter more so than for other techniques and will, for the most part, not be repeated.

The archaeometric literature where cluster analysis of artefact compositions is used is especially voluminous and largely responsible for the views outlined above. Scientific papers (i.e. archaeometric as opposed to non-archaeometric) seem often to be written to a fairly standard format and, particularly in conference proceedings, limited in length. Statistical considerations are rarely the main focus of a paper and often get short shrift, being dealt with – if at all – by reference to previous work. This has resulted in the development of 'standard procedures' that are too easy to apply uncritically.

Exceptions to this generalisation include Pollard (1983, 1986) who provides good and critical reviews of practice in the early 1980s and, more recently, Bishop and Neff's (1989) similarly critical review of practice. Earlier influential papers include those by Harbottle (1976) and Bieber et al. (1976). Papers co-authored by these individuals, or emanating from the laboratories where they work, often have more than the average amount of statistical interest. Other papers that devote more than usual care to the discussion of statistical issues include Hart et al. (1987); Leese et al. (1986); Mertz et al. (1979); Picon (1984), and the early paper by Prag et al. (1974). This is *not* intended as an exhaustive list!

A majority of the studies listed in the bibliography use either Ward's method or average linkage as the clustering technique with some form of Euclidean distance as the measure of similarity. There is, I think, a tendency for British authors to favour the former approach and American authors the latter. If this is a valid observation it may be because the much-used CLUSTAN package, in which Ward's method is a recommended technique, is of British origin, whereas influential American work, emanating from the Brookhaven Laboratory, has tended to favour average linkage.

Other techniques than Ward's method and average linkage have been and are used. Some of these other applications are early (i.e. 1970s) and clearly occurred when cluster analysis in archaeometry was at an experimental stage. It is also obvious that the choice of some authors has been constrained by the packages available to them, while others are happy to accept results that are 'interpretable' regardless of the method used. The consensus would seem to be that Ward's method and/or average linkage produce interpretable results on a reasonably regular basis (e.g., Pollard, 1986). Explicit and *a priori* justifications for the use of other approaches (other than that they have been used before and are available) are difficult to find.

Examples of the use of single linkage include Kuleff et al. (1985) and Schubert (1986), while complete linkage is used by Mirti et al. (1990) and Rauret et al. (1987). Studies that have used both single linkage and average linkage include Blasius et al. (1983); Djingova and Kuleff (1992), and Tubb et al. (1980). Cox and Pollard (1981); Salazar et al. (1986), and Stewart et al. (1990) are among those that use Ward's method and average linkage. Other papers that report the results of using several techniques include Attas et al. (1977); Bower et al. (1988); Cabral et al. (1983); Lambert et al. (1978); Matthers et al. (1983), and Rice and Saffer (1982).

Some of these papers report the results of k-means methods (including relocation) which are finding increasing use and deserve to be even more widely applied. Other examples include Burmester (1983b); Herz and Doumas (1991); Hughes (1991); Hughes and Vince (1986); Knapp et al. (1988); Liddy (1988; 1989); an early application by Ottoway (1974); Rehman et al. (1991), and Woolf et al. (1986). Other examples of the use of non-hierarchical approaches include applications of mode analysis — a method that seeks natural clusters — in Krywonos et al. (1980) and Matthers et al. (1983). Hughes and Vince (1986) report a use of the NORMIX routine within CLUSTAN, a model-based approach that uses maximum likelihood estimation, and note that it confirms the results of a Ward's method analysis with relocation. Wisseman et al. (1987), in a comparative study, include a rare application of a hierarchical divisive procedure.

Other paraphernalia available with a cluster analysis, such as the use of co-phenetic correlations or testing of the number of clusters, is much less widely reported. Most commonly, principal component analysis or discriminant analysis (Chapters 9, 10) are used to present or validate a cluster analysis. Examples of PCA used in this way include Cracknell (1982); Cox and Pollard (1981); Hatcher et al. (1980), and Sheridan (1989). Recent examples of the use of discriminant analysis, which currently seems to be the more popular supplementary technique, include Djingova and Kuleff (1992); Hart et al. (1987); Pernicka et al. (1990); Pollard and Hatcher (1986), and Stevenson and Curry (1991).

Recent applications of cluster analysis to typological study and assemblage comparison are harder to locate than examples of the study of artefact compositions. The literature in these areas is more diverse than the archaeometric literature, so I may have been looking in the wrong places; another possibility is that greater restraint is now exercised in the use of the cluster analysis in these areas than was once the case. It will be obvious from what follows that I have found it difficult to generalise usefully about the 'state of the art' or to identify trends in the literature. One reason is that while there is much greater choice in how to carry out a cluster analysis than, say, a PCA or discriminant analysis, this is not reflected in more extensive methodological discussion in much of the literature. A second reason is that much of what is published is, technically, no different from applications dating from the mid-1970s when cluster analysis packages became widely available.

In the fields of both typology and assemblage comparison, the early experimental work of Hodson and his co-workers was influential. Hodson et al. (1966) and Hodson (1970, 1971) are all examples of typological study. The first two of these papers used a (subsequently) much-analysed data set of thirty bronze La Tene fibulae. The 1966 paper used attribute data and a simple matching coefficient to compare single and average linkage, concluding that the latter but not the former produced 'classifications of demonstrable archaeological significance' (p. 311). The 1970 paper used quantitative rather than qualitative data (e.g., lengths, angles and counts) and contrasted single-link, average-link and k-means procedures as well as a little used 'double-link' method. Among the conclusions it was doubted (p. 317) if single link would 'ever be very useful to archaeologists' while the potential usefulness of k-means procedures with Euclidean distance was noted. Hodson (1971) is a further exploration of the k-means method illustrated on handaxe data using metric variables and their ratios, subsequently log transformed.

In summary, the conclusions of this work are that single linkage is of limited practical use; that average linkage is one of the potentially more useful hierarchical techniques; and that the non-hierarchical k-means method, when Euclidean distance is an appropriate measure of similarity, is even better. Except that, with the advent of CLUSTAN in the 1970s, Ward's method competes with average linkage as a suitable hierarchical procedure, many archaeological users of cluster analysis would probably not dissent from these conclusions.

Another way of viewing this is that, over twenty years on, there have been few technical developments (other than increased computer-package availability) that have seriously influenced the way in which archaeologists have used cluster analysis, if the method is used at all. Where developments have occurred is in the way in which archaeologists have viewed the usefulness of cluster analysis as a tool for typology. For contributions to this debate the reader is referred to the references given in the introduction to Chapter 7.

In this context it is of interest to note the comment in Hodson (1977, pp. 398–9) that he chose 'to rely on perception rather than quantification in setting up a general typology for Hallstatt artifacts' and that 'this is not really a matter of choice but is dictated by the wish to achieve comprehensive practical results in a finite time'. The paper is otherwise concerned with the cluster analysis of graves and functional types within them using single linkage in conjunction with Jaccard's coefficient. Hodson's (1969) earlier pioneering paper, concerned with looking for useful methods rather than results, used average linkage and Euclidean distance on the percentage of tool types within assemblages. The potential problems with such fully compositional data are noted elsewhere (Section 4.6). Hodson's solution – to use an arc-sin transformation – has been emulated by others (Close 1977, 1978; Callow, 1986c), but does not address the compositional problem directly. The much later publication by Hodson (1990) does not, in its use of cluster analysis, differ from Hodson (1977) in any important respect.

It was noted in Chapters 4, 7 and 8 that correspondence analysis is possibly supplanting PCA as an approach to analysing assemblage data expressed in the form of counts. In the context of cluster analysis an approach sometimes used is to carry out an analysis on 'factors' or 'components' determined by an initial correspondence analysis of the data (e.g., Diaz-Andreu and Fernandez-Miranda, 1991). This and the analogous use of cluster analysis on principal components has been discussed in Section 8.5.

Hodson's (1977) paper is an example of a polythetic cluster analysis applied to burial assemblages. The choice between polythetic and monothetic procedures, and in the latter case the choice between association analysis and use of the information statistic, was discussed in Section 8.7. Whallon (1972), in a study of pottery typology, advocated the use of association analysis, but it does not seem to have been widely used for reasons discussed in earlier sections.

If an area where there have been recent developments and which is currently popular can be identified, it is that of 'spatial' analysis using k-means (pure locational) clustering or unconstrained clustering. Applications date from the pioneering work of Kintigh and Ammerman (1982) and Whallon (1984), with critical comparisons of the approaches available in Blankholm (1991); Gregg et al. (1991), and Kintigh (1990). Other applications have been noted in Section 8.6. A possible reason for this current popularity is that the methodology has been developed by archaeologists and is perceived as 'congruent' with archaeological needs. The lack of congruence in other applications of cluster analysis, except possibly to mortuary studies, may serve to explain the relative lack of recent published applications.

On the subject of validating clusters or using objective methods to determine an appropriate number of clusters the archaeological literature is comparatively silent. Aldenderfer (1982) and Whallon (1990) are honourable exceptions. In Chappell's (1987) study of stone axe morphology, Ward's method is applied to principal components weighted according to their eigenvalues and generally producing groups based on axe size. It is stated that this procedure was based on extensive experimentation on the basis that it gave 'the most intuitively acceptable results' (p. 130). Cavalloro and Shimada (1988), in their analysis of brick size in Sican marked adobes, choose complete linkage 'because it had produced the most reliable and interpretable results with other data sets' (p. 97). Jones (1980), who compares different clustering techniques on burial assemblages, remarks that 'the only basis on which to choose how many clusters seems ... to be largely empirical'.

There is clearly an element of subjectivity in such approaches that begs the question of what is meant by terms such as 'intuitively acceptable' and 'reliable'; however, this kind of approach is (implicitly) in common use. Magne and Klassen (1991), who study anthropomorphs on the basis of presence or absence of attributes using average link applied to Jaccard's coefficient, choose their method on the grounds that it gives the best co-phenetic correlation coefficient. This is a more objective approach but presupposes that a tree structure is appropriate to the data to be completely justified.

9

Discriminant Analysis – The Main Ideas

9.1 Introduction

Discriminant analysis starts from the presumption that a set of objects are known to belong to one of two or more groups. Two aspects are commonly distinguished – that of discrimination, where new variables are defined that in some sense best distinguish between the known groups; and that of allocation or classification, where objects are assigned to existing groups on the basis of their characteristics. Discriminant analysis is not really an exploratory technique and thus stands apart from the methods discussed in earlier chapters. In principle it presumes that a 'model' for the data exists – namely, that the group to which an individual belongs is known. In archaeological practice usage can be of a more exploratory nature in that the methodology is used as a convenient way of displaying results obtained via a cluster analysis, for example.

There are many different approaches to discriminant analysis (Fatti et al., 1982; Hand, 1981; Mardia et al., 1979; Seber, 1984; Krzanowski, 1988). Archaeological practice is overwhelmingly dominated by one of these – Fisher's linear discriminant analysis – because of its availability in many of the more popular computer packages. This chapter concentrates on this approach.

Section 9.2 provides some examples using the SPSS-X package. The ideas introduced there are developed in Section 9.3. An appreciation of Mahalanobis distance is useful and readers should re-read Section 4.9 if necessary. Matters of detail and more specialised topics are reserved for Chapter 10.

9.2 Discriminant analysis

In PCA new variables are defined as linear combinations of the (possibly transformed) original variables. They 'explain' as much of the variation in the data as possible and the hope is that a low-dimensional plot based on the new variables will show any structure that exists in the data. No prior assumption of structure exists; the method provides a tool for exploring whether there is interesting structure.

In discriminant analysis, by contrast, it is assumed that distinct and

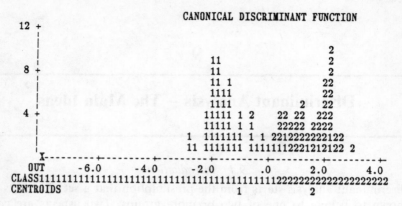

Figure 9.1: SPSS-X *discriminant analysis of the data in Table A1 – two groups*

Note: The figure is of edited SPSS-X output for the data in Table A1 where the two groups correspond to the two sites.

known groups exist in the data and that the data include all possible groups. Linear combinations of the variables are sought that display the difference between the presumed groups as clearly as possible. The aim may be to confirm that the presumed groups are indeed distinct; to characterise groups on the basis of the coefficients associated with the linear combinations of the variables; to identify individuals that do not readily fit into their presumed group; to identify those variables that best discriminate between groups, or to provide a criterion for allocating unclassified individuals to a group. Thus in Table A1 or A2 it might be assumed that specimens from the two sites are chemically distinct, and a single linear combination of variables is sought that maximises the difference between the sites. In this example it would be hoped that scores are quite distinct for specimens from the two sites.

To illustrate, some output from an SPSS-X discriminant analysis is shown in Figure 9.1 for the two-group case referred to above, using the data of Table A1. A particularly simple form of discriminant analysis – linear discriminant analysis – is being illustrated here. The raw data in Table A1 have been standardised in a manner that is discussed in Section 9.3, and a linear combination – the 'standardised canonical discriminant function' in SPSS-X terminology – of the form

$$F = .55Al - .59Fe - .19Mg - .07Ca + .16Na + .14K - .74Ti$$
$$+ .69P + .38Mn + .46Sb + .24Pb$$

obtained. Figure 9.1 is the histogram obtained on substituting the standardised values of the variables into this, a 1 representing specimens from Leicester and 2 Mancetter. A score of just slightly more than 0 on F is the

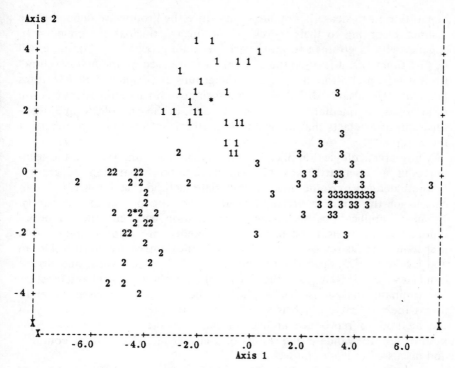

Figure 9.2: SPSS-X *discriminant analysis of the data in Table A1 – three groups*

Note: The same data as in Figure 9.1 are used but with three groups determined by previous PCA/ cluster analyses.

boundary between the two groups. Standardisations other than that used here are possible (Cooley and Lohnes, 1971; Leese (pers. comm.)) that occur in packages used in the literature, so that the same data analysed in different environments will not necessarily give the same numerical results.

It can be seen that the two presumed groups are reasonably well separated. One Mancetter specimen is misclassified in the Leicester group, and seven Leicester specimens are misclassified in the Mancetter group. Another way of viewing this is that 97 of 105 or 92% of specimens are correctly allocated, although we shall see later that this is an overoptimistic assessment.

With three groups there are two discriminant functions, and plots of scores on these functions will display group separation. For illustration the three groups suggested in the PCA of Figure 3.8(a), based on an analysis of Tables A1 and A2 and omitting specimen 12, are taken as given. Figure 9.2 shows the outcome of a discriminant analysis in which group 3 is distinct from the other two; groups 1 and 2 appear to be separate although the point in group 2 nearest to group 1 is actually allocated to group 1.

Separation in this case is not surprising since the groups are defined to be distinct according to their scores on the linear principal components. In general, with G groups and more variables than groups (p > G), there will be G-1 functions; it is often the case that a plot based on the first two functions will be useful for showing that the groups are distinct. Other features of Figure 9.2, of a kind that is often informative, include the identification of artefacts intermediate between groups and, in the case of group 3, identification of artefacts that appear to be outliers not obviously members of the group.

Many archaeological applications of discriminant analysis use the technique in a descriptive way to demonstrate the separation of groups, possibly defined by some other multivariate method (Section 8.10). Applications of this kind in archaeological science are particularly common. Biometric applications, involving discrimination between different populations on the basis of skeletal measurements, are also common. Two archaeological (as opposed to archaeometric) examples are given in Doran and Hodson (1975, pp. 237–46) who cluster Hallstatt C swords and British handaxes respectively, on the basis of morphological characteristics. Discriminant analysis is then applied to the groups obtained in order to assess their validity, which is also judged with reference to the ability of the method to reproduce archaeologically accepted classifications. This kind of use, to 'validate' the results of a cluster analysis, is quite common and discussed at length in Section 10.8.

9.3 The main ideas

The main ideas underlying linear discriminant analysis are discussed, mainly with reference to the two-group case, in this section. A more detailed exposition is given in Section 10.2.

The usual considerations concerning data transformation apply, and it can help if the data within groups are normally distributed. Subsequent standardisation is not essential as the graphical output is not scale-dependent, but a form of standardisation to be discussed later in this section does aid interpretation. If obvious outliers are detected in preliminary data analysis then it will often be sensible to omit them from the analysis. One assumption of two-group linear discriminant analysis is that there are, in fact, just two groups in the data and outliers may be an indication that this assumption is violated. Contrary to what is often stated a (multivariate) normal distribution of data within groups is not an essential requirement for the method to be used descriptively. There are, however, some advantages in having normally-distributed data that are discussed in Section 10.4. It is also assumed that the two groups are of a similar geometrical size and shape in multivariate space.

Given two groups whose members are (presumed to be) known, a discriminant analysis results in a function of the form

Discriminant Analysis – Main Ideas

$$F = a_0 + a_1Y_1 + a_2Y_2 + \ldots + a_pY_p \quad (9.1)$$

where the Y_i may be transformed data. For each row of the data matrix a score, f_i, can be calculated from (9.1) and the average value of f_i in each group, $\bar{f}_1$ and $\bar{f}_2$, and variances within each group, s_1^2 and s_2^2, may be calculated. For good discrimination we would ideally like the means to be widely separated with small variances within groups so that individuals are clustered tightly about the group centroid. The coefficients, a_i, in (9.1) are defined to try and achieve this.

If s_1^2 and s_2^2 do not differ significantly they can be averaged to get a common within-group variance, s_w^2. The appropriate averaging formula, which takes account of the different group sizes, n_1 and n_2, is

$$s_w^2 = [(n_1 - 1)s_1^2 + (n_2 - 1)s_2^2]/(n_1 + n_2 - 2) \quad (9.2).$$

(Readers familiar with the independent two-sample t-test (Fletcher and Lock, 1991, pp. 86–8) will recognise this as the same averaging formula used there.) In an ideal world (9.2) will be 'small'. The distance between group means should, by contrast, be 'large'. Treating the means as single points, and taking into account sample sizes, the between-group variance, s_b^2, which can be used as measure of group separation is

$$s_b^2 = n_1(\bar{f}_1 - \bar{f})^2 + n_2(\bar{f}_2 - \bar{f})^2 \quad (9.3)$$

where $\bar{f}$ is the mean of all the discriminant scores. Since, ideally, (9.3) should be 'large' and (9.2) 'small' it makes sense to define the discriminant function coefficients to maximise the between- to within-group variance

$$s_b^2/s_w^2 = (s_w^2)^{-1}s_b^2 \quad (9.4)$$

where the right hand side of the equation is written as it is to emphasise an analogy with the case of three or more groups discussed later. The idea involved here, of defining the a_i's to separate the two groups as much as possible while keeping them compact, is (once you have it) a simple and natural one due to Fisher (1936). While the idea is simple, the mathematical details are less so and beyond the scope of this book, but need not concern users of the method.

The foregoing development did not involve any assumption about the distribution of the data within groups and is 'distribution-free' in this sense. The main assumption involved is that the spread of scores about the group mean is similar in both groups. The question of assumptions is addressed in more detail in Sections 10.3 to 10.5.

It is possible to report the outcome of a discriminant analysis in a variety of ways which will be illustrated with a small example based on the elements Fe, Ti and P from Table A1. Summary information on the means of these elements for the two sites is given in Table 9.1.

Terminology varies from book to book and package to package and in

Table 9.1: Element means for three variables from the two sites in Table A1

	Element		
	Fe	Ti	P
Leicester	.696	.100	.117
Mancetter	.476	.080	.139

Table 9.2: Discriminant analysis output for three variables from Table A1

Element	Standardised CDF	Unstandardised CDF	Fisher's LDF 1	Fisher's LDF 2
Fe	.59	4.14	1.892	−6.673
Ti	.22	14.96	478.403	447.437
P	−.64	−39.097	459.169	540.062
Constant			−262.994	−275.759

the following exposition that, associated with the SPSS-X package, will be used. If variables are used in their raw form then the 'unstandardised canonical discriminant function' is obtained. This is given in Table 9.2 where it can be seen that the coefficients are of different orders of magnitude and are not comparable. More usefully, for interpretive purposes, the variables can be standardised to obtain the 'standardised canonical discriminant function' in which the coefficients can legitimately compared. This is also shown in Table 9.2 and suggests that Fe and P are of about equal importance in discriminating between the two groups, with Ti somewhat less so. The standardisation that is appropriate here differs from that assumed elsewhere in the book and must now be explained.

In equations (9.2) and (9.3) within- and between-group variances were defined for the discriminant scores. Means and variances can equally well be defined for individual variables within groups and combined in a similar fashion. Additionally, covariances between elements within groups can be defined and averaged to get between- and within-group covariances. The information on variances and covariances in the groups can be collected in p by p matrices, S_1 and S_2 say. These are then averaged similarly to (9.2) to give the within-groups covariance matrix

$$S_w = [(n_1 - 1)S_1 + (n_2 - 1)S_2]/(n_1 + n_2 - 2) \qquad (9.5).$$

If the i'th diagonal element of this, the estimated within-group variance of the i'th variable, is denoted by s_{wi}^2 then the standardisation needed to obtain the standardised canonical discriminant function is

$$Y_i = (Z_i - \bar{Z}_i)/s_{wi} \qquad (9.6).$$

An assumption made above is that S_1 and S_2 are not significantly different

Discriminant Analysis – Main Ideas

and may be averaged as in (9.5). This is not obviously correct in the present case and the issue will be considered further later in this section.

To illustrate some of the details, S_1, S_2 and S_w are given below for the three-variable example (actual values have been multiplied by 10^4 and rounded to two decimal places).

$$S_1 = \begin{pmatrix} 294.82 & 16.05 & -4.59 \\ 16.05 & 2.24 & -.71 \\ -4.59 & -.71 & 2.05 \end{pmatrix} \quad S_2 = \begin{pmatrix} 90.38 & 11.08 & 5.58 \\ 11.08 & 2.04 & .63 \\ 5.58 & .63 & 3.48 \end{pmatrix}$$

$$S_w = \begin{pmatrix} 205.50 & 13.88 & -.15 \\ 13.88 & 2.15 & -.12 \\ -.15 & -.12 & 2.67 \end{pmatrix}$$

where, for example, in S_w $205.50 = (58 \times 294.82 + 45 \times 90.38)/103$ etc. and sample sizes are $n_1 = 59$ and $n_2 = 46$. In Table 9.2 the standardised canonical discriminant function is simply obtained from the unstandardised version by multiplying the coefficients by s_{wi} (e.g., for Fe $.59 = 4.14 \times \sqrt{.02055}$).

Table 9.1 showed the group means for the elements used in the analysis. It can be shown that the Mahalanobis distance between these means, D say, is just the Euclidean distance between the mean discriminant scores for the two groups. In the definition of Mahalanobis distance in equation (4.14) the z's correspond to the group means and S to the within-group covariance matrix. For our three-variable example SPSS-X output gives $D^2 = 4.28$, or $D = 2.07$ and this is just the distance between the group centroids of the standardised canonical discriminant functions, .91 and −1.16.

Equation (9.1) can be used to allocate individuals to groups. The simplest allocation procedure is to compute an individual's score and allocate it to the group whose centroid is nearest as measured by Mahalanobis distance. This can be shown to be equivalent to a procedure whereby linear functions are obtained for each group, scores on the functions are obtained for an individual, and allocation is to the group for which the score is highest. The functions in question are 'Fisher's linear discriminant functions' (Section 10.2) and are also shown in Table 9.2. For the two-group case the unstandardised canonical discriminant function can be obtained as the difference in coefficients of Fisher's functions divided by D (e.g. for Ti, $14.96 = (478.03 - 447.437)/2.07$).

The analyses reported here have assumed that S_1 and S_2 are not significantly different, rather glossing over the fact that inspection of their form suggests this is an assumption of some dubiety. Formal tests of the assumption are available and are noted in Section 10.5. One can be obtained in SPSS-X. In archaeological practice these tests are often ignored; since they are very sensitive to the often questionable assumption of normality and are prone to give 'significant' results, there is some justification for this. Alternatives to linear discriminant analysis exist (Section 10.5, 10.6) but can be demanding of data or are not implemented in the packages most used in the archaeological literature. Where the question of assumptions is addressed, the impression is often given that linear discriminant analysis is believed to be very robust to violation of assumptions. While open to debate, this may help explain the widespread use of a single approach to discrimination. It is also the case that the technique is often used for descriptive purposes where group separation is so clear that there is no need to agonise about the integrity of assumptions implicit in the methodology.

The remaining paragraphs briefly note the extension of the methodology to three or more groups, chiefly to establish notation used in Chapter 10. The within-group covariance matrix summarises information on the size and shape of groups, assuming these to be essentially similar. The definition given in equation (9.5) extends naturally to three or more groups. A between-groups covariance matrix, S_b, that summarises information on the configuration of group centroids can also be defined (analogously to (9.3) but taking into account covariation as well). With p variables and $p > G$ groups $(G - 1)$ discriminant functions can be defined; a plot of scores on these functions, if possible, would ideally show well-separated groups with individuals tightly located about their group centroid. In practice the first two functions are usually used (though see Capannesi et al. (1991) for a three-dimensional example). A simple equivalent of (9.4) no longer exists, but it is noted in Section 10.2 that analysis is based on the analogous matrix $S_w^{-1} S_b$, and extracts linear functions from this that aim to maximise between to within-group scatter. In defining S_w and S_b division by $(N - G)$ and $(G - 1)$ is involved, where N is the total number of observations, so that we can write $S_w = W/(N - G)$ and $S_b = B/(G - 1)$ where W and B have an interpretation as within- and between-groups scatter matrices. Some treatments of discriminant analysis, and other topics, are based on W and B rather than the covariance matrices, and in particular the relationship

$$T = W + B \qquad (9.7)$$

is used which decomposes the total 'scatter' T into between- and within-group scatter. Ideally W will be 'small' and B 'large' in some sense. In Section 8.8 it was noted that in some methods of cluster analysis, where the groups are unknown, groups are defined to try and achieve this state of affairs.

10

Further Aspects of Discriminant Analysis

10.1 Introduction

In Section 10.2 some of the mathematics of two-group discriminant analysis and the extension to three or more groups is given in more detail. Section 10.3 considers some aspects of statistical inference. Some readers may wish to omit these and go directly to Sections 10.4 and 10.5 which discuss the assumptions of normality and equal covariances respectively. An alternative to linear discriminant analysis that makes fewer assumptions is quadratic discriminant analysis and this is introduced in Section 10.5. It is not widely used, partly because of sample-size considerations which are discussed more generally in Section 10.6. The question of assessing how good discrimination is, including allocation procedures, is addressed in Section 10.7. These procedures are often used to 'validate' clusters determined by a cluster analysis, and this is discussed critically in Section 10.8. Section 10.9 examines methods of variable selection; examples to illustrate several of the procedures discussed are given in Section 10.10; and Section 10.11 concludes with a review of the discriminant analysis literature in archaeology.

10.2 The mathematics of discriminant analysis

Results pertinent to the two-group case discussed in Section 9.3 will be summarised. Assume that there are p variables and let $\bar{z}_i$ be the (px1) vector of means in group i. For two groups a discriminant function

$$F = a_0 + a_1 Z_1 + a_2 Z_2 + \ldots + a_p Z_p \tag{10.1}$$

is defined to maximise the between- to within-group variance

$$s_b^2 / s_w^2 = [n_1(\bar{f}_1 - \bar{f})^2 + n_2(\bar{f}_2 - \bar{f})^2]/s_w^2 \tag{10.2}$$

where $\bar{f}$ is the mean of the discriminant scores; $\bar{f}_i$ is the mean of the scores within group i; and s_w^2 is the estimated within-group variance of the scores. It can be shown that this leads to a solution in which the vector of coefficients $\mathbf{a}$, where $\mathbf{a}' = (a_1 a_2 \ldots a_p)$, is proportional to

$$S_w^{-1}(\bar{z}_1 - \bar{z}_2) \tag{10.3}$$

where S_w is the (pxp) within group covariance matrix. A consequence of this is that the distance between group means for the discriminant scores is (Lachenbruch, 1975, p. 10)

$$(\bar{f}_1 - \bar{f}_2) = (\bar{z}_1 - \bar{z}_2)' S_w^{-1} (\bar{z}_1 - \bar{z}_2) \tag{10.4}$$

which is just Mahalanobis's D^2 between group centroids on the original scale. If $\bar{z}_i$ is the (px1) vector of observations for the i'th individual, the Mahalanobis distance to the centroid of group 1 is

$$D_i^2 = (z_i - \bar{z}_1)' S_w^{-1} (\bar{z}_i - \bar{z}_1) \tag{10.5}$$

which can be rewritten as

$$D_i^2 = -2[\bar{z}_1' S_w^{-1} \bar{z}_i - c_1] + c \tag{10.6}$$

where c_1 is a constant specific to group 1 and c is a constant not dependent on the group. The bracketed quantity is a linear function of the Z_i specific to group 1 and is Fisher's linear discriminant function for that group; a similar function for group 2 is defined in the same way. Specimens can be allocated to the group for which D_i is smallest or, equivalently, for which Fisher's linear discriminant function is largest.

These ideas carry over to the three and more group case but the mathematics is more complicated. With G groups (G > 2) the discriminant functions can be defined to maximise between- to within-group separation but this concept must now be formulated in matrix terms. Assuming p > G there are, at most, (G − 1) discriminant functions and it can be shown that their coefficients are based on the eigenvectors of the matrix $S_w^{-1} S_b$ with the eigenvalues providing information on their importance (e.g., Krzanowski, 1988, pp. 295–6). The matrix S_b is the between-groups covariance matrix discussed in Section 9.3. The derived functions are often called canonical variates. In practice the first two functions are often used to display separation between groups and this will work well if the first two eigenvalues dominate the analysis.

10.3 Statistical inference

In this section some aspects of statistical inference in relation to discriminant analysis are noted. An appreciation of these is helpful in understanding computer-package output and some aspects of the literature, although it will often be the case that the assumptions necessary for the following results to be strictly valid will rarely apply.

If formal tests of the significance of between-group differences are needed, several possibilities, all depending on the normality assumption, are available. For the two-group case Hotellings T^2 statistic, a multivariate generalisation of the univariate Student's t-statistic, is most commonly used. If D^2 is the Mahalanobis distance between-group centroids, (10.4), then T^2 is defined as

$$T^2 = [n_1 n_2 / (n_1 + n_2)] D^2 \tag{10.7}$$

and

$$F = T^2(n_1 + n_2 - p - 1)/(n_1 + n_2 - 2)p \qquad (10.8)$$

has the F-distribution with p and $(n_1 + n_2 - p - 1)$ degrees of freedom if the population means are the same (Lachenbruch, 1975, p. 25; Krzanowski, 1988, p. 327). Chatfield and Collins (1980, p. 125) note that with similar sample sizes this test is not sensitive to the equal covariance matrix assumption and is the only test in common use. It may be noted, as with the tests for three and more groups, that a statistically significant result does not mean that the discriminant rule will necessarily classify well.

For the multiple group case Wilk's lambda (Λ) statistic is often used. This, as with other possible tests, is a function of the eigenvalues of $S_w^{-1} S_b$ and can be written as

$$\Lambda_p = [(1 + \lambda_1)(1 + \lambda_2) \ldots (1 + \lambda_p)]^{-1} \qquad (10.9)$$

where the λ_i are the eigenvalues of $S_w^{-1} S_b$ and all p variables are used. Small values of Λ tend to indicate large group separation. To test hypotheses about the significance of group separation one possibility, among several, is to transform (10.9) as

$$-[n + 1 - (p + G)/2]\log(\Lambda) \qquad (10.10)$$

which follows, approximately, a chi-squared distribution with $p(G-1)$ degrees of freedom if there are no group differences in the population. If, given the first k discriminant functions, a test of the significance of the remainder is needed, the test is readily adapted. Equation (10.9) is defined for the subset of interest only and (10.10) is tested with $(p-k)(G-k-1)$ degrees of freedom (e.g., Klecka, 1980, p. 39).

Wilk's lambda is also commonly used in stepwise selection procedures in which the discriminant function is 'built up' a step at a time, culminating in a function using only a subset of the variables (Sections 10.9, 10.10). The first variable to be entered into the function is that with the smallest Λ_1. Variables continue to be added so long as their contribution, additionally to those already in the model, is significant. Given that k variables have been entered, an approximate F-test for the significance of an additional variable can be based on

$$F = [(n - G - k)/(G - 1)][(\Lambda_k/\Lambda_{k+1}) - 1] \qquad (10.11)$$

Selection involves adding the variable that maximises F or, equivalently, minimises Λ_{k+1}. At any stage after two variables are entered, a variable previously in the model may be deleted if it has ceased to be significant, given later entries.

Other criteria than Λ may be used; SPSS-X, for example, offers five which are discussed in Klecka (1980, pp. 52–8) or one of the SPSS-X manuals (Norusis, 1985). These differ primarily in the criterion for group separation to be optimised at any stage. Although the use of Λ is common it is not recommended by Seber (1984, p. 341). One problem is that the performance in classifying individuals can deteriorate as the function is built up. This is because the method tends to increase the separation of well-separated groups rather than of poorly-separated groups. An example is noted in Section 10.10.

In many archaeological applications formal tests of the kind discussed here are not undertaken. Where they are, tests of the significance of group differences tend to perform 'too well' and lead to the conclusion that differences are highly significant. This occurs particularly when some other multivariate procedure, such as cluster analysis, is first used to define groups, as this 'loads the dice' in favour of there being statistically significant group differences. For many analyses the formal statistical significance can be taken for granted and may be of limited interest. Often informal methods of assessment may be more useful.

These informal methods may include inspection of the discriminant analysis plot and of the values of D_i^2 for each group (Section 10.7). While an observation may be allocated to the group for which this is smallest, it does not imply that the observation is close to the group. For large groups with a multivariate normal distribution, D_i^2 should be distributed approximately as chi-squared with p degrees of freedom, so that unusual values that do not really belong to a group might be detected by reference to this distribution.

10.4 The normality assumption

Papers using linear discriminant analysis sometimes give the impression that a (multivariate) normal distribution within groups is an essential requirement of the method. This is not so, as was observed in Section 9.3. Nevertheless there are advantages to having normally-distributed data and it is needed for the inferential aspects of discriminant analysis discussed in Section 10.3. The importance of normality is examined in more detail in this section. The two-group problem is assumed until further notice.

The general idea in two-group discrimination is to partition (multivariate) space into two regions. Individuals are classified according to which region they fall into. The regions are defined to coincide, as far as possible, with the space occupied by the two groups. It is not always possible to separate the groups perfectly so that some individuals will fall into the wrong region and are hence misclassified relative to their assumed group. The success of a discrimination rule can be assessed on the basis of the proportion of misclassifications.

If the probability distribution of individuals within a group is known, then for any pair of regions the probability of misclassification can be evaluated. Given a knowledge of the distributions, the regions (or discriminant rule) can be defined to minimise the probability of misclassification or error rate.

Aspects of Discriminant Analysis

In the case where observations are multivariate normally distributed with known means and known and equal covariances, the optimal rule leads to a 'true' discriminant function whose sample analogue is just (10.2). The actual sample discriminant function is

$$a_0 + \mathbf{z}'\mathbf{S}_w^{-1}(\mathbf{z}_1 - \mathbf{z}_2) \tag{10.12}$$

where a_0 is a constant.

In other words, the distribution-free procedure described in Section 9.3 leads to a function that is the sample analogue of a procedure that is optimal for normally-distributed data. This is one reason for wanting the data to be normally distributed when using the linear discriminant functions of the last chapter.

This means that if the data are not normal, then use of (10.2) is unlikely to be optimal. Is this a serious concern? Seber (1984, p. 299), summarising other studies, suggests that the linear discriminant function is robust with mild skewness in the data or with symmetric distributions having longer tails than the normal. With very skewed data, which may include the lognormal distribution, the linear discriminant function can perform very badly. Thus there are clear advantages in transforming to normality in such cases which may include, for example, compositional studies of artefacts where trace elements are log-normally distributed.

When some or all of the variables are binary, the joint distribution of variables in a group clearly cannot be normal. Nevertheless the linear discriminant function will sometimes, but not always, perform well. Seber (1984, pp. 297–9) summarises research on this matter. Problems may arise if there are moderate to large positive correlations among the binary variables, or when the correlations between binary variables or between binary and continuous variables differ markedly between groups.

10.5 The equal covariance matrices assumption

The results of Chapter 9 are based on the assumption that the averaging of group covariance matrices into a single within-groups covariance matrix, $\mathbf{S}_w$, is legitimate. This requires the assumption of equal covariance matrices within the populations from which the samples are drawn. A similar assumption is required for the optimality result based on the use of the normal distribution noted in the last section.

This supposes that groups may have different means but are otherwise of similar geometrical size and shape in multivariate space. This is not usually something that can be inspected visually; however, a simple but real example is shown in Figure 10.1. This shows the logged concentrations of two elements (Al and Fe) for a sample of 177 specimens of mortaria that divides clearly into two groups associated with two different sites of origin.

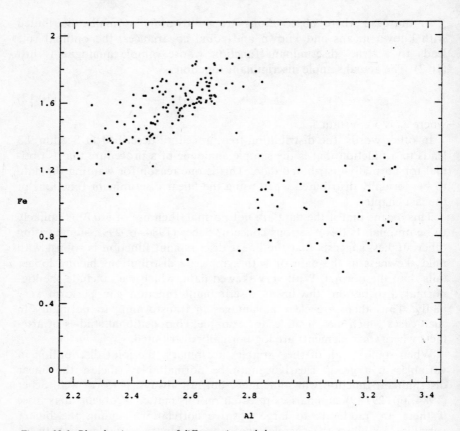

Figure 10.1: Plot showing groups of different size and shape

Note: Linear discriminant analysis assumes that groups are of similar size and shape. The example above, based on real pottery data, for two variables shows how the assumption can be violated.

The two groups appear to be of different shapes, this raises a number of questions including whether or not such groups are seriously different in shape; whether or not the linear discriminant analysis is seriously affected in any event; and what alternatives are available if the linear discriminant approach is abandoned.

One alternative is to use the method of quadratic discriminant analysis. Working within a similar theoretical framework based on the normality assumption – but relaxing the assumption of equal covariance matrices – results in the derivation of a discriminant function quadratic rather than linear in the Z_i.

$$a_0 - [\mathbf{z}'(\mathbf{S}_1^{-1} - \mathbf{S}_2^{-1})\mathbf{z}' - 2\mathbf{z}'(\mathbf{S}_1^{-1}\mathbf{z}_1 - \mathbf{S}_2^{-1}\mathbf{z}_2)]/2 \qquad (10.13).$$

Lachenbruch (1975, pp. 46–7) and Seber (1984, pp. 299–300) summarise work that has been done on the choice between linear and quadratic discrimination. If the data are normal, then the chief advantages of using quadratic discrimination arise when there are large covariance differences and large sample sizes. The linear discriminant function is tolerant to small covariance differences and generally preferable for small sample sizes (sample size considerations are discussed explicitly in Section 10.6). Since the linear but not the quadratic discriminant function is robust to certain kinds of departure from normality noted in Section 10.4, the general message is that for many practical situations use of the linear discriminant function is preferable. In the context of predictive modelling of archaeological site location Kvamme (1990, p. 275) reports work where quadratic discrimination was so strongly influenced by outliers as to offset any theoretical advantages. Krzanowski (1988, p. 340) notes that quadratic discrimination 'is not one that tends to be used very often in practical applications'.

If necessary a formal test of the hypothesis that the population covariance matrices, Σ_1 and Σ_2, are equal is possible. This assumes multivariate normality and is based on asymptotic (large sample) maximum likelihood theory. Details are given in Krzanowski (1988, p. 327) for example, and are given below for reference. Use is made of matrix notation and determinants and some readers may wish to omit this, though it should be noted that SPSS-X can provide the relevant numerical information.

In the notation established in the previous chapter the statistic

$$\mathbf{M} = (n_1 + n_2)\log|c\mathbf{S}_w| - n_1\log|c_1\mathbf{S}_1| - n_2\log|c_2\mathbf{S}_2| \qquad (10.14)$$

where $c = (n_1 + n_2 - 2)/(n_1 + n_2)$, $c_1 = (n_1 - 1)/n_1$ and $c_2 = (n_2 - 1)/n_2$, has an asymptotic chi-squared distribution with $p(p+1)/2$ degrees of freedom if the population covariances are equal. To test for the equality of more than two covariance matrices, equation (10.14) generalises in an obvious way (Krzanowski, 1988, p. 370) and leads to a large sample chi-squared test with $p(p-1)(G-1)/2$ degrees of freedom. Seber (1984, pp. 449–50) notes two rather complex modifications of the statistic due to Box (1949), one of which leads to an approximate F-test. In SPSS-X, Box's M statistic can be obtained if requested and it is briefly noted in the examples of Section 10.10. The test will tend to give statistically significant results for large samples and is sensitive to violations of the multivariate normality assumption.

Explicit reference to the issues and use of the statistics alluded to in this and the preceding two sections in the archaeological literature is uncommon. Exceptions include Vitali and Franklin's (1986a, p. 167) paper on the classification of ceramics on the basis of elemental compositions where the use of raw rather than logarithmically transformed data and use of the equal covariance matrix assumption are explicitly justified. Other work, emanating from the British Museum Research Laboratory on the sourcing of flint axes, has used quadratic discrimination (Craddock et al., 1983), but in general such applications are rare.

The assumption of equal covariance matrices often goes unexamined and is almost certainly violated in many archaeological applications. One reason may be that the results obtained (among those published) often demonstrate the clear separation between groups that is sought without the need to worry about the strict validity of assumptions. Wynn and Tierson (1990, p. 77), in a study of the morphology of later Acheulean handaxes, also suggest that discriminant analysis 'is known to be extremely robust both to inequality of group covariance matrices and other assumptions'. While the statistical literature suggests that this may be a little optimistic, a belief in the veracity of this view may help explain the widespread use of linear discriminant analysis in archaeological practice without formal statistical testing. Approaches that avoid some of the assumptions of linear discriminant analysis, such as quadratic discrimination and non-parametric methods, are either very demanding of the data (next section) and/or not implemented in the statistics packages (often SPSS-X or BMDP) that tend to be most used.

10.6 Sample size

The choice between using the linear or quadratic discriminant function depends, to some extent, on the size of the sample. Sample-size questions are, in general, difficult to answer since they usually depend on a variety of factors and may require the specification of information that is unknown. Here some recommendations culled from the literature are collected.

A linear discriminant analysis can be carried out if the number of observations, n, exceeds the number of variables, p. If, however, n is close to p, then results may be unstable. Harbottle (1976), citing other work, suggests that to be on the safe side the n/p ratio should be between 3 and 5. Lachenbruch (1975; p. 16) provides a table that may be helpful in certain cases. This assumes normality and equal sample sizes and depends both on the number of variables and group separation. A crude guide is that for large group separation one needs about 2p observations in order to be within .05 of the optimum error rate, and for smaller group separation about 3.5p observations. This last figure is within the bounds suggested by Harbottle (1976).

The choice between linear and quadratic discrimination is discussed by Seber (1984, pp. 299–300), also assuming multivariate normality. In general quadratic discrimination is much more demanding of data. Quadratic discrimination performs poorly for small sample sizes even when the covariances are different. It is suggested that if both groups have less than 25 observations then linear discriminant analysis is to be preferred, though it may perform unacceptably badly with large p and large covariance differences. For 'big' sample sizes and $p > 6$ with large covariance differences, quadratic discrimination is best – 'big' is interpreted to mean a sample size of 25 in each group if $p = 4$, and an additional 25 observations in each group for each additional two dimensions (e.g. with $p = 10$ or six extra dimensions $3 \times 25 = 75$ extra observations are suggested for each group or $2 \times 100 = 200$ observations in total).

Given that any optimal properties of quadratic discrimination depend heavily on the normality assumption and on having lots of data, even in the two-group case, its potential for application to archaeological problems seems limited. Thus, linear discriminant analysis is likely to remain the workhorse in applications. Since linear discrimination can perform very badly with very different covariances, particularly when groups are close to each other, the equal covariance assumption should ideally be checked and results treated with caution where it is badly violated.

Notwithstanding the above remarks, Craddock et al. (1983) report work on the classification of flints from different mines where quadratic discrimination and linear discrimination work equally well even though sample sizes are small. Success is judged in terms of the ability to classify the flints to the correct mine. This suggests a more pragmatic approach – when in doubt about choice of method, rather than agonising about it, use both and select (for future use in classification) that which performs best. Krzanowski (1988, pp. 346–7) makes a similar suggestion. This begs the question of how the success of a discrimination rule should be judged and this is considered in the next section.

10.7 Assessing how good discrimination is

One outcome of a discriminant analysis is a rule for classifying or allocating individuals to a group. In practice one wishes to assess how good a particular rule is. There is an extensive theoretical literature on this topic, much of which is too complicated to apply routinely in practice (Krzanowski, 1988, p. 361). Simpler, data-based procedures are required for routine practical use and some are now discussed.

Perhaps the most common approach is the resubstitution method. The same individuals used to define a classification rule are allocated to one or other group on the basis of that rule. Because individuals influence the rule subsequently used to classify them, this procedure typically produces an overoptimistic assessment of the success of a procedure. To avoid this and produce a more realistic assessment, various forms of sample-splitting have been proposed. An attractive option is the 'cross-validation' or 'leave one out' approach. Each individual is classified using a rule based on the (n-1) observations that omit that individual. The overall success of classification is typically somewhat lower, and more realistic, than for the resubstitution method.

Cross-validation would seem to require carrying out n (rather than 1) discriminant analyses; however, mathematical relationships between the results for n and n-1 observations can be exploited to ease the computational burden (Lachenbruch, 1975, pp. 36–7). The MINITAB package allows both resubstitution and cross-validation as does the BMDP package (calling it 'jack-knifing').

An illustrative example will be given in Section 10.10. In the archaeological literature, although the benefits of some form of cross-validation have been appreciated for some time (e.g., Benfer and Benfer, 1981, p. 382), examples of its use are limited. One reason may be the absence, at the time of writing, of a cross-validatory procedure in the popular SPSS-X package.

Studies that have involved some form of cross-validation include Craddock et al. (1983), D'Altroy and Bishop (1990) and Mello, Monna and Oddoe (1988) on the chemical compositions of polished flint axes, Inca ceramics and Mediterranean marbles respectively. Morey's (1986) study of Amerindian canid remains and Baxter's (1988) analysis of the morphology of post-Medieval wine bottles may also be noted. The majority of references to linear discriminant analysis in the bibliography do not, however, discuss the topic and use the resubstitution method, if any.

Individuals are inevitably classified into some group even if they are not close to one. Figure 10.2 is an artificially-constructed two-variable example (but based on real data) in which there are two groups and two outliers that belong to neither. That to the extreme right is much closer to group 2 than to group 1 and will be classified into group 2; that at the bottom of the graph is almost equidistant from either. We shall suppose that the two outliers are initially considered to belong to groups 2 and 1 respectively.

Global measures of the success of a classification, such as those discussed earlier, will not reveal the existence of such aberrant points (although plots of the kind illustrated in Figures 9.1 or 9.2 may). A more sensitive analysis can be obtained by examining the Mahalanobis distance of individuals from the groups to which they are allocated. This, in turn,

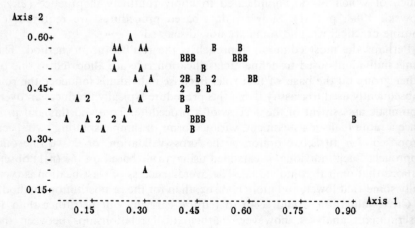

Figure 10.2: Artificial data showing two groups and two outliers

Aspects of Discriminant Analysis

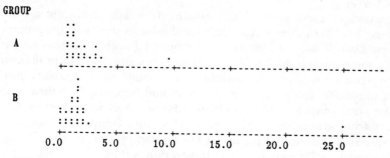

Figure 10.3: Dot-plots of the Mahalanobis distance from the group centroids of the data shown in Figure 10.2

can be done on the basis of analyses with or without cross-validation. Using output generated using MINITAB, Figure 10.3 shows the distribution of distances for each of the groups for the analysis done without cross-validation. The two outliers are clearly revealed. With cross-validation (not shown), even clearer results were obtained. The success of classification in both cases was 100%. Although such inspection of distances is advocated by Pollard (1986), for example, published applications are limited.

It has previously been noted (Section 10.3) that if data are (multivariate) normally-distributed and group sizes are large, then the distribution of the D_i^2 in Figure 10.3 should follow a chi-squared distribution in the absence of problems. Further useful diagnostic statistics can be generated given the normality assumptions. With two groups, assuming equal probabilities of group membership in advance of analysis, we have

$$P(g/D) = P(D/g)/[P(D/1) + P(D/2)] \qquad (10.15)$$

where $P(g/D)$ is the probability of belonging to group g given a discriminant score, D, and $P(D/g)$ is the probability of having a particular discriminant score given membership of group g. For the case of normally-distributed data with equal covariances (10.15) reduces to

$$P(g/D) = \exp(-D_g^2/2)/[\exp(-D_1^2/2) + \exp(-D_2^2/2)] \qquad (10.16)$$

where D_g^2 is the Mahalanobis distance of an individual from the centroid of group g.

As a simple illustrative example the outlier at the bottom of Figure 10.2 has $D_g^2 = 9.58$ and 12.783 for g = 1 and 2 respectively. It is thus closer to group 1 than 2 and is allocated to group 1. The analysis in Figure 10.3 showed, nevertheless, that it was not particularly close to that group. Substitution into (10.16) gives $P(1/D)=0.83$ and $P(2/D)=.17$. This shows it is about five times as likely to belong to group 1 (.83/.17) if it belongs to either.

Suitably scaled values of the $\exp(-D_g^2/2)$ will give the $P(D/g)$ of (10.15).

Inspection of these probabilities, which are available within SPSS-X, will indicate how likely it is that an individual belongs to any of the groups.

This kind of use is advocated by Vitali and Franklin (1986a) who place it in a more general mathematical setting. They suggest that, in the context of provenance studies, 'the usefulness of conditional probability has not been previously recognised', though it should be noted that their usage is in the same spirit as Pollard's (1986) advocation of inspection of the distances from groups discussed earlier. Vitali and Franklin (1986a, p. 168) provide an illustration of the use of $P(g/D)$ and $P(D/g)$.

A summary of the discussion in this section is that:

(a) Cross-validation rather than the resubstitution method provides a more suitable global assessment of the success of an analysis.

(b) Inspection of the $P(g/D)$ provides an indication of the relative probabilities of group membership. Values near .5 in the two-group case indicate considerable doubt about which group an individual belongs to. In the context of provenance studies Pollard (1986) suggests that values in excess of .95 are desirable.

(c) Inspection of values of $P(D/g)$ or, analogously, distances of individuals to group centroids will identify observations that do not conform to the groups to which they are supposed to belong. If such individuals are identified it may well be sensible to rerun an analysis after omitting or reassigning them (Krzanowski, 1988).

(d) Of the packages with which I am familiar, MINITAB implements cross-validation and can produce information on the Mahalanobis distances of individuals to group centroids as well as values of $P(g/D)$. SPSS-X, in addition to this latter quantity, also provides information on $P(D/g)$ for the group to which an individual is allocated. Use of the statistics is illustrated in examples in Sections 10.8 and 10.10.

10.8 Use for validating clusters

A common use of discriminant analysis in applications is to 'validate' the results of a cluster analysis (Aldenderfer, 1982). The two applications in Doran and Hodson (1975) referenced in Section 9.2 were of this kind, as are many of the references to discriminant analysis in the bibliography. Discriminant analysis is applied to the groups determined by a cluster analysis. If most individuals are successfully classified into the correct group the cluster analysis is considered to be 'validated'. Negative results would be informative in these circumstances; if many individuals are misclassified or not clearly classified, then this would cast doubt on the usefulness of the cluster analysis. Unfortunately a 'successful' discriminant analysis does not, by itself, 'validate' a cluster analysis if the term 'validate' is understood to mean that the clusters are shown to be genuinely distinct rather than merely representing a sensible partition of the data.

To illustrate this point a simple artificial example is used that also illustrates some of the methods of the previous section. Figure 8.1(a) illustrated a plot based on thirty randomly-generated observations for two standardised normal random variables. The labelling corresponds to the two clusters obtained using Ward's method (without relocation) shown in Figure 8.1(b). This would often be interpreted as showing two clear clusters; identical results are obtained using average linkage cluster analysis.

In this case it is known, by construction, that the two clusters are not distinct, but simply represent a partition of the sample. Two other arbitrary partitions can be created by dividing the data into two equal-sized groups on the basis of scores on the X axis and Y axis respectively. These, together with the cluster analysis, give three partitions of the data.

Discriminant analysis was applied with all three partitions; the success in each case, using the resubstitution method, was 87% for the partition based on the X axis; 93% for the partition based on the Y axis; and 97% for the cluster analysis partition (cross-validation did not affect the overall picture). The last two partitions, though differing in the initial assignment of about a quarter of the observations, are almost equally successfully 'validated'.

For the partition based on the cluster analysis values of $P(g/D)$ (see Section 10.7) were calculated for the group to which an observation was assigned; 25/30 were in excess of .98 with the lowest being .59 for observation 25 and the next lowest .76. Number 25, which is the only observation misclassified, is on the boundary between the two groups in Figure 8.1(a) and is moved to the other group if relocation is used, leading to a 100% success in the discriminant analysis. Discriminant analysis is occasionally used in this way to 'correct' a cluster-analysis-based classification.

Values of $P(D/g)$ were also calculated for each observation, for the group to which they were allocated, and ranged from .08 to .92. There is nothing untoward here; with a sample of size thirty, some values as low as .08 are to be expected. Inspection of Mahalanobis distances of observations to group centroids similarly highlighted no peculiarities.

These results are perfectly reasonable since the partition of the sample achieved by cluster analysis is a sensible one. The point to emphasise, as others have done (Pollard, 1986), is that discriminant analysis performs too well in discriminating between arbitrary but sensible partitions of the data. Cluster analysis will produce clusters, even if distinct groups do not exist. Other partitions of the data are often possible that discriminant analysis would 'validate' equally well, and the implication is that discriminant analysis should be used for validating clusters with extreme caution and may be most useful when negative results are obtained.

For illustrative purposes an artificial example has been used, but it is possible to achieve a similar demonstration with real data. For example, it is possible to partition the data of Table A1 in two distinct ways (using scores on the second and third principal components) such that 100%

discrimination is achieved with cross-validation in both cases. (This is left as an exercise for those interested!) Once again the partitions are sensible but arbitrary; they do not form externally separated clusters, but discriminant analysis provides no indication of this.

10.9 Variable selection

Rather than using all the available variables in a discriminant analysis, it is quite common to attempt to select a subset of variables that 'best' discriminate between groups in some sense. There are potential pitfalls in the way such selection procedures are used, and the aim of this section is to discuss and illustrate these with reference to the two-group case. The same principles apply to more than two groups (see the example in Section 10.10).

A good reason for wanting to use a subset of variables, rather than the full set, is that it can actually improve discrimination. This may appear paradoxical; in effect, variables that carry no discriminating information can blur the effect of those that do and, for descriptive purposes, the use of a good subset may be better.

Problems that occur in using selection procedures in practice include:

(a) 'best' can be defined in different ways and different definitions do not always give rise to the same results;

(b) in practice variables are usually selected in a stepwise fashion, adding to or subtracting from a subset one at a time; given an agreed definition of 'best', such methods are not guaranteed to find the best model;

(c) with many variables it may often be the case that several subsets (of equal size) are almost equally good at discrimination.

Thus, 'best' subsets are not necessarily best nor, in any useful sense, unique. Attempts to interpret a 'best' subset in terms of substantive meaning, or to use it for future work, are beset with dangers if this is not recognised.

This point will be illustrated shortly, exploiting a relationship between two-group discrimination and multiple regression analysis that must first be explained. Suppose observations are coded 0 or 1 according to which group they are in and denote the resulting 0–1 variable by Y. It is well known that there is a formal relationship between two-group discriminant analysis and the multiple regression model

$$Y = b_0 + b_1 Z_1 + b_2 Z_2 + \ldots + b_p Z_p \qquad (10.17)$$

Specifically, the estimated coefficients are the same as those of the discriminant function apart from a constant factor. This means that problems in two-group discrimination can be approached using well-understood regression methods. Full details are given in Flury and Reidwyl (1988, pp. 94–114).

In particular, methods of model selection used in multiple regression can be applied to obtain 'best' subsets whose fits may be compared using the coefficient of determination, R^2 (Shennan, 1988, pp. 129–31). Flury and Reidwyl (1988) exploit the method of backward elimination in which, starting from the full set, variables are deleted one at a time until all are significant according to some preselected criterion. Forward selection starts with no variables and adds them one at a time until no additions are significant; stepwise regression combines this with backward elimination when, after each entry, the importance of variables already in the model is reviewed – they may be deleted. Other approaches attempt to compare all possible models for different-sized subsets. A full understanding of the details is not needed here and text books such as Draper and Smith (1981) or Rawlings (1988) can be consulted if these are of interest.

The two important points to bear in mind are that the model selected may depend on both the method of choice and the criterion for selection or deletion of a variable. This will be illustrated by applying different methods to a set of vessel-glass compositional data similar to that in Tables A1 and A2. Using cluster analysis, the data divided clearly into two groups which were shown by the use of h-plots to be genuinely distinct in their compositions (Baxter and Heyworth, 1990); these were taken as given for illustrative purposes and twenty variables used. Stepwise selection (which gives the same results as forward selection) and backward elimination were applied using the default options from the MINITAB package where the criterion for inclusion is significance at the 5% level (approximately). Additionally, the package allows one to select the best five models, judged by R^2, for fixed subset sizes.

The first two procedures initially select three- and six-variable models respectively containing the elements (Fe, Cr, Sr) and (Mg, Ca, Ti, Mn, V, Zn) with R^2 of 80.4 and 84.0. If backward elimination is forced to proceed until only three variables remain, those selected (Mg, Ca, Ti) are completely different from those obtained by stepwise regression, and give a marginally better model judged by the R^2 of 81.6. Using just the nine elements identified above, backward elimination gives the best three-variable model available. The best five three-variable models all include Mg and Ti with Ca, Sr, Cr, Zn and V respectively, with R^2 ranging from 81.6 to 80.7. Although there is little to choose between them, the model selected by stepwise regression is not even among the best five available! Note that, because of its ready availability in packages such as SPSS-X, stepwise selection is often used in practice.

These results can be understood by reference to the h-plot for the nine standardised variables in Figure 10.4. The high degree of correlation between many of the variables is evident (this is because they tend to have high values in one group and low values in another). Thus many of the variables carry similar discriminating information and can be interchanged with little effect on the efficiency of discrimination.

Flury and Reidwyl (1988) show that R^2 is related to the distance

208   EXPLORATORY MULTIVARIATE ANALYSIS IN ARCHAEOLOGY

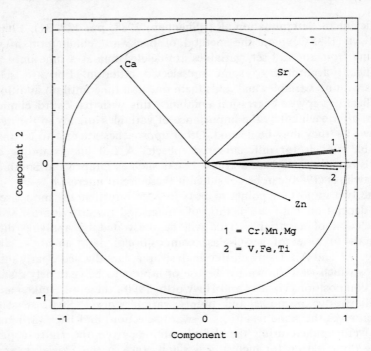

Figure 10.4: An h-plot of Winchester vessel glass compositional data

between groups so that all the subsets identified (and others) will discriminate between the groups almost equally well. If all that is needed is a demonstration that the predefined groups can, in fact, be separated successfully, then any of the subsets could be used. Using the methods of Section 10.7 discrimination is almost perfect in every case, and can be made 100% successful by increasing the number of variables used.

Where the danger lies is in attempting to 'explain' separation in terms of a single supposedly 'best' subset without allowing for the correlations within the full suite of variables. In the context of archaeological glass, for example, there is an expectation that there will be correlated suites of elements that characterise the glass (Henderson, 1985). To imply that single elements within these suites are responsible for separation – as studies that use stepwise methods sometimes do – is misleading and oversimplified (Baxter and Heyworth, 1990).

In practice SPSS-X and BMDP rather than MINITAB (which is only suited to the two-group case) are used, but can exhibit similar phenomena to those noted above. Although both packages allow considerable control over the manner of entry or deletion of variables, it is the default options that tend to be used in the archaeological literature. It is worth examining the consequences of this in a little more detail.

Attention will be confined to the case of forward selection where variables are entered into the model one at a time until the candidate for entry fails to satisfy some preset criterion. At any stage an F statistic such as that in (10.11) is calculated for all variables not in the model, and the variable with the largest F is considered for entry to the model. The default in SPSS-X is to enter the variable if F exceeds 1, and in BMDP to enter it if F exceeds 4. A consequence is that, with the same data, SPSS-X will typically select a larger model than BMDP if the defaults are used.

Even with a more stringent entry criterion than the SPSS-X default, my experience is that a larger model than is necessary will often be selected. It is not uncommon for the success of a classification (whether judged by resubstitution or cross-validation) to increase to a peak as variables are added, and then remain stable or even decline as further variables are entered. Thus, judged by the success of classification, a final eight-variable model may actually be worse than the four-variable model selected en route to it. It is also often possible, by judicious experimentation, to select better and/or smaller models that include variables not picked out by the selection procedure. These observations, which reinforce previous comments about caution in interpreting a selected model as best, are based on unpublished research on artefact compositional data sets with about twenty variables.

10.10 Examples

As a small illustrative example, the Bronze Age cup data of Table A4, previously examined via PCA in Section 4.12, will be used. This is a two-group discrimination problem, between early and late cups, and ratios of rim diameter to the remaining four variables will be used – these will be denoted ND, SD, H and NH.

Preliminary inspection of dot-plots of variables in the two groups did not show any signs of clear outliers or non-normality, so no further transformation was used. The sample sizes n_1 and n_2 are 20 and 40, and p, the number of variables, is four. Using SPSS-X Box's M statistic, based on (10.14), suggested that the difference between S_1 and S_2 was significant at about the 2% level. There is thus some evidence, though not completely overwhelming, that the covariance matrices for the two groups differ.

Proceeding as if the assumption is satisfied, the unstandardised canonical discriminant function has the form

$$F = 10.03 + 1.84ND + 5.89SD + 5.05H - .86NH \qquad (10.18)$$

and the standardised form is

$$F = .10ND + .48SD + .70H - .06NH \qquad (10.19).$$

Equation (10.19) suggests that, if the variables are capable of discriminating between early and late cups, then height and (to a lesser extent) shoulder diameter, expressed as a ratio of rim diameter, are the best discriminators. In passing it may be noted that if early cups are coded as 0 and late cups as 1 then, as discussed in Section 10.9, a regression analysis gives similar results, except that the coefficients of the variables in (10.18) are about 12.3 times the regression coefficients.

Assessing the best discriminating factors in the present case is, in fact, rather academic, since there is little evidence of any difference in shape between the two periods. Using the resubstitution method, 60% of the cups are classified correctly, which is little better than would be achieved by random allocation, and using cross-validation (in MINITAB) only 53% are correctly allocated. Calculations of P(g/D), as in (10.16), mostly give values in the range .4 to .6 for $g=1$ or 2, showing that there is no real basis for allocating cups to the early or late group on the basis of shape characteristics. This conclusion is confirmed by the stacked histogram of Figure 10.5 which shows no discrimination between the groups.

For a second example the data of Tables A1 and A2 will be used. In the first instance the two sites from which the specimens come will be used to define two groups. Following this an analysis based on the three groups suggested by PCA and cluster analysis will be undertaken. In both cases specimen 12 was omitted on the basis of analyses reported in earlier chapters.

For the two-group case, inspection of dot-plots for variables within groups showed that the (marginal) distributions looked reasonably normal in many cases, but were quite clearly non-normal in others. Overall the normality assumption is clearly not satisfied. Transformation to normality in all cases is not, however, possible since the departures from normality tend to take the form of bi-modality (e.g., Fe and Ca for Mancetter) or modes at an extreme of the data (e.g., Sb and Pb for Mancetter) which do not lend themselves to this. There are also clear outliers within groups for

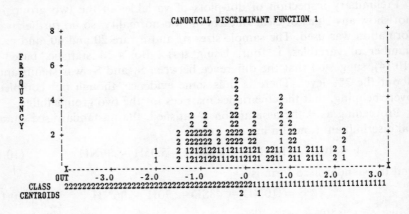

Figure 10.5: Discriminant analysis of the Bronze Age cup data in Table A4

Note: The diagram is edited SPSS-X output

Aspects of Discriminant Analysis

some variables (e.g., Cr has one outlier within each site; Li has one outlier for Leicester).

The bi-modality suggests that the assumption of two chemically distinct groups, co-incident with the sites, may be incorrect, and the PCAS reported in Chapter 3 suggest three rather than two groups. Nevertheless we will proceed on the assumption of two groups, using untransformed data and including within-group outliers, for illustrative purposes. The equal covariance matrix assumption is also clearly violated as judged by (10.14) and statistics based on it (though these depend on the normality assumption for their strict validity). In summary, the ideal assumptions for the analysis are violated and this should be borne in mind in interpretation.

Despite all this, the analysis appears reasonably successful. The success rate is 93% using the resubstitution method and 87% using cross-validation. The figures change to 96% and 82% if quadratic discrimination is used, so there is no real reason for preferring the latter approach. For linear discriminant analysis with resubstitution, the misclassified specimens are 2, 3, 41, 54, 58, 80 and 101 with 9, 23, 48, 83, 85, 86 and 102 additionally being misclassified if cross-validation is used. In the case of cross-validation, 91% of Mancetter and 81% of Leicester specimens are classified correctly.

If, rather than using two groups defined by site of origin, the three chemically-distinct groups suggested by other multivariate analyses are used then Figure 9.2 results. This figure shows fairly clear separation between the groups apart from two points intermediate between groups 1 and 2 (both labelled as belonging to group 2) and one point intermediate between groups 1 and 3 (labelled 3). One point to the right of group 3 is clearly distinct from it and two or three other members of group 3 are distinct from the main group. This generally satisfactory separation is to be expected since the PCA, on the basis of which the groups were defined, also showed fairly clear separation apart from one or two points intermediate between groups. It was suggested in Section 10.8 that this kind of application of discriminant analysis, which is one of the more common uses in archaeology, is often relatively uninformative. It confirms the separation evident from the PCA but, without appeal to the PCA, does not demonstrate the distinctiveness of the groups in terms of external isolation.

Using the resubstitution method specimens 22 and 101 are misclassified from groups 2 and 3 into group 1. This may simply arise as a result of an original misclassification on the basis of visual interpretation of Figure 3.8 and the specimens might be reallocated accordingly. This kind of approach is sometimes used as an alternative to the use of relocation methods in cluster analysis. If cross-validation is used the success rate is 93%, as opposed to 98%, with a further five specimens (3, 19, 30, 48 and 67) being misallocated. All but one of these involve groups 1 and 2; Figure 3.8 (rather than 9.2) suggests that this is because the boundary between the groups is not well defined.

Rather than using global misclassification rates to 'validate' results obtained by other methods of multivariate analysis – as is most usual – a more sensitive analysis of the kind suggested by Pollard (1986) or Vitali and Franklin (1986a) is possible. Pollard suggests that it is desirable that $P(g/D)$ should exceed .95 for group membership to be secure (however groups are defined). For the resubstitution method this is so for all but three specimens (22, 30, 101) and in fact $P(g/D)$ exceeds .99 in nearly all other cases. This would often be taken as indicative of a successful analysis in the sense that with three exceptions specimens are securely allocated to their presumed group. With cross-validation the picture changes to some degree. Of the five specimens allocated to a different group if cross-validation rather than resubstitution is used, one (number 19) has a $P(g/D)$ of 1.0 of being in group 2 in the former case, and of .99 of being in group 3 in the latter case. Overall, with cross-validation, seven specimens have a $P(g/D)$ of less than .95. Though not bad, this is, typically, less impressive than results for the resubstitution method.

Following Vitali and Franklin (1986a) or Pollard (1986), inspection of $P(D/g)$ or distances of specimens from group centroids is also informative. Using the resubstitution method the specimen to the extreme right of Figure 9.2, number 55, is securely allocated to group 3, judged by $P(g/D)$ whatever method is used. In fact, the distance from the group centroid is 72.6, corresponding to a $P(D/g)$ of .002, suggesting it is not a true member of the group. The next most extreme are 70.7 and 38.2 for D and .003 and .010 for $P(D/g)$ for specimens 19 and 26, from group 3, to the extreme south-west and north of the group centroid. Together with the PCAs of Chapters 3 and 4, these results, suggest that there are three reasonably distinct chemical groups in the data to which about 90% of the specimens can be allocated with reasonable confidence. The remaining specimens appear to be either intermediate between groups or, in the case of 19 and 55, belong to none of them (along with 12 which was omitted from the start).

The larger of the groups is based mainly, though not exclusively, on Leicester material, with the other two based mainly on Mancetter material. Within the Mancetter material the grouping did not correlate with any of the recorded archaeological variables (e.g., position within site). It can be shown that the 'recipes' used in the glass making are typical of other Roman glass of the period (Jackson et al., 1991). While there are clear differences in the average composition at the two sites, it is not possible to separate them clearly as was originally hoped.

If the separation into three broadly different groups is accepted, then it is of interest to ask how they differ and which variables are important discriminators. A simple and effective way of doing this would be to calculate element means for each group (or site) and present them in a tabular or graphical form (Jackson et al., 1991). This will not necessarily show

how variables are interrelated or which are important discriminators, and for these purposes multivariate methods are useful. One possibility, in the present case, would be to present the data in the form of a biplot after eliminating specimens of uncertain allocation and outliers. For illustrative purposes, however, the groups will be retained as they are and variable selection procedures within SPSS-X will be used.

Calculating Wilk's Λ as defined in (10.9) for $p = 1$ for each variable separately shows that fifteen of the twenty-two are very highly significant; the most significant, with a value of .21, is Sb, and this is first entered into the model as part of a stepwise selection procedure. Given the correlations that exist between Sb and other variables once Sb is in the model, other apparently important discriminating variables may not be needed since they contain similar information to Sb. In fact, given Sb and using the selection procedure described in Section 10.3 following equation (10.10), the second variable selected is Co which would be the fourteenth best discriminator if only a single variable were used.

The particular test statistic used in SPSS-X has an approximate F distribution assuming normality of the data, and the default value that has to be exceeded for a variable to be added to the model is $F = 1$. This can be modified by the user and $F = 3$, corresponding approximately to a 10% level of significance for a single preselected variable, has been used here. Proceeding with sequential entry (no deletions being needed) until the candidate for entry is not significant gives an eight-variable model based on (Sb, Co, Pb, Fe, Cr, P, Mg, Mn). The success of this model at discriminating between groups is 97% or 96% according to whether resubstitution or cross-validation is used. This last figure is larger than obtained using all the variables, and illustrates one of the potential advantages of variable selection, omitting variables with no additional discriminating power.

Some miscellaneous points about this analysis are now mentioned. Firstly, the overall significance of the model is unknown; use of $F = 3$ as the criterion for selection – as with other possible choices – is essentially arbitrary, particularly since the data are non-normal and any significance level is notional. A less stringent criterion would simply result in a larger model. Secondly, the results were identical to those obtained using the other variable selection procedures in SPSS-X, although the order of entry and deletion varied from method to method. Essentially, the methods differ in the definition of group separation that is optimised at each stage (Klecka, 1980, pp. 52–8). Thirdly, with a little experimentation in MINITAB, results as good as those obtained by the variable selection procedures can be produced using largely different subsets. Thus, with only Sb and Co in common, many of the models containing eight variables including six of the trace elements taken from (Cu, Li, Ni, Sr, V, Y, Zn, Zr) will produce similar results to the variable selection procedure. The point to reiterate here is that variable selection procedures are not guaranteed to produce subsets that are 'best' or 'unique' in any useful sense of these terms, and to go beyond the limited objective of data display and interpret the selected subset is probably misguided in many instances. Finally, in terms of the success of classification, it is actually possible to do as well or better using a subset of the eight 'best' variables. Thus, omitting Mn and P from this subset does not affect the classification success; additionally, omitting Pb actually improves the classification success using the resubstitution method (though not cross-validation) to 98%.

10.11 Discriminant analysis in the archaeological literature

The vast majority of the papers listed in the bibliography are to applications of linear discriminant analysis; exceptions will be noted in the course of the review. Benfer and Benfer (1981, p. 388) in a brief historical review note that McKern and Munro (1959) 'published the first modern example of discriminant function analysis with physical anthropological data used to classify archaeological sites by phase'. Most applications, however, date from the early 1970s and later, with Graham's (1970) paper on discrimination between Palaeolithic handaxe groups a much-cited early example. De Bruin et al. (1972) and Sieveking et al. (1972) are early archaeometric examples, both appearing in the journal *Archaeometry*. The use of discriminant analysis in physical anthropology dates back to the 1920s (Benfer and Benfer, 1981, p. 388); related archaeological applications are to be found in early issues of *World Archaeology* and the *Journal of Archaeological Science* such as Brothwell (1972) and Brothwell and Krzanowski (1974).

As with PCA and cluster analysis, applications to artefact and material compositions determined by methods such as neutron activation analysis, X-ray fluorescence and inductively coupled plasma spectroscopy are common. Ceramics and clays are the most studied materials, but analyses of garnets (Bimson et al., 1982); lacquers (Burmester, 1983b); mortars (Cappanesi et al., 1991); Roman coins (Carter and Frurip, 1985); flint axes (Craddock et al., 1983); Medieval glass (Djingova and Kuleff, 1992); marble (Germann et al., 1988), and obsidian (Williams-Thorpe et al., 1984) can all be noted. Similar compositional data, but for faunal remains, have been studied by Francalacci (1989). The use of lead isotope data is discussed in Gale (1989); Gale and Stos-Gale (1989), and Stos-Gale (1989), with marble isotope data used in Leese (1988).

Studies divide broadly into those where the groups used in the analysis are determined by archaeological or contextual information and those where the groups are determined by some other statistical method such as cluster analysis. As examples of the former, Carter and Frurip (1985) group their Augustan quadrantes by date; Hedges and Salter (1979) group iron currency bars according to site of origin. The use of discriminant analysis for validating a cluster analysis is discussed in Section 10.8, and examples occur in Attas et al. (1977); Birgül et al. (1979); Burmester (1983b); Hart et al. (1987); King et al. (1986); Kuleff et al. (1989); Newton et al. (1988); Pernicka et al. (1990); Pollard and Hatcher (1986), and White et al. (1983).

Typological biological applications that attempt to discriminate between predefined groups on the basis of metric measurements of bone material – often skulls – are also common. Human populations have been studied by Bartel (1979); Brothwell (1972); Brothwell and Krzanowski (1974); Brown (1981, 1987); Collier (1989), and Freedman and Lofgren (1981). Dogs and related species have also attracted attention (Beneke, 1987; Higham et al., 1980; Morey, 1986, and Walker and Frison, 1982) as have cattle and similar animals (Browne, 1983; Higham et al., 1981) and horses (Beneke, 1990).

Archaeological, as opposed to archaeometric or biometric, applications tend to be more varied in kind. Discrimination between predefined assemblages of lithic material on the basis, usually, of metric measurements on the individual artefacts is quite common. Examples include Benfer and Benfer (1981) on projectile points; Burton (1980), waste flakes; Graham (1970), Paleolithic handaxes; Montet-White and Johnson (1976), end scrapers; Wenban-Smith (1989), bifaces; Whittaker (1987), projectile points; and Wynn and Tierson (1990), Acheulean handaxes. Other types of artefact studied in a similar spirit include post-Medieval wine bottles (Baxter, 1988); Bronze Age cups (Lukesh and Howe, 1978); pots (Smith, 1988), and Norwegian spearheads (Solberg, 1986). Occasionally the assemblages compared are defined by a cluster analysis (e.g., Aldenderfer, 1982; Doran and Hodson, 1975), but this does not seem to be common in this kind of study.

Note that in my terminology studies of the above kind are classified as typological/morphological even though they involve the comparison of assemblages. This is because the basic unit of analysis is the artefact rather than the assemblage. If a discriminant analysis shows assemblages to be distinct, this will be because they tend to be composed of artefacts of different types.

Discriminant analyses where the assemblage is the basic unit of analysis and where, for example, assemblages are grouped by period are less common. Examples include Callow and Webb (1981) and Pitts and Jacobi (1979) on lithic assemblages, and Tomber (1988) on pottery assemblages.

Some studies use discriminant analysis for the purpose of site classification and can be viewed as either assemblage comparisons or typological studies. In the former category is the work of Bettinger (1979) and Schlanger and Orcutt (1986). Sites are classified in groups and then discriminant analysis is used to see whether the assemblages of material within sites differ between groups. Martin (1989), by contrast, is a typological study in which sites are grouped by age and characterised by variables the distance to the nearest island, etc.

Bettinger (1979) characterises assemblages at sites by the frequency of variables such as the number of floors or projectile points, but also reports analyses where these variables are treated on a presence/absence (0-1) basis. Other studies, all typological, that use discriminant analysis on 0-1 data are Bartel's (1981) study of Neolithic figurines, grouped in twelve sites/periods, characterised by the presence or absence of 134 attributes; Grebinger and Adam's (1974) study of pottery styles based on the presence or absence of decorative motifs, and Bain's (1985) similar study of Maori rock drawings. The view is expressed in this last paper that discriminant analysis is 'ideally suited to coping with presence or absence data' (p. 50) but without supporting evidence. This issue is noted briefly at the end of Section 10.4.

Turning now to issues which cut across specific areas of application, the comments at the end of the last paragraph are related to the issue of data normality, which is desirable though not essential (Section 10.4), and data transformation. To improve the distribution of variables, logarithmic transformation of some or all of them is quite common in archaeometric studies (e.g., Craddock et al., 1983; De Bruin et al., 1972; Hughes et al., 1983; Mello, Monna and Oddoe, 1988; White et al., 1983). Ideally such decisions should be informed by preliminary data inspection which may militate against transformation (e.g., Bimson et al., 1982; Vitali and Franklin, 1986a).

The paper by Bimson et al. (1982) uses element ratios (to silicon). In another context Higham et al. (1980), in a study of canid remains from Thailand, report analyses of ratios of measurements divided by condylobasal length in order to eliminate size effects. Miscellaneous applications of other transformations include Callow and Webb's (1981) use of the arc-sin transformation to try and avoid problems posed by the compositional (summing to 100%) nature of their data, and Baxter's (1988) use of logged ratios to avoid a similar problem caused by expressing dimensional measures as a proportion of their sum.

The assumption of equal covariance matrices seems rarely to be tested or commented on. Leese's (1983, p. 31) analysis of coin compositions notes that the Box test (Section 10.4) shows 'evidence for variation in the dispersion matrices' and attributes this to five outliers; Hart et al. (1987, p. 583) note the unequal covariance problem that 'cannot be remedied'; and Smith (1988, p. 913) notes that both the normality and equal covariance assumptions are violated, but that 'findings are plausible when theoretically interpreted'. The use of quadratic discriminant analysis to avoid the problem is rare; the analysis of flint axe compositions by Craddock et al. (1983) is one example.

The general impression gained from many of the references is that relatively little attention is given to verifying the normality and equal-covariance assumptions in many applications. Wynn and Tierson (1990,

p. 77) state that 'the technique of discriminant analysis is known to be extremely robust both to inequality of group covariance matrices and other assumptions'. This belief, whether true or not, coupled with a lack of widely-available software for alternative procedures, perhaps explains the neglect.

Stepwise methods are used in about one-fifth of the references. Discussion of the reasons why is typically minimal and often amounts to little more than a statement that BMDP program P7M was used, or something similar. If SPSS-X is used, then the WILKS routine is usually chosen – possibly because it is listed first in the options available in the package. The 'best' discriminating variables are sometimes listed in applications; potential dangers in interpreting such a set are discussed in Section 10.9.

In Section 10.7 it was suggested that to assess the global success of a discriminant analysis some form of cross-validation, rather than the resubstitution method, is desirable. Although this has been advocated for some time (Benfer and Benfer, 1981), few studies use such an approach. Of those in the bibliography D'Altroy and Bishop (1990); Hughes et al. (1983); Mello et al. (1988); Mello, Monna and Oddoe (1988), and Needham et al. (1989) are examples from the analysis of artefact compositional data. In typological studies Baxter (1988); Beneke (1990); Knutsson et al. (1988) and Morey (1986) may be cited. The study by Knutsson et al. presents results that show a 16% difference in classification rates, based on use-wear characteristics of flint flakes, if cross-validation (75%) rather than resubstitution (91%) is used.

More detailed analysis of a classification based on inspection of quantities such as $P(g/D)$, $P(D/g)$ or D^2, the Mahalanobis distance from a group centroid, is also relatively unusual. Apart from the exposition by Vitali and Franklin (1986a), a good illustration of such use is given in Pollard and Hatcher's (1986) analysis of oriental greenware compositions. Other studies that make some use of these quantities – tangentially in some cases – include Henderson and Warren (1983); King et al. (1986), and Kohl et al. (1979).

Formal significance tests, either of the significance of group differences or of individual functions based on (10.8), (10.9) or similar statistics, are sometimes used, though often *en passant*. Examples include Boyd and Boyd (1991); Higham et al. (1980), and Leese (1983).

Finally, two papers out of the ordinary run and not yet noted are Luedtke (1979) and Leach and Manly (1982). The first of these, on the identification of sources of chert artefacts, contains a useful discussion on the types of error that may occur in a discriminant analysis, along with a comparative study that illustrates the use of variable transformation, the use of different variable subsets, and the use of D^2 for assessing a classification. The paper by Leach and Manly (1982), on the sourcing of lithic artefacts, contains a more than usually detailed discussion of the problems involved in using linear discriminant analysis. They note that 'the assumption of common variance has little justification in sourcing studies' (p. 80) and devote several pages to discussing this and other problems, such as those caused by missing data. They then describe and illustrate a series of algorithms designed to overcome the problems that they perceive.

The view of the robustness of linear discriminant analysis to the violation of assumptions in Leach and Manly (1982) is in marked contrast to the optimistic assessment of Wynn and Tierson (1990) quoted previously in this section. Alternative forms of discriminant analysis to the linear version exist and deserve exploration, but this will probably not happen until they are made widely available through general-purpose packages.

11

The Final Chapter

> Archaeologists are particularly ill-trained for, and unused to, the rigorous and logical thought necessary for the informed use of quantitative methods, while the rare statisticians who have tried their hands at archaeology typically have understood the nature of archaeological data, questions, and models only partially, vaguely, or incorrectly, so that their efforts are usually no better than the archaeologist's own.
>
> (Whallon, 1984, p. 243)

11.1 Introduction

As one of the 'rare statisticians' referred to above, I am acutely aware of the problems referred to by Whallon. Archaeological collaborators have saved me in the past from making a fool of myself by trying to publish 'interesting' statistical analyses of archaeological data: analyses that were, in fact, complete nonsense. One instance occurred when statistical results of apparent importance (concerning the technology of Roman glass manufacture) were rejected out of hand by my archaeological collaborators. They, of course, were correct to do so. The statistical analyses were clear-cut, but it was obvious (though not to me) that they made little sense in terms of what was known about Roman glass technology. The problem proved to be non-obvious measurement error in a subset of the data which, when accounted for, led to rather different and less exciting conclusions.

Archaeology is an eclectic discipline; where it calls on scientific and statistical knowledge, few individuals combine all the necessary archaeological, scientific and statistical skills at a high level. Ideally, collaboration should take place between archaeologists and statisticians in those areas where a statistical input is of potential interest. Unfortunately, not all archaeologists regard statisticians as useful creatures and there are, in any case, not enough interested statisticians to go round.

While a book is no substitute for collaborative involvement, it is hoped, nevertheless, that this one does something to reveal both the potential and the pitfalls of some widely-used multivariate methods in archaeology. An obvious limitation is that I am primarily a statistician, so that my qualifications for writing on applications in archaeology could be queried. The impetus came from two sources. Firstly, in consulting or working with archaeologists on a range of problems, it has been clear that approaches and techniques which I (and other statisticians) take for granted have seemed novel, interesting and useful to my collaborators. Secondly, the reporting of statistical uses in the literature often leaves much to be desired and can be of limited assistance to those who wish to emulate previous practice. If at least some potential users find this book of assistance in understanding what has been or can be done with multivariate statistics in archaeology, it will have served its purpose.

As noted elsewhere, many of the techniques discussed in the book could be considered 'old-fashioned', though I believe they will continue to find widespread application. Some alternative, and in some cases newer, approaches are noted briefly in the next section. Some personal views (or prejudices) concerning statistical writing in the archaeological literature are aired in the final section. This includes a brief account of why I think the reporting of many applications is unsatisfactory.

11.2 Developments in the statistical literature and archaeology

11.2.1 Exploratory data analysis

Exploratory data analysis (EDA) is not an especially new subject (Tukey, 1977), but its 'philosophy' has been greeted with approval by some quantitative archaeologists over the last decade or so. Thus Voorrips (1990, p. 3) includes EDA among techniques which 'in many instances are more useful to the archaeologist than the "classical", methods for descriptive statistics and multivariate analysis'. (He also notes that 'until now few archaeologists have actually been applying these methods' (p. 8).)

Exploratory data analysis is sometimes opposed to confirmatory data analysis which, in turn, is associated with hypothesis-testing procedures. These latter are sometimes seen as peripheral to archaeological requirements (witness the space given to them in Doran and Hodson (1975) or Shennan (1988)), and EDA, conceived of as 'an inductive approach to recognizing patterns in a data set' (Carr, 1987, p. 220), has obvious attractions.

In fact, statisticians are by no means agreed on what is to be understood by 'exploratory' as opposed to 'confirmatory' data analysis; or on the distinction, if any, between data analysis and statistics. This is clear from the discussion in Gifi (1990, pp. 21–31) which liberally quotes from statisticians who have pronounced on the topic. So far as its protagonists are concerned, EDA seems not just to be a collection of techniques, but also an attitude to data analysis. This 'attitude', elevated to the status of a 'philosophy', appears to have an appeal to those who feel the need to apply statistical methods within some unifying framework.

Archaeologists have adapted EDA ideas to their own needs. Thus, Carr (1987, pp. 220–1) sees inductive EDA methods as opposed to deductive 'confirmatory' methods. In archaeological application, if I understand his argument, he sees a need for EDA to 'articulate' with archaeological theory/reasoning and proposes 'constrained' EDA as 'an inductive middle-step within a stepwise analytic design that has an overall deductive orientation and that is begun with deduction'. If this is to be read as saying that statistical analysis should serve a specified archaeological goal, there is in fact no especial need to limit oneself to a particular statistical 'philosophy' even if, in practice, EDA methods turn out to be widely useful.

My view is that EDA and other simple descriptive techniques should be more widely used than they seem to be, and certainly in the initial stages of any data analysis. The 'correct' attitude, however conceived (e.g., that one is engaged in an inductive pattern-seeking activity), can be adopted at this stage without ruling out other approaches. Wider and more confident use of simple statistics might well obviate the need for multivariate analysis in some cases, but it is unhelpful to see the different approaches as being in some kind of opposition. A certain mystique surrounds the use of multivariate methodology; the intent – if not the mathematics – of many multivariate methods is easy to understand, and many methods are easy to apply with current software. It is perfectly legitimate, and useful, to use them for data exploration or data screening in a spirit identical to the way in which EDA is often used.

11.2.2 Miscellaneous multivariate methods

The techniques described in this book are old-fashioned but useful. In this section a pot-pourri of newer approaches that may have an impact on archaeological usage in the future are noted. At present, routine application of many of these techniques is not feasible, often because of a lack of accessible software, but also for other reasons that will be mentioned as appropriate.

Principal component analysis can be thought of as a particular example of a 'projection pursuit' technique. The term 'projection pursuit' was introduced by Friedman and Tukey (1974) and the ideas and methodology expounded in a paper by Jones and Sibson (1987). These last authors describe PCA as 'something of a blunt instrument' (p. 2); new variables are defined to explain as much of the variation in the data as possible, but its success as an exploratory tool requires that large variation also be 'interestingly structured variation', and this is not a logical necessity. Roughly speaking, in projection pursuit, new variables are defined to display 'interesting' structure in the data, such as clustering. 'Interesting' can be defined in different ways, and the methodology described by Jones and Sibson (1987) leads to a series of different views of the data which can be interesting in different ways, or not at all. I know of no archaeological uses of the method, and the consensus in the published discussion of the paper seemed to be that much was to be learned about it before routine application became possible.

The same is almost certainly true of shape analysis, at least as applied to archaeological problems. This is a topic attracting much current research interest among statisticians (Bookstein, 1986; Kendall, 1989; Goodall, 1991). The subject is mathematically complex; although archaeology has been mentioned as an applied area of potential interest (Goodall, 1991, p. 1), convincing published examples are needed.

It was hinted, in the discussion of EDA, that some scholars may only be comfortable in their use of statistics if it can be embedded within a coherent 'philosophical' framework. Bayesian methodology has an obvious appeal for such scholars and over the years has excited considerable debate within statistics about its merits. A detailed account of Bayesian ideas is not possible here; Barnett (1982) can be consulted for a comparative discussion. The Bayesian view, to provide a rather simplistic account, is to use probability to represent degrees of belief. This allows the combination of 'prior' information with observed data to produce a 'posterior distribution' about the phenomenon of interest that summarises how the data has modified one's initial beliefs. This mimics the way some people think science ought to work, and has a seductive semantic appeal that I suspect attracts non-specialists unfamiliar with the technicalities of the approach. Buck et al. (1991) give a more detailed account written with an archaeological audience in mind.

In this last reference it is suggested (p. 812) that the works by Doran and Hodson (1975) and Orton (1980) contain material that indicates 'clearly how much this approach has to offer archaeologists'. Ruggles (1986) also stresses the potential of the Bayesian approach. Although some examples of Bayesian analysis in archaeology are cited in Buck et al., they are reasonably rare. Orton (1992, p. 139) notes that he has advocated the use of Bayesian statistics in archaeology for many years, but also that he has done little about it.

The enthusiasm for Bayesian methods in some quarters, coupled with a lack of applications, is interesting, and reasons for it may be suggested. First of all, the mathematics is not easy. Thus, of three papers on the interpretation of radio-carbon dates by Buck et al. (1991), Litton and Leese (1991) and Naylor and Smith (1988), the first two were written, in part, to help explain the third. These papers all involve the collaboration of mathematicians and statisticians from the University of Nottingham, UK, who have done much to make such analyses numerically tractable (if not yet widely accessible). This points to a second factor influencing the absence of applications and that is the computational difficulty of implementing the methodology. While Litton and Leese (1991) note recent developments that overcome such problems, the need for active collaboration with an informed practitioner of Bayesian methods is probably essential at present, and this may limit the use of the methodology.

Buck and Litton (1991) is one of few applications of Bayesian methods to archaeological problems of multivariate analysis. They undertake a cluster analysis of clay pipes based on dimensional measurements. The example used assumes multivariate normality of the data and, among other things, that prior information on the variation and covariation of the measurements is available. The paper exemplifies a particular modelling approach to a common archaeological problem. It is not dependent on the normality assumption, though in practice this, or similar specific distributional assumptions, is needed for the approach to work. The authors suggest that their examples 'demonstrate the wide ranging applicability' of their techniques and that what is needed – in line with what was suggested above – 'is real collaboration between archaeologists and statisticians in order to develop explicit models for the underlying processes which give rise to other types of archaeological data' (p. 97).

The practical difficulties of applying Bayesian methodology are likely to face other 'modelling' approaches to the statistical analysis of archaeological data for the simple reason that each data set and problem must be treated 'on its own merits', rather than being subjected to the application of some standard technique. This is not in principle a bad thing, but is demanding of intellectual effort and time. For this reason, though I hope to be proved wrong, I would expect modelling-based methods to attract increasing interest in the future, but for 'exploratory' and accessible methods of the kind discussed in the book to remain the bread-and-butter of multivariate applications.

11.3 Statistics in the archaeological literature

This final section presents some personal views on statistical publication in the archaeological literature. The generalities in Section 11.3.1 are essentially a statement of my prejudices and can be ignored by the scholastically sensitive. Section 11.3.2, on the sins of statistical reportage in the archaeological literature, could more readily be justified, though I have refrained from citing specific instances of what I consider to be bad practice. Statistical publication in non-statistical journals is a subject that constantly exercises statisticians, and most of the observations made are not specific to archaeological or archaeometric publication.

11.3.1 Some generalities

While questions about the philosophy and foundations of statistics have excited considerable, often vitriolic debate, many applied statisticians are simple (some would say simple-minded) creatures. They like to do statistics and only worry about foundational and philosophical issues, if at all, in the privacy of their bathtub (to quote an eminent statistician from my post-graduate days). This way things get done, but possibly inappropriately if the statistician strays from his or her field of expertise.

Quantitatively-minded archaeologists, by contrast, are much given to debating foundational issues in print (e.g., Aldenderfer, 1987; Barcelo, 1991; Carr, 1985; Whallon and Brown, 1982; Voorrips, 1990a). Articles of the kind I have in mind, often concerned with the philosophy and 'congruence' or 'concordance' of statistical applications in archaeology, can be long on theory and short on practice. They can also make difficult reading (though this is my problem). I suspect their influence on practice to be costive, if any. The next two paragraphs contain a gross caricature of aspects of the debate on quantitative methodology that sometimes takes place in the social science literature (not just archaeology), and must not be taken too seriously by those with an interest in and knowledge of such matters.

Two unhelpful extremes exist in the literature. The first associates quantitative methodology with some ism (Xism, say) which, after deep thought, is rejected in favour of the ism of the day (Yism). Neophyte scholars ill disposed to matters quantitative then read the definitive treatise on Yism to which they subscribe, reject Xism and, by association and without having to try it, also reject quantitative methodology. This last step is unsound, but allows statistical methodology to be rejected from a position of ignorance, on high-sounding but spurious 'philosophical' grounds.

The other extreme is represented by Xist texts and papers. These are written by scholars sympathetic to quantitative methodology, but are sometimes couched in terms that seem designed to alienate all but committed fellow Xists. Apart from polysyllabic profusion, such texts contain much matter on topics such as the philosophy of scientific discovery and the logic of induction with little reference to works of statistics and few or no practical examples. Often a reading of such Xist texts – never mind Yist rebuttals – will be sufficient to deter our hypothetical neophyte from the use of statistics. To be convincing, 'theory' needs to be leavened with a good dose of practical application.

I can live comfortably with the idea that many practising archaeologists do a good job of work without resorting to statistical methodology. This absence of use should, however, stem from the nature of the problems confronted *and* a knowledge of that which specific methods do or don't have to offer. Blanket rejection of statistical methodology on generalised and abstract grounds should not be an excuse for non-numeracy. The converse also applies, in the sense that statistical methods should be used because they are (potentially) useful, and not simply because they are there or because they are believed to impart objectivity – often spurious – to an analysis.

11.3.2 Statistics in archaeological publications

The quality of reporting of statistical techniques and results in archaeological publications is highly variable but often poor. An obvious reason for this is pressure on space. In archaeometric conference proceedings limited to ten pages, for example, the statistics often comes a poor third to the science and archaeology in terms of space allotted. The marginalisation of statistical material to appendices in archaeology journals, sometimes printed in a reduced typeface, is also to be regretted.

In the publication of scientific results it is considered good practice to report the experimental protocol in sufficient detail that interested parties can reproduce the experiment if desired. An analogous criterion could be applied to the publication of statistical results, and many papers are lacking from this perspective.

An original intention was to produce a detailed analysis of the way in which particular techniques have been used in archaeology, based on a content analysis of papers in the bibliography. This proved difficult because often the detail is not there. It is not adequate simply to state that a PCA was carried out, for example, since this does not uniquely define the approach used. Have the data been transformed or standardised, for example? The answers are frequently no and yes respectively, and can sometimes be deduced from internal clues in the papers, but it is best that they be stated explicitly. Similarly in a cluster analysis, as an absolute minimum, the use of any data transformation, the choice of (dis-)similarity coefficient and clustering method used should be explicitly stated, as should the basis for choice of the final number of clusters. It is surprising how often such information is omitted.

A particularly pernicious practice is the referencing of computer packages as a substitute for methodological discussion. That a PCA 'was implemented using the FACTOR routine in SPSS-X with VARIMAX rotation' is a hypothetical but quite typical formulation. I want to know this, but in addition to, rather than instead of, a proper discussion. Why PCA was chosen; what form of PCA was used; and why rotation and VARIMAX in particular were used are all questions deserving of an explicit answer. In cluster analysis to state – again hypothetically but typically – that 'Ward's method in CLUSTAN was used' is not a sufficient discussion. Why was this method chosen? Typically the data will have been standardised, and squared Euclidean distance used as the dis-similarity coefficient (since these are the recommended practices in CLUSTAN), but this should be stated explicitly.

It is, in fact, often obvious that a particular approach has been used because it is the default option in an available computer package. The popularity of PCA of the correlation matrix; 'principal component factor analysis' with VARIMAX rotation; Ward's method of cluster analysis; and stepwise discriminant analysis using Wilk's lambda is, in part, attributable to this factor. While the choice will often be a sensible one, it should be explicitly justified and users should be aware of alternatives.

In other cases the justification for a particular approach, and its minimal documentation, seems to be tradition. The origins of the tradition, while not necessarily lost in the mists of antiquity, are on close inspection sometimes found to be rather inaccessible. It is frustrating to be referred to an author's unpublished Ph.D. thesis or a ten-year-old internal technical report for non-obvious detail. Either sufficient technical detail should be provided, or else more accessible references given. Other traditions are associated with an author's academic environment, and some scholars can be 'provenanced' on the basis of the way in which they treat their statistical material. Approaches to statistical analysis are passed on through generations of Ph.D. students, and the original basis for choosing a technique sometimes gets lost in the process.

Even when space is limited, adequate referencing and specification of the precise form of a particular technique used are minimum requirements that should be satisfied. The level of technical detail to provide is less easy to advise on since it will depend on the intended audience. Cluster analysis, PCA and discriminant analysis *are* standard techniques and could often be dealt with by providing a suitable reference so long as the exact version used is specified. For unusual techniques or non-standard uses of a technique, the details ought to be spelled out. The correspondence analysis literature is poor in this respect, with discussion often limited to a statement along the lines that a correspondence analysis was carried out with presentation of the resulting graph.

If statistical methods are worth using at all, then they are worth using and reporting properly. This may require closer and more sympathetic collaboration among archaeologists, scientists and statisticians than is presently the case, and also greater willingness on the part of journal editors to accord space to statistical aspects of archaeological studies.

APPENDIX A
Data Sets

The data in Tables A1 and A2 were made available by Caroline Jackson and were collected in the course of work on her Ph. D. thesis at Bradford University. The two tables contain twenty-two elemental concentrations for 105 specimens of Romano-British waste glass found at two furnace sites, Leicester and Mancetter, in England

The data in Table A3 were originally given in Mellars (1976), the subset extracted being that used in Pitts (1978).

Table A4 is adapted from data given in Lukesh and Howe (1978); in particular the subset used is that for which full information was available.

The data in Table A5 were taken from Bølviken et al. (1982) and show the number of occurrences of sixteen types at forty-three Early Stone Age sites in Northern Norway.

Tables A6 and A7 are based on articles by Nielsen and Englestad in the collection edited by Madsen (1988a), while Table A8 is Table A7 recoded in a manner described in the text. A particular merit of this collection is that the data are given for many of the analyses described, and a subset from Madsen's own article (Madsen, 1988b, p. 18), while not given here, provides the basis for an illustrative example described in Chapter 6.

Table A1: Major and minor element values for two sites and 105 specimens of waste Romano-British glass

	Al	Fe	Mg	Ca	Na	K	Ti	P	Mn	Sb	Pb	Site
1	2.51	.53	.56	6.98	17.44	.73	.09	.15	.58	.12	.03	2
2	2.36	.49	.53	6.71	17.69	.68	.09	.13	.40	.23	.04	2
3	2.30	.36	.49	8.10	15.94	.68	.07	.13	.77	.00	.01	2
4	2.42	.52	.56	6.93	17.59	.72	.09	.14	.47	.18	.02	2
5	2.32	.37	.51	7.51	16.27	.69	.07	.13	.21	.00	.02	2
6	2.34	.56	.52	6.10	18.61	.69	.10	.11	.30	.32	.03	2
7	2.50	.46	.50	6.83	17.46	.79	.08	.15	.40	.06	.02	2
8	2.47	.53	.55	6.55	18.55	.75	.09	.12	.35	.23	.04	2
9	2.41	.67	.62	6.18	18.33	.81	.12	.14	.52	.31	.07	2
10	2.64	.50	.63	7.76	15.66	.63	.08	.16	.21	.00	.01	2
11	2.77	.58	.50	7.33	16.10	.68	.08	.14	.57	.00	.01	2
12	2.43	.69	.72	6.27	17.84	.98	.12	.22	.63	.13	.04	2
13	2.50	.36	.53	8.51	15.46	.60	.07	.16	.45	.00	.01	2
14	2.63	.46	.47	7.25	16.26	.59	.07	.12	.30	.00	.01	2
15	2.66	.41	.50	7.35	17.12	.63	.07	.15	.11	.00	.01	2
16	2.43	.62	.52	6.89	17.17	.69	.08	.13	.44	.18	.05	2
17	2.55	.53	.52	7.91	16.20	.62	.07	.15	.38	.00	.01	2
18	2.44	.54	.56	6.65	17.68	.97	.10	.12	.40	.25	.03	2
19	2.22	.34	.46	7.08	16.14	.63	.06	.15	.12	.00	.01	2
20	2.59	.37	.46	7.57	15.71	.56	.07	.16	.07	.00	.01	2
21	2.45	.48	.55	6.84	17.73	.76	.09	.14	.62	.14	.05	2
22	2.42	.49	.51	7.00	16.32	.93	.08	.14	.42	.10	.03	2
23	2.27	.38	.48	7.88	16.28	.52	.07	.14	.26	.00	.01	2
24	2.48	.55	.55	6.64	18.76	.75	.09	.12	.36	.24	.04	2
25	2.27	.32	.39	6.75	17.95	.75	.07	.12	.18	.00	.01	1
26	2.32	.84	.55	6.19	19.78	.70	.10	.11	.24	.37	.08	1
27	2.46	.49	.54	6.82	18.07	.75	.08	.13	.60	.13	.04	2
28	2.67	.34	.49	6.94	18.04	.54	.06	.11	.44	.00	.01	2
29	2.47	.42	.51	7.57	17.94	.76	.07	.14	.41	.00	.01	2
30	2.40	.45	.54	7.62	17.76	.64	.08	.13	.40	.11	.02	2
31	2.41	.36	.54	8.15	16.65	.54	.07	.13	.44	.04	.01	2
32	2.68	.38	.59	8.47	16.14	1.54	.07	.14	.42	.00	.01	2
33	2.41	.63	.53	6.84	17.77	.76	.08	.16	.45	.09	.03	2
34	2.38	.55	.55	6.73	17.37	.76	.08	.16	.44	.11	.04	2
35	2.50	.78	.56	6.40	18.35	.73	.11	.11	.27	.26	.03	1
36	2.38	.84	.54	6.17	18.05	.70	.10	.11	.26	.30	.03	1
37	2.50	.54	.58	7.21	16.86	1.05	.09	.14	.57	.15	.04	2
38	2.35	.43	.51	8.02	17.52	.56	.07	.14	.29	.00	.01	2
39	2.31	.74	.54	6.26	18.59	.69	.10	.10	.25	.31	.03	1
40	2.42	.36	.47	7.31	17.76	.62	.07	.13	.24	.04	.01	2
41	2.34	.54	.54	6.76	17.62	.68	.09	.13	.42	.19	.04	1
42	2.21	.85	.56	6.21	19.64	.71	.09	.11	.25	.33	.03	1
43	2.17	.56	.56	6.22	20.03	.69	.10	.10	.23	.43	.04	1
44	2.40	.54	.54	7.14	16.87	.79	.08	.13	.56	.23	.05	2
45	2.58	.37	.49	7.36	16.58	.65	.07	.13	.47	.00	.01	2
46	2.45	.89	.55	6.19	18.30	.71	.11	.12	.26	.29	.03	1
47	2.24	.52	.52	6.36	18.69	.60	.09	.11	.29	.30	.03	1
48	2.49	.48	.55	7.32	18.14	1.00	.08	.14	.40	.11	.02	2
49	2.40	.50	.54	6.70	18.85	.70	.09	.12	.39	.23	.04	2
50	2.27	.75	.55	6.24	19.53	.67	.09	.11	.25	.31	.02	1
51	2.27	.87	.56	6.39	18.98	.68	.09	.11	.29	.27	.03	1
52	2.34	.43	.58	9.42	15.72	.59	.08	.13	.14	.00	.01	1
53	2.49	.85	.54	6.36	18.01	.73	.11	.11	.27	.26	.03	1

continued

Table A1 (continued)

	Al	Fe	Mg	Ca	Na	K	Ti	P	Mn	Sb	Pb	Site
54	2.43	.44	.50	6.77	17.70	.74	.08	.15	.48	.06	.02	1
55	2.25	.59	.56	5.52	20.55	1.01	.12	.09	.26	.41	.05	1
56	2.33	.37	.52	7.31	16.75	.49	.06	.11	.90	.00	.01	2
57	2.46	.47	.52	7.03	17.48	.67	.08	.14	.49	.13	.04	2
58	2.55	.56	.58	7.17	17.34	.72	.09	.14	.69	.16	.04	1
59	2.38	.64	.61	5.99	19.63	.79	.12	.14	.50	.30	.07	2
60	2.38	.47	.52	6.79	17.36	.67	.08	.13	.45	.14	.05	2
61	2.71	.37	.47	7.50	16.57	.48	.07	.13	.21	.00	.01	2
62	2.23	.73	.54	6.07	18.58	.64	.10	.10	.23	.30	.02	1
63	2.45	.77	.56	6.41	19.07	.73	.11	.11	.28	.30	.03	1
64	2.58	.37	.54	7.57	16.11	.61	.07	.14	.14	.00	.01	2
65	2.46	.35	.51	7.72	16.51	.56	.07	.12	.17	.00	.01	1
66	2.17	.54	.57	6.23	19.98	.67	.10	.10	.21	.43	.04	1
67	2.59	.58	.56	7.61	16.74	.68	.08	.17	.50	.03	.01	2
68	2.22	.48	.52	6.44	18.66	.62	.09	.11	.31	.28	.03	1
69	2.52	.86	.56	6.45	18.32	.74	.12	.12	.26	.27	.03	1
70	2.34	.78	.58	6.37	19.34	.73	.10	.11	.26	.31	.03	1
71	2.64	1.11	.59	7.89	17.78	.75	.12	.15	.26	.23	.03	1
72	2.32	.64	.58	5.66	20.08	.79	.13	.12	.31	.41	.05	1
73	2.73	.74	.55	6.12	18.83	.77	.11	.10	.29	.31	.06	1
74	2.51	.78	.55	6.44	18.30	.73	.11	.12	.26	.26	.03	1
75	2.37	.81	.55	6.38	19.03	.70	.10	.11	.24	.32	.03	1
76	2.31	.88	.57	6.42	18.90	.76	.10	.12	.28	.30	.03	1
77	2.50	.78	.56	6.46	18.57	.73	.11	.12	.26	.27	.03	1
78	2.57	.80	.56	6.43	18.41	.75	.12	.12	.26	.26	.03	1
79	2.24	.84	.56	6.26	19.49	.73	.09	.12	.23	.32	.03	1
80	2.37	.44	.50	6.78	17.15	.70	.08	.15	.45	.06	.02	1
81	2.48	.77	.55	6.36	18.30	.73	.11	.12	.26	.26	.03	1
82	2.26	.58	.61	6.16	19.47	.74	.10	.11	.21	.43	.03	1
83	2.59	.48	.60	8.76	14.50	.51	.07	.13	.27	.00	.01	1
84	2.25	.66	.52	6.20	18.06	.64	.09	.11	.24	.31	.04	1
85	2.43	.48	.56	7.60	15.57	.62	.08	.16	.49	.00	.01	1
86	2.49	.93	.55	6.18	16.54	1.10	.12	.13	.25	.28	.03	1
87	2.46	.76	.55	6.37	17.95	.72	.11	.12	.26	.27	.03	1
88	2.47	1.05	.56	7.62	17.02	.70	.11	.14	.26	.23	.03	1
89	2.16	.74	.53	6.09	17.25	.65	.09	.11	.25	.32	.03	1
90	2.26	.58	.52	6.41	17.28	.67	.09	.13	.28	.25	.03	1
91	2.29	.78	.56	6.24	18.45	.70	.10	.11	.26	.32	.03	1
92	2.30	.78	.53	6.28	18.20	.65	.10	.11	.25	.31	.03	1
93	2.52	.65	.55	6.16	18.69	.74	.11	.10	.29	.33	.06	1
94	2.28	.68	.55	6.37	18.60	.68	.10	.11	.24	.32	.03	1
95	2.25	.62	.56	5.55	19.47	.74	.13	.11	.31	.42	.05	1
96	2.32	.80	.54	6.34	18.25	.66	.10	.11	.25	.32	.03	1
97	2.35	.74	.55	6.54	18.44	.71	.10	.11	.26	.29	.03	1
98	2.45	.42	.61	9.79	16.22	.62	.08	.13	.14	.00	.01	1
99	2.19	.84	.54	6.13	17.99	.69	.10	.11	.24	.33	.04	1
100	2.62	.82	.54	6.25	17.79	.73	.11	.09	.29	.32	.04	1
101	2.35	.65	.54	6.73	17.91	.72	.10	.12	.28	.25	.06	1
102	2.44	.35	.51	7.70	16.27	.62	.07	.13	.16	.00	.01	1
103	2.42	.68	.53	6.15	17.19	.77	.13	.11	.27	.26	.03	1
104	2.52	.79	.56	6.37	18.11	.74	.12	.11	.26	.27	.03	1
105	2.37	.75	.55	6.33	18.55	.69	.10	.11	.25	.32	.03	1

Note: Values are given as the % of the oxide form of the element in the glass. Site 1 is Leicester and site 2 Mancetter.

Table A2: Trace element values for two sites and 105 specimens of waste Romano-British glass

	Ba	Co	Cr	Cu	Li	Ni	Sr	V	Y	Zn	Zr	Site
1	246	9	20	60	20	21	432	19	9	26	242	2
2	217	7	19	78	18	14	409	17	8	26	231	2
3	242	8	19	13	17	20	487	22	9	25	43	2
4	232	8	19	70	17	18	425	19	8	31	236	2
5	208	7	19	18	15	17	413	12	9	28	140	2
6	205	7	22	62	21	21	390	16	8	29	264	2
7	229	12	18	232	17	16	388	15	8	26	223	2
8	225	7	18	61	20	15	411	17	8	28	238	2
9	230	8	21	111	19	18	426	22	8	35	300	2
10	239	8	20	33	15	15	418	14	8	19	171	2
11	248	16	19	31	16	19	440	16	8	27	138	2
12	237	15	22	264	18	19	398	21	7	41	257	2
13	238	6	19	9	15	17	482	15	8	22	184	2
14	222	5	16	6	15	17	426	11	8	23	117	2
15	221	4	16	6	15	10	389	11	8	18	195	2
16	232	9	18	99	16	15	417	16	7	29	234	2
17	232	5	17	11	15	17	430	15	8	24	190	2
18	235	7	18	54	18	17	418	18	7	31	268	2
19	197	5	91	16	29	16	369	11	7	33	100	2
20	212	5	17	5	13	12	397	9	8	18	145	2
21	243	8	19	81	15	20	432	18	9	29	245	2
22	230	7	18	37	15	17	414	17	8	23	229	2
23	230	9	18	31	15	15	421	16	9	19	117	2
24	225	7	19	64	19	18	411	17	8	30	226	2
25	212	4	17	7	16	12	356	10	8	15	93	1
26	196	7	19	195	20	15	387	15	8	33	241	1
27	240	8	20	68	18	20	430	18	8	27	187	2
28	258	8	19	17	14	13	414	14	8	18	105	2
29	224	9	26	38	15	18	431	13	8	27	179	2
30	233	8	24	43	17	20	440	17	9	29	206	2
31	236	9	22	13	14	19	462	16	9	22	169	2
32	274	8	21	17	16	20	495	16	9	17	158	2
33	229	11	23	165	18	19	405	15	8	29	216	2
34	227	12	23	142	16	18	398	15	8	28	212	2
35	207	7	26	99	24	16	395	17	9	43	290	1
36	200	7	25	78	23	16	378	17	8	45	280	1
37	255	8	25	82	21	21	442	20	9	32	202	2
38	242	10	22	33	13	14	442	17	9	19	176	2
39	198	8	21	85	21	20	378	15	8	32	254	1
40	234	7	20	22	16	12	413	13	8	19	176	2
41	223	7	24	119	19	18	410	17	8	30	212	1
42	196	7	23	87	21	19	392	14	8	39	243	1
43	186	6	24	52	18	15	414	14	8	29	249	1
44	242	11	25	72	24	20	438	19	8	25	197	2
45	234	7	23	8	15	19	450	14	8	24	71	2
46	206	7	28	80	24	20	378	17	8	51	275	1
47	196	6	24	48	20	17	391	15	8	25	224	1
48	115	8	26	43	20	18	432	16	8	25	209	2
49	225	7	25	62	20	18	419	16	8	27	224	2
50	201	6	25	72	21	16	390	15	8	33	249	1
51	208	7	25	89	23	18	397	15	8	32	239	1
52	196	5	15	5	13	13	482	11	9	17	142	1
53	202	7	21	94	20	18	388	17	9	48	290	1

continued

Data Sets

Table A2 (continued)

	Ba	Co	Cr	Cu	Li	Ni	Sr	V	Y	Zn	Zr	Site
54	227	12	20	111	15	14	403	15	8	24	211	1
55	187	6	99	65	19	17	379	16	8	31	311	1
56	271	8	20	15	15	21	460	22	8	22	161	2
57	235	7	20	70	17	18	423	16	8	30	202	2
58	265	8	26	50	17	20	457	27	9	30	221	1
59	227	8	27	112	16	18	424	23	8	32	332	2
60	226	8	25	82	16	16	412	17	8	29	211	2
61	219	7	25	18	12	13	398	12	8	18	142	2
62	192	6	25	54	19	16	374	16	8	32	253	1
63	209	7	26	83	21	15	398	18	8	49	274	1
64	218	7	24	11	13	14	409	12	8	18	155	2
65	220	7	23	13	12	14	407	12	9	15	57	1
66	183	6	23	72	15	14	414	15	8	30	249	1
67	271	12	25	31	13	19	438	24	9	22	186	2
68	206	6	23	46	16	16	405	15	8	24	213	1
69	205	7	26	97	18	18	393	18	9	48	280	1
70	206	7	19	76	19	15	401	16	8	34	221	1
71	227	7	22	77	19	18	406	20	10	37	295	1
72	200	7	26	66	16	16	392	19	8	40	276	1
73	212	7	28	103	23	17	391	18	9	32	249	1
74	208	7	28	101	22	18	396	18	9	44	258	1
75	202	7	26	82	18	17	391	17	8	36	239	1
76	211	7	22	73	22	16	402	17	8	48	216	1
77	208	6	26	106	21	15	401	17	9	47	247	1
78	208	6	22	100	21	17	397	18	9	45	260	1
79	196	6	21	85	17	14	395	15	8	41	208	1
80	219	12	25	104	14	16	391	16	8	25	169	1
81	205	7	27	100	21	16	392	18	9	42	238	1
82	193	5	24	73	19	15	393	15	8	31	221	1
83	240	5	16	5	12	12	519	16	9	16	65	1
84	190	6	25	70	16	16	382	15	8	34	252	1
85	251	5	29	15	13	13	408	21	8	21	219	1
86	209	7	24	82	101	18	371	19	9	57	323	1
87	205	7	26	100	21	15	393	18	9	43	306	1
88	218	7	26	74	19	17	393	19	9	34	328	1
89	189	6	25	69	17	17	366	16	8	32	268	1
90	203	6	25	62	21	15	385	16	8	46	257	1
91	201	6	20	73	18	14	390	16	8	33	268	1
92	199	6	21	76	19	15	386	17	8	43	268	1
93	207	6	25	79	20	16	390	18	8	30	279	1
94	199	6	20	113	19	14	391	16	8	87	264	1
95	194	6	21	65	15	16	386	19	8	34	332	1
96	200	6	21	79	19	15	387	17	8	45	269	1
97	204	6	20	75	22	16	401	16	8	43	264	1
98	209	5	16	6	11	12	518	12	10	19	148	1
99	189	6	19	108	16	15	374	15	7	40	249	1
100	210	6	21	68	19	15	382	17	8	31	269	1
101	210	6	21	48	16	15	400	15	7	37	254	1
102	223	8	13	12	11	13	410	12	8	17	88	1
103	222	7	19	64	18	14	378	17	8	30	368	1
104	207	6	22	100	20	19	394	18	9	43	311	1
105	205	6	20	86	18	15	394	17	8	43	279	1

Note: Values are given as parts per million of the element in the glass. Site codes are 1 for Leicester and 2 for Mancetter.

Table A3: Numbers of tools present in thirty-three Mesolithic assemblages

	Tool type (number present)					
	A	B	C	D	E	Total
1	97	27	2	1	0	127
2	73	15	1	0	0	89
3	5	57	1	0	0	63
4	37	4	0	0	0	41
5	25	28	2	2	0	57
6	53	5	1	0	0	59
7	68	37	8	0	1	114
8	136	95	3	10	43	287
9	41	0	3	0	0	44
10	690	181	26	15	1	913
11	78	165	19	2	0	264
12	108	10	1	0	0	119
13	41	3	0	0	0	44
14	1283	217	18	0	0	1518
15	44	31	2	0	0	77
16	30	21	14	0	0	65
17	28	12	1	0	1	42
18	226	518	101	0	0	845
19	1281	1052	1	1	444	2779
20	1458	1927	0	2	512	3899
21	28	0	2	0	0	30
22	69	88	0	10	8	175
23	31	29	1	0	0	61
24	65	15	0	0	0	80
25	30	2	1	0	0	33
26	130	40	2	0	11	183
27	248	326	334	7	5	920
28	285	129	57	10	19	500
29	63	0	4	0	0	67
30	60	32	5	0	2	99
31	21	12	1	0	0	34
32	71	6	0	0	0	77
33	33	24	8	0	0	65
Total	6936	5108	619	60	1047	13770

Notes: A = Microliths; B = Scrapers; C = Burins; D = Axes/adzes; E = Saws. The source of the data is Mellars (1976) and is based on the subset used in analysis by Pitts (1979).

Table A4: Dimensions of Bronze Age cups from Italy

RD	ND	SD	H	NH	Type
11.1	10.0	10.3	5.5	2.5	2
9.5	9.2	9.8	4.8	2.0	2
20.8	20.9	22.0	9.5	3.8	2
19.5	18.2	19.5	8.8	2.7	2
15.5	15.5	18.8	9.8	3.2	2
11.7	11.1	11.5	3.8	1.4	2
10.8	10.7	10.8	3.5	1.7	2
15.0	16.1	16.4	11.8	3.5	2
18.5	16.4	18.0	10.5	4.8	2
11.0	8.9	9.5	5.8	3.7	2
9.0	8.0	9.5	5.8	3.0	2
9.0	7.2	8.8	7.4	4.0	2
12.1	9.6	11.0	6.4	4.7	2
10.7	9.0	10.8	6.3	4.0	2
19.5	18.0	18.5	10.8	4.7	2
20.0	19.0	19.5	10.6	5.0	2
18.0	17.0	17.5	9.0	4.4	2
19.0	16.2	17.0	8.0	4.2	2
24.0	22.0	23.0	7.7	3.0	2
15.2	14.0	15.0	3.5	3.0	2
13.2	12.5	12.5	3.5	1.9	2
29.5	28.0	28.0	9.5	3.5	2
12.0	10.4	12.0	4.8	2.2	2
10.1	10.1	10.1	4.3	1.8	2
22.0	20.0	21.0	8.3	3.0	2
29.0	28.0	28.0	10.8	3.2	2
9.8	8.2	9.0	4.0	2.5	2
8.7	8.0	8.7	6.0	2.5	2
19.1	16.5	20.0	11.3	5.0	2
10.0	9.2	10.8	6.4	2.5	2
10.5	10.1	11.2	6.0	3.2	2
9.0	8.5	10.3	5.0	2.0	2
18.5	17.0	19.0	9.8	4.5	2
19.0	17.0	17.2	8.5	5.3	2
18.3	17.0	18.0	8.3	4.0	2
18.5	18.2	18.5	8.5	4.5	2
17.2	16.2	17.2	8.3	4.3	2
9.5	9.1	10.0	6.3	3.2	2
9.0	8.8	9.9	6.0	2.5	2
19.5	18.0	19.5	9.1	3.3	2
11.0	10.1	12.0	4.3	3.0	1
15.0	14.0	17.0	5.5	2.0	1
8.5	8.2	9.4	3.8	2.5	1
19.0	19.0	20.0	14.3	5.0	1
10.1	9.8	11.8	5.3	1.7	1
11.5	10.5	9.9	3.3	2.0	1
9.0	8.4	8.4	3.6	1.5	1

continued

Table A4: (continued)

RD	ND	SD	H	NH	Type
10.5	9.3	9.5	5.8	2.8	1
11.0	10.2	11.1	5.4	2.5	1
19.0	18.0	20.5	10.8	5.0	1
8.8	7.8	8.0	3.8	2.0	1
9.5	8.0	8.5	4.0	2.3	1
13.3	13.3	14.5	8.3	3.5	1
8.0	7.2	9.6	6.0	3.0	1
8.0	8.0	8.0	6.6	1.6	1
15.8	15.0	17.4	11.1	5.0	1
8.4	7.2	7.4	5.3	2.9	1
12.9	12.0	12.3	5.7	2.1	1
13.0	12.4	13.4	8.2	2.5	1
6.6	6.2	7.0	4.1	2.2	1

Note: RD = Rim diameter; ND = Neck diameter; SD = Shoulder diameter; H = Height; NH = Neck height.
Source: Lukesh and Howe (1978)

Table A5: Assemblages by tool type

							Tool type								
A	B	C	D	E	F	G	H	I	J	K	L	M	N	O	P
2	0	0	0	1	0	0	0	12	0	0	4	0	2	0	0
0	0	0	0	0	1	0	0	2	0	8	0	0	0	0	0
1	0	0	0	0	0	0	0	3	0	0	0	0	0	0	0
0	0	0	0	0	0	0	0	3	0	0	1	0	1	0	0
0	0	0	0	0	0	0	0	2	0	0	3	0	0	0	0
0	0	2	1	0	0	1	0	0	0	1	1	0	0	0	0
0	0	0	1	0	0	1	0	2	0	0	1	0	0	0	0
0	0	0	0	0	0	0	0	1	0	0	0	1	0	0	0
2	0	0	0	0	1	0	0	4	0	1	6	0	0	0	0
4	2	0	0	0	0	0	0	2	1	1	0	0	0	0	0
2	2	1	0	1	0	0	0	3	1	5	1	0	0	2	1
10	0	7	0	4	6	1	0	5	3	1	7	1	0	0	0
1	0	1	0	2	1	1	0	2	0	1	2	0	0	0	3
1	0	0	0	0	4	0	0	0	0	6	4	0	0	0	4
1	0	0	0	1	0	0	0	0	0	1	0	1	0	0	0
0	0	1	0	2	0	0	0	7	0	0	0	0	0	0	2
1	0	2	0	2	1	1	0	5	0	2	1	0	0	0	0
2	0	0	0	1	0	0	0	3	0	0	1	0	0	0	0
4	0	1	0	0	2	0	0	16	7	0	3	0	4	0	0
0	0	0	0	0	0	0	0	18	2	1	5	0	3	0	0
1	0	1	1	0	1	3	0	5	2	0	5	0	0	0	0
0	0	2	0	0	2	0	0	2	0	0	0	0	0	0	0
0	0	0	0	1	1	0	0	2	0	1	2	0	0	0	0
0	0	1	1	0	0	0	0	0	0	0	0	0	0	0	0
4	8	4	4	0	3	0	0	13	0	0	8	2	1	0	0
1	2	8	4	0	7	5	2	21	4	1	9	1	0	0	0
0	0	0	0	0	2	0	0	1	0	0	1	2	0	0	0
0	0	2	0	0	2	0	0	5	2	0	0	0	0	0	0
0	0	0	0	0	3	2	0	3	0	0	1	0	0	0	0
0	0	0	0	0	1	0	0	5	0	0	2	0	0	0	0
0	0	0	0	0	1	0	0	3	0	0	0	0	0	0	0
1	0	0	1	0	1	1	0	0	0	0	3	0	0	0	0
0	0	1	0	0	1	0	0	0	0	0	1	0	0	0	0
0	0	0	1	1	0	1	0	0	0	0	0	0	0	0	0
0	0	1	0	1	0	1	0	2	0	0	1	0	0	0	0
0	0	4	1	0	0	0	0	3	0	0	1	1	0	0	0
0	0	2	0	0	0	0	0	3	0	0	0	0	0	0	0
0	0	0	0	0	0	0	0	7	0	5	0	0	0	0	0
5	0	3	0	1	0	1	0	6	0	3	3	0	0	0	1
24	0	5	0	6	9	2	0	12	2	1	13	1	0	0	0
16	1	10	0	9	25	11	0	32	2	3	19	5	0	0	3
2	0	0	0	0	0	0	0	1	1	3	1	0	0	0	0
18	0	8	0	14	26	11	0	30	2	8	26	4	0	0	4

Tool types: A = tanged arrows; B = blade arrows; C = transverse and oblique arrows; D = atypical arrows; E = microliths; F = flake knives; G = blade knives; H = notched knives; I = core and flake scrapers; J = blade scrapers; K = disc scrapers; L = burins; M = axes; N = chisels; O = slate axes; P = perforators.
Source: Bølviken et al. (1982, pp. 50–1)

Table A6: Scrambled grave by type presence/absence matrix

Data Sets

Table A6: (continued)

```
                                                                    Site
. . 1 . . . . . . 1 1 . . . . . . . . . . . . 1 . . . . 1 . . . . . . . . 1 . . .  48
. . . . . . . . . . . 1 . . . . . . . . . . . . . . . . . . . . . . . 1 . 1 . . .  49
. . . . . . . . . 1 1 . . . . . . . 1 . . . . . . . . . . . . . . . . 1 . 1 . . .  50
. . . . . . . . . . . 1 . . . . . . . . . 1 . . . . . . . . . . . . . . . . . . .  51
. . 1 . . . . . . 1 1 . . . . . . . 1 . . . . . . . . . . . . . . . . 1 . 1 . . .  52
. . . . . 1 . 1 . . . . . . . . . . . . . 1 1 . . . . . . . . . . . . . . . . . 1  53
. . . . . . . . . 1 . . . . . . . . 1 . . . . . . . . . 1 . . . . . . . . . . . .  54
. . . . . . . . 1 1 . . . . . . . . . . . . . . . . 1 . . 1 . 1 . . . . . . . . .  55
1 . . . . . . . . . 1 . . . . . . . . 1 . . . . . . . . . . . . . . . . . . . . .  56
. . . . . 1 . . . . . . . . . . . . . . . . . . . . . . . . . . . . . . . 1 . . .  57
. . . . . . . . 1 . 1 . . . . . . . . 1 . . . . . . . . . . . . . . . . . 1 . . .  58
. . . . . . . . . . . . . . . . . . . . . . . . . . . . . . 1 . . . 1 1 . .  59
. . . . . . 1 . . . . . . . . . . 1 . . 1 . . . . . . . . . . . . . . . . 1 . . .  60
. . . 1 . . . . . . . . . . . . . . . . 1 . . 1 . . . . . . . . . . . . . . . . .  61
. . . 1 . . . . . . . . . . . . . 1 . . . . . 1 . . . . 1 . . . . . . . . . . . .  62
. . . . . . . . 1 . . . . . . 1 . . . . . . . . . . . . . . . . . . . . . . . . .  63
. . . . . . . . . 1 1 . . . 1 . . . . . . 1 1 . . 1 . . 1 . . 1 . . . . . . . . .  64
. . . . 1 . . . . . . . . . . . . . . . . 1 . . . . . . . . . . . . . . . . . . .  65
. . . . . . . . 1 1 . . . . . . . . 1 . . . . . . . . . . . . . . . . . 1 . . . .  66
. . . . . . . . 1 1 . . . . 1 . . . 1 . . . . 1 . . . . . . . . . . . . . . . . .  67
. . . . . 1 . 1 . . . . . . . . . . . . . . . . . . . . . . . . . . 1 . . . . . .  68
. . . 1 . . . . . . . . . . . . 1 . . . . 1 . . . . . . . . . . 1 . . . . . . . .  69
. . . . . . . . . . . . . . . . . . . . . 1 . . . . . . . . . 1 . . . . . . . . .  70
1 1 . . . . . . . 1 1 . . . . . . 1 . . . 1 . . . . . . . . . . . . . . . . . . .  71
1 . . . . . . . . . . . . . . . . . . . . . 1 . . . . . . . . . . . . . . . . . .  72
. . . . . . . . . 1 1 . . . . . . . . . . . . . . 1 . . . 1 . . . . . . . . . . .  73
. . . . . . . . . . 1 . . . . . . . . . . . . . . . . . . . . . . . . . . . . 1 .  74
. . . . . 1 . . . . . . . . . . . . . . . . . . . . . . . 1 . . . . . 1 . . . . .  75
. . . . 1 . . . . . . . . . . . . . . . . . . . . . . . . . . 1 . . . . . . . . .  76
. . . . . . . . . 1 . . . . . . . . . . . . . . . . . . . . . . . . . . . 1 . . .  77
```

Source: Adapted from Nielsen (1988, p. 48) whose site numbering is used. Columns have been randomly ordered – they are not identified as this is not needed for the illustration in the text. Types are of ornaments found in female graves, a 1 showing the presence of the ornament type in the grave.

Table A7: Variables characterising the morphology of pit houses from Arctic Norway

\multicolumn{6}{c}{Variables}					
1	2	3	4	5	6
3	1	2	2	1	3
2	1	1	2	1	3
3	1	1	2	1	3
1	2	1	2	1	1
3	2	2	2	1	2
1	2	2	2	1	2
4	2	2	2	1	1
2	1	2	2	1	3
2	1	1	2	1	3
2	2	1	2	2	1
2	1	1	2	1	3
4	2	2	2	1	3
4	2	3	2	1	3
3	1	3	2	1	3
1	1	1	1	1	1
2	1	2	1	1	3
4	1	1	2	1	1
3	1	2	2	1	3
2	2	2	1	1	1
1	2	3	2	1	3
3	2	2	2	1	3
3	1	2	2	1	3
4	2	2	1	2	3
1	1	2	1	1	3
3	2	2	2	1	3
3	2	2	2	1	3
1	2	2	2	2	3
4	2	2	1	1	3
3	1	2	2	1	3
3	1	2	1	1	3
3	1	1	2	1	3
3	1	1	2	1	1
1	1	1	2	1	3
4	1	1	2	2	3
4	2	2	2	1	3
2	1	1	2	1	3
3	2	2	2	1	3
4	1	1	2	1	3
4	1	1	2	1	3
3	1	1	2	1	3
2	1	1	2	1	3
4	1	1	2	2	3
4	1	1	2	1	3
4	2	1	2	1	3
4	1	1	2	2	3

Note: Variables: 1 = Hearths; 2 = Depth; 3 = Size; 4 = Form; 5 = Orientation; 6 = Entrances/depressions in walls.
Source: Engelstad (1988) – see Table A8 for variable categories.

Data Sets

Table A8: Indicator matrix for the data of Table A7

						Variable									
A	B	C	D	E	F	G	H	I	J	K	L	M	N	O	P
0	0	1	0	1	0	0	1	0	0	1	1	0	0	0	1
0	1	0	0	1	0	1	0	0	0	1	1	0	0	0	1
0	0	1	0	1	0	1	0	0	0	1	1	0	0	0	1
1	0	0	0	0	1	1	0	0	0	1	1	0	1	0	0
0	0	1	0	0	1	0	1	0	0	1	1	0	0	1	0
1	0	0	0	0	1	0	1	0	0	1	1	0	0	1	0
0	0	0	1	0	1	0	1	0	0	1	1	0	1	0	0
0	1	0	0	1	0	0	1	0	0	1	1	0	0	0	1
0	1	0	0	1	0	1	0	0	0	1	1	0	0	0	1
0	1	0	0	0	1	1	0	0	0	1	0	1	1	0	0
0	1	0	0	1	0	1	0	0	0	1	1	0	0	0	1
0	0	0	1	0	1	0	1	0	0	1	1	0	0	0	1
0	0	0	1	0	1	0	0	1	0	1	1	0	0	0	1
0	0	1	0	1	0	0	0	1	0	1	1	0	0	0	1
1	0	0	0	1	0	1	0	0	1	0	1	0	1	0	0
0	1	0	0	1	0	0	1	0	1	0	1	0	0	0	1
0	0	0	1	1	0	1	0	0	0	1	1	0	1	0	0
0	0	1	0	1	0	0	1	0	0	1	1	0	0	0	1
0	1	0	0	0	1	0	1	0	1	0	1	0	1	0	0
1	0	0	0	0	1	0	0	1	0	1	1	0	0	0	1
0	0	1	0	0	1	0	1	0	0	1	1	0	0	0	1
0	0	1	0	1	0	0	1	0	0	1	1	0	0	0	1
0	0	0	1	0	1	0	1	0	1	0	0	1	0	0	1
1	0	0	0	1	0	0	1	0	1	0	1	0	0	0	1
0	0	1	0	0	1	0	1	0	0	1	1	0	0	0	1
0	0	1	0	0	1	0	1	0	0	1	1	0	0	0	1
1	0	0	0	0	1	0	1	0	0	1	0	1	0	0	1
0	0	0	1	0	1	0	1	0	1	0	1	0	0	0	1
0	0	1	0	1	0	0	1	0	0	1	1	0	0	0	1
0	0	1	0	1	0	0	1	0	1	0	1	0	0	0	1
0	0	1	0	1	0	1	0	0	0	1	1	0	0	0	1
0	0	1	0	1	0	1	0	0	0	1	1	0	1	0	0
1	0	0	0	1	0	1	0	0	0	1	1	0	0	0	1
0	0	0	1	1	0	1	0	0	0	1	0	1	0	0	1
0	0	0	1	0	1	0	1	0	0	1	1	0	0	0	1
0	1	0	0	1	0	1	0	0	0	1	1	0	0	0	1
0	0	1	0	0	1	0	1	0	0	1	1	0	0	0	1
0	0	0	1	1	0	1	0	0	0	1	1	0	0	0	1
0	0	0	1	1	0	1	0	0	0	1	1	0	0	0	1
0	0	1	0	1	0	1	0	0	0	1	1	0	0	0	1
0	1	0	0	1	0	1	0	0	0	1	1	0	0	0	1
0	0	0	1	1	0	1	0	0	0	1	0	1	0	0	1
0	0	0	1	1	0	1	0	0	0	1	1	0	0	0	1
0	0	0	1	0	1	1	0	0	0	1	1	0	0	0	1
0	0	0	1	1	0	1	0	0	0	1	0	1	0	0	1

Note: A, B, C and D indicate no, one, or two hearths or a charcoal concentration; E, F indicate deep or shallow depth; G, H, I indicate small, medium or large size; J, K indicate oval or rectangular form; L, M are two different orientations; N, O, P indicate the entrances/depressions in walls on one side, front and one side or none.
Source: Engelstad (1988)

APPENDIX B
Matrix Algebra – An Informal Guide

Introduction

The aim of this appendix is to provide a brief guide to the language and results of matrix algebra that are used in parts of the text. A matrix is treated as a table of numbers and matrix algebra as a way of manipulating tables to get other tables having useful properties. The treatment is non-rigorous and none of the analogies used should be believed. Mathematicians should stop reading at this point; those wanting a more respectable treatment can turn to Healy (1986) or similar texts.

A *matrix* of order n by p is simply a table of numbers having n rows and p columns. Matrices will be denoted by **A, B, C, ... X, Y, Z**. The number in the i'th row and j'th column of a matrix, **X** say, will be denoted x_{ij}; the location itself is the (i,j) *cell*.

For illustrative purposes a small example based on the first five rows and three columns of Table .A1 will be used; this matrix, **X**, is shown below.

$$\mathbf{X} = \begin{array}{ccc} Al & Fe & Mg \\ 2.51 & .53 & .56 \\ 2.36 & .49 & .53 \\ 2.30 & .36 & .49 \\ 2.42 & .52 & .56 \\ 2.32 & .37 & .51 \end{array}$$

This is of order 5 by 3 and, for example, $x_{23} = .53$, $x_{41} = 2.42$ etc. From **X** other matrices may be defined; let **X̄** be the 5 by 3 matrix whose columns hold the means of the columns of **X**

$$\mathbf{\bar{X}} = \begin{array}{ccc} 2.382 & .454 & .530 \\ 2.382 & .454 & .530 \\ 2.382 & .454 & .530 \\ 2.382 & .454 & .530 \\ 2.382 & .454 & .530 \end{array}$$

Guide to Matrix Algebra

and define, for reasons that will be clear later, $\mathbf{Y} = (\mathbf{X} - \mathbf{\bar{X}})/2$ to get

$$\mathbf{Y} = \begin{matrix} .064 & .038 & .015 \\ -.011 & .018 & .000 \\ -.041 & -.047 & -.020 \\ .019 & .033 & .015 \\ -.031 & -.042 & -.010 \end{matrix}$$

where $y_{12} = .038 = (.530 - .454)/2$, for example.

This illustrates that matrices of the same order can be added and subtracted together like ordinary numbers, and that matrices can be multiplied by a number (1/2 in this instance) by multiplying all the elements by that number. Multiplication of two matrices is sometimes possible but trickier; for $\mathbf{C} = \mathbf{AB}$ to be defined the number of columns in $\mathbf{A}$ must equal the number of rows in $\mathbf{B}$. If $\mathbf{A}$ is of order n by p and $\mathbf{B}$ of order p by q then $\mathbf{C}$ is of order n by q. It follows from this that unless $p = q$ the matrix $\mathbf{BA}$ is not well defined; if $p = q$ then $\mathbf{AB}$ and $\mathbf{BA}$ are both defined but will usually differ. Where $\mathbf{C} = \mathbf{AB}$ is defined $\mathbf{A}$ is said to be *post-multiplied* by $\mathbf{B}$ while $\mathbf{B}$ is *pre-multiplied* by $\mathbf{A}$. Some more technical detail with examples will be given later.

Unlike ordinary numbers, matrix division is not simply defined, and to discuss the analogy of division some terminology needs to be established. If the rows and columns of an n by p matrix, $\mathbf{A}$, are interchanged to get a p by n matrix, the result is called the *transpose* of $\mathbf{A}$ and written $\mathbf{A}'$ or $\mathbf{A}^T$. The usefulness of the transpose will be seen shortly. The transpose of $\mathbf{Y}$ is given below.

$$\mathbf{Y}' = \begin{matrix} .064 & -.011 & -.041 & .019 & -.031 \\ .038 & .018 & -.047 & .033 & -.042 \\ .015 & .000 & -.020 & .015 & -.010 \end{matrix}$$

From the earlier discussion $\mathbf{Y}'$ is of order 3 by 5 and $\mathbf{Y}$ is of order 5 by 3 so the matrix $\mathbf{C} = \mathbf{Y}'\mathbf{Y}$ is well defined and of order 3 by 3. In general, for $\mathbf{C} = \mathbf{AB}$ (where this is defined) c_{ij} is obtained by multiplying the k'th element of row i in $\mathbf{A}$ by the k'th element of column j in $\mathbf{B}$, for all k, and summing the results. Symbolically this gives

$$c_{ij} = a_{i1}b_{1j} + a_{i2}b_{2j} + \ldots + a_{ip}b_{pj} = \sum_{k=1}^{p} a_{ik}b_{kj} \quad (1).$$

In the special case where $\mathbf{A} = \mathbf{Y}'$ and $\mathbf{B} = \mathbf{Y}$ what results is

$$c_{ij} = y_{1i}y_{1j} + y_{2i}y_{2j} + \ldots + y_{ni}y_{nj} = \sum_{k=1}^{n} y_{ki}y_{kj} \quad (2).$$

(Here the a's and b's are replaced by y's; the (i,j) cell of $\mathbf{A}$ is the (i,j) cell of $\mathbf{Y}'$ which, by the definition of a transpose, is the (j,i) cell of $\mathbf{Y}$. Hence

$a_{i1} = y'_{i1} = y_{1i}$ etc.) Thus c_{ij} is obtained by multiplying each element in row i of **Y** by the corresponding element in row j and summing the results. For example, c_{12} is the first column of **Y** 'multiplied' by its second column, or

$$(.064 \times .038) + (-.011 \times .018) + (-.041 \times -.047)$$
$$+ (.019 \times .033) + (-.031 \times -.042)$$
$$= .002432 - .000198 + .001927 + .000627 + .001302$$
$$= .00609$$

Obtaining the elements of $\mathbf{C} = \mathbf{Y}'\mathbf{Y}$ in this way gives

$$\mathbf{C} = \mathbf{Y}'\mathbf{Y} = \begin{matrix} .007220 & .006090 & .002375 \\ .006090 & .006830 & .002425 \\ .002375 & .002425 & .000950 \end{matrix}$$

In this particular example **Y** is a centred data matrix whose elements are $c(x_{ij} - \bar{x}_j)$ where $c = 1/2$. From (2), the (i,j) element of $\mathbf{Y}'\mathbf{Y}$ is just

$$c^2 \sum_{k=1}^{n} (x_{ki} - \bar{x}_i)(x_{kj} - \bar{x}_j) \tag{3}$$

which, with $c = (n-1)^{-1/2} = 1/2$, is just the estimated covariance of variables i and j if i differs from j, and is the estimated variance if $i = j$. Thus $\mathbf{Y}'\mathbf{Y}$ is the estimated covariance matrix of the data. This is why **Y** is defined as it was earlier. Had the data been standardised so that $y_{ij} = c(x_{ij} - \bar{x}_j)/s_j$ then $\mathbf{Y}'\mathbf{Y}$ would be the correlation matrix.

If a matrix has the same number of rows as columns it is said to be *square*. If, additionally, a square matrix **S** has $s_{ij} = s_{ji}$ it is said to be *symmetric*. The matrix **C** is an example of a square, 3 by 3, symmetric matrix. Square symmetric matrices play an important role in multivariate statistics.

The sum of the diagonal elements of a square matrix is called its *trace*. For **C** the trace, written as $\text{tr}(\mathbf{C})$, is $(.00722 + .00683 + .00095)$ or $.015$. Note that the diagonal elements are also the variances of the variables, so that the trace of **C** is the sum of the variances, or total variance, of the data. From (3) the (i,i) cell of **C** is just the sum of squares of the elements in column i of **Y** so that the trace is the sum of squares of all elements of **Y**. This has an interpretation as variance if the constant c is included in (3) and as a variation otherwise. This interpretation is used in Section 6.4.

If all the off-diagonal elements of a square matrix are zero it is said to be *diagonal*. If, additionally, all the diagonal elements are 1's the matrix is called the *identity matrix* and is denoted by the symbol **I**, which is analogous to 1 in the ordinary number system.

Suppose **S** is a square matrix and we can find another matrix, **T** say, of the same order such that $\mathbf{ST} = \mathbf{I}$. In this case **T** is called the *inverse* of **S** and is written as S^{-1} (i.e. $SS^{-1} = I$). This is loosely analogous to dividing S by itself (with ordinary numbers $2/2 = 1$, for instance; the inverse of 2 is 1/2 and 2 multiplied by 1/2 is equal to 1). It is possible, if you are not a mathematician, to think of a matrix product of the form AB^{-1} as analogous to 'division' of A by B (so long as the orders are compatible).

The inverse of C is

$$C^{-1} = \begin{matrix} 780.4 & -33.5 & -1865.5 \\ -33.5 & 1564.3 & -3909.1 \\ -1865.5 & -3909.1 & 15695.0 \end{matrix}$$

It can be verified that the diagonal elements of CC^{-1} are equal to 1 (e.g. the (1,1) cell is $(.00722 \times 780.4 - .00609 \times 33.5 - .002375 \times 1865.5) = 1$, allowing for rounding in the inverse) and off-diagonal elements are equal to 0 (e.g. the (1,2) cell is $(-.00722 \times 33.5 + .00609 \times 1564.3 - .002375 \times 3909.1) = 0$).

In ordinary algebra division by zero (e.g. 2/0) is not defined, and similar problems arise with matrices. It is possible to convert a square matrix, B say, to a single number (using complicated products of the b_{ij} that we will not enter into) called the *determinant*. The determinant is denoted $|B|$. For two matrices of comparable order the determinant can sometimes be thought of loosely as measuring the 'size' of the matrix. If the 'size' gets too small, funny things can start to happen in manipulating tables; in particular if $|B| = 0$ then B is said to be *singular* and the inverse B^{-1} and AB^{-1} are not defined. This last example is like trying to divide by zero.

A digression on vectors

Sometimes a matrix consists of a single row and is called a *row vector*. A matrix consisting of a single column is called a *column vector*. It is sometimes useful to think of an n by p table as composed of either a collection of n row vectors or, equivalently, of p column vectors. Lower-case letters can be used to highlight the fact that we are dealing with vectors; thus x will be used to indicate a column vector and x' a row vector.

It is possible to think of the n row vectors as defining n points in p-dimensional space or to think of the p column vectors as defining p points in n-dimensional space. These ideas may sound like something out of science fiction but are fundamental to an understanding of PCA, correspondence and discriminant analysis. To make matters simple, suppose that p=2 so that the i'th row of the matrix A is just $a' = (a_{i1} a_{i2})$. It is then possible to plot a_{i1} against a_{i2} as a single point on an ordinary piece of graph paper. In other words, the row vector has been displayed as a single point in a two-dimensional space, and the n rows could be displayed as n such points. With p=3 we have $a = (a_{i1}\ a_{i2}\ a_{i3})$ and this can be plotted as a point in the ordinary three-dimensional space in which you sit as you read these words. With n rows a cloud of such points is defined. With p bigger than 3, plotting of the points is not possible and mathematics must be resorted to, but it is still possible to talk of a cloud of points in p-dimensional space.

In PCA, correspondence and discriminant analysis as often used, a data matrix, **Y**, is converted to some other matrix, **A** say, and the first two or three columns of **A** are plotted to get a picture of the data. The form of conversion in each case is designed to emphasise some aspect of the data that is of special interest.

To get some idea of what is happening, imagine you are sitting at an empty desk in a room with a small insect hovering before you. Its position could be measured using three co-ordinates (y_{i1} y_{i2} y_{i3}) defined by reference to one corner of the room and the two walls that define it (taking their meeting place on the floor as the (0 0 0) point). The insect irritates you and you squash it on to the surface of your desk. It now exists (albeit in a deceased state) on a two-dimensional surface and its co-ordinates (a_{i1} a_{i2}) can be measured with respect to two adjacent sides of the desk. In mathematical language what is involved is a *projection* of the insect from a three-dimensional space to a two-dimensional *sub-space* represented by the surface of the desk.

More improbably, imagine that you are now being irritated by a cloud of hovering insects that you simultaneously flatten onto your desk before they have time to move. You can now take out a ruler and measure the distance between bodies. Flattening them onto the desk will have changed the distances between pairs of insects but, if you do the job efficiently, the distances may well be a good approximation to the distances between them when they were still hovering. Principal components analysis does this sort of thing; it often involves a projection from p to two- or three-dimensions such that the distances are preserved as well as possible.

To pursue this just a little further, suppose your desk is covered with a rubber (and highly elastic) tablecloth. Having squashed the insects, you can stretch the tablecloth in various ways to see what effect it has on the pattern of bodies. This will distort the distances between insects. In correspondence and discriminant analysis, such 'distortions' are undertaken in a controlled mathematical fashion in order to endow the new distances with desirable properties.

Finally, suppose that the resting spots of two insects are marked with drawing pins, with a third pin inserted at the centroid of the scatter, and the two pins joined to the centroid pin with string. The two pieces of string represent the vectors from the centroid to the points, and their length could be physically measured. If the strings are at right angles to each other they are said to be *orthogonal*.

More formally if **a** is a column vector such that $\mathbf{a}' = (a_1 a_2 \ldots a_n)$ the squared *length* of **a** is defined to be $\mathbf{a}'\mathbf{a}$ or

$$a_1^2 + a_2^2 + \ldots + a_n^2. \tag{4}$$

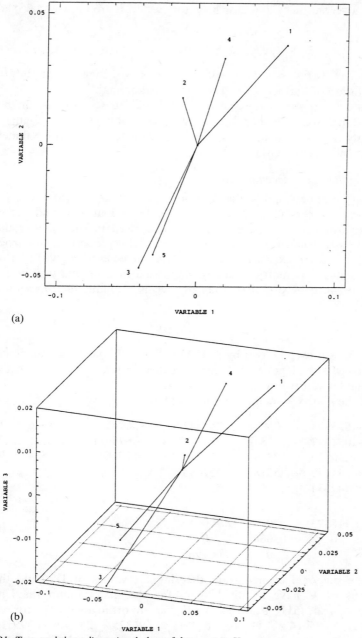

Figure B1: Two- and three-dimensional plots of data matrix Y

Note: Diagram (a) is based on the first two columns of the data matrix and diagram (b) on all three columns.

Given a second n by 1 vector **b** vectors **a** and **b** are orthogonal if $\mathbf{a'b} = 0$ where

$$\mathbf{a'b} = a_1b_1 + a_2b_2 + \ldots + a_nb_n \qquad (5).$$

If a matrix, **U** say, has the property that $\mathbf{U'U = I}$ then this implies that all columns are of unit length and mutually orthogonal.

Figure B1(a) shows the five rows of **Y** depicted as vectors in the two-dimensional space defined by the first two columns of **Y**. Figure B1(b) attempts to represent the same five rows depicted in the three dimensions defined by the three columns of **Y**.

The singular value decomposition

Ordinary numbers, unless they are prime, can be factorised as the product of other numbers (e.g., $24 = 2 \times 3 \times 4$). Data matrices, and correlation and covariance matrices, can also be factorised in such a way that the matrices which constitute the factors have useful properties. In particular a remarkable result exists, the *singular value decomposition* (SVD) of a matrix, which states that matrices can be factorised in a unique way as follows. Let **Y** be an n by p matrix; then the SVD is a unique factorisation of the form

$$\mathbf{Y = UDV'} \qquad (6)$$

Where **U** is n by p; **V** is p by p; $\mathbf{U'U = V'V = I}$; and **D** is a diagonal p by p matrix whose diagonal elements are called *singular values*. The squares of the singular values are called *eigenvalues*.

For the matrix **Y**, defined earlier, we have

$$\mathbf{U} = \begin{matrix} -.623923 & .575096 & .255842 \\ -.038097 & -.663849 & .414960 \\ .550960 & .178297 & .414716 \\ -.331475 & -.355185 & -.552093 \\ .442534 & .265641 & -.533433 \end{matrix} \qquad \mathbf{V} = \begin{matrix} -.694014 & .711366 & .110916 \\ -.674511 & -.696325 & .245179 \\ -.251646 & -.095539 & -.963112 \end{matrix}$$

$$\mathbf{D} = \begin{matrix} .118323 & 0 & 0 \\ 0 & .030668 & 0 \\ 0 & 0 & .007965 \end{matrix}$$

and the diagonal elements are the singular values of **D**. Note that the columns of **V** have length 1 (e.g., for the first column the length is $(.694014^2 + .674511^2 + .251646^2) = 1$) and are orthogonal (e.g., for columns 1 and 2 $(-.694014 \times .711366 + .674511 \times .696325 + .251646 \times .095539) = 0$). Similar properties hold for **U**. A MINITAB macro for obtaining a singular value decomposition is given in Appendix C.

In practice a computer package would be used to obtain the SVD. For the example being used here let $\mathbf{P} = \mathbf{UD}$ then, applying the rules discussed earlier

$p_{11} = (-.623923 \times .118323 + .575096 \times 0 + .255842 \times 0) = -.073824$
$p_{12} = (-.623923 \times 0 + .575096 \times .030668 + .255842 \times 0) = .017637$
$p_{13} = (-.623923 \times 0 + .575096 \times 0 + .255842 \times .007965) = .002038$

etc. (the full matrix $\mathbf{P}$ is given after equation (14)). Thus $\mathbf{Y} = \mathbf{PV}'$ and, for example,

$y_{11} = (-.073824 \times -.694014 + .017637 \times .711366 + .002038 \times .110916) = .064$

as required (those following through this calculation should bear in mind that some rounding error is involved – which is why computers are useful – and that it is the transpose of $\mathbf{V}$ that is being used).

Before looking at why the components of the SVD are useful we will engage in some purely algebraic manipulation. It can be shown that for any three matrices $\mathbf{A}$, $\mathbf{B}$ and $\mathbf{C}$ for which $\mathbf{ABC}$ is defined $(\mathbf{ABC})' = \mathbf{C}'\mathbf{B}'\mathbf{A}'$; also $(\mathbf{A}')' = \mathbf{A}$. Using these facts and noting that $\mathbf{D} = \mathbf{D}'$ we get from (6)

$$\mathbf{Y}'\mathbf{Y} = (\mathbf{UDV}')'\mathbf{UDV}' = \mathbf{VDU}'\mathbf{UDV}' = \mathbf{VD}^2\mathbf{V}' \qquad (7)$$

since $\mathbf{U}'\mathbf{U} = \mathbf{I}$. If both sides are (post-)multiplied by $\mathbf{V}$

$$\mathbf{Y}'\mathbf{YV} = \mathbf{VD}^2 \qquad (8).$$

Similar manipulations give

$$\mathbf{YY}' = \mathbf{UD}^2\mathbf{U}' \qquad (9)$$

and

$$\mathbf{YY}'\mathbf{U} = \mathbf{UD}^2 \qquad (10).$$

The columns of $\mathbf{V}$ are called the *eigenvectors* of $\mathbf{Y}'\mathbf{Y}$ and the columns of $\mathbf{U}$ are the eigenvectors of $\mathbf{YY}'$. The diagonal elements of $\mathbf{D}^2$, which we will assume are ordered by size, are the eigenvalues of both $\mathbf{Y}'\mathbf{Y}$ and $\mathbf{YY}'$.

So far this is all just algebraic manipulation and definition. To see how some of the algebraic quantities can be given statistical meaning we will use the fact that $\mathrm{tr}(\mathbf{AB}) = \mathrm{tr}(\mathbf{BA})$ where $\mathrm{tr}(\ )$ is the trace and the matrix products are assumed to be well defined. Let $\mathbf{Y}'\mathbf{Y}$ be the covariance matrix for a set of data and equate $\mathbf{A}$ with $\mathbf{VD}^2$ and $\mathbf{B}$ with $\mathbf{V}'$ in (7) to get

$$\mathrm{tr}(\mathbf{Y}'\mathbf{Y}) = \mathrm{tr}(\mathbf{VD}^2\mathbf{V}') = \mathrm{tr}(\mathbf{V}'\mathbf{VD}^2) = \mathrm{tr}(\mathbf{D}^2) \qquad (11).$$

From earlier results $\mathrm{tr}(\mathbf{Y}'\mathbf{Y})$ was shown to be the total variance in the data, .015. $\mathrm{Tr}(\mathbf{D}^2)$ is just the sum of the eigenvalues $.118823^2 + .030668^2 + .007695^2 = .0140003 + .00094505 + .0000592 = .015$, where .118823 etc. are the singular values or diagonal elements of $\mathbf{D}$ given earlier. This shows that the total variance in the data is just the sum of the eigenvalues. If $\mathbf{Y}'\mathbf{Y}$

is a correlation matrix then the traces in (11) are just equal to p, the number of variables.

Now define new variables

$$\mathbf{P} = \mathbf{YV} = \mathbf{UD} \tag{12}$$

where from (1)

$$p_{i1} = y_{i1}v_{11} + y_{i2}v_{21} + \ldots + y_{ip}v_{p1}$$

$$p_{i2} = y_{i1}v_{12} + y_{i2}v_{22} + \ldots + y_{ip}y_{p2} \tag{13}.$$

and, more generally, p_{ij} is similarly defined for $j = 1 \ldots p$. These new variables $P_1, P_2, \ldots, P_p$ are defined to be the *principal components* of $\mathbf{Y}$. Observed scores on P_1 are $(p_{11}\ p_{21} \ldots p_{n1})$ and on P_2 are $(p_{12}\ p_{22} \ldots p_{n2})$.

Given that $y_{ij} = c(x_{ij} - \bar{x}_j)$ the P_j have zero mean. Given this, in the same way that the variances of $y_1, \ldots y_p$ were shown to be the diagonal terms of $\mathbf{Y'Y}$, the variances of the P_j are given by the diagonal elements of $\mathbf{P'P}$ that sum to the trace. Now

$$\mathbf{P'P} = \mathbf{V'Y'YV} = \mathbf{V'VDU'UDV'V} = \mathbf{D}^2 \tag{14}$$

since $\mathbf{U'U} = \mathbf{V'V} = \mathbf{I}$. Since $\mathbf{D}^2$ is diagonal this shows that the P_j are orthogonal or uncorrelated; if the j'th eigenvalue (diagonal element) of $\mathbf{D}^2$ is written as λ_j – this also shows that the variance of p_j is λ_j.

In the present case $\mathbf{P}$ is given by

$$\mathbf{P} = \begin{matrix} -.0738 & .0176 & .0020 \\ -.0045 & -.0204 & .0032 \\ .0652 & .0055 & .0032 \\ -.0392 & -.0109 & .0042 \\ .0524 & .0081 & .0041 \end{matrix}$$

and a plot of the first against the second component is given in Figure B2(a). The eigenvalues are $\lambda_1 = .014$, $\lambda_2 = .00941$ and $\lambda_3 = .000059$ so that the first two components account for 99.6% of the variation in the data. Comparing this plot with those in Figure B1 it will be seen that the configuration of points is similar.

Figure B2(b) shows a plot based on the first two rows of $\mathbf{V}$ and together B2(a) and B2(b) constitute a form of biplot. The fact that the vector for Mg is somewhat shorter than those of the other two elements suggests it is of limited importance in influencing the configuration in Figure B2(a). This is because the variance of Mg, the (3,3) cell in $\mathbf{C}$, is much smaller than that of the other two.

In MINITAB and other packages similar results can be obtained directly using PCA on the covariance matrix. It should be noted that this will produce an identical configuration to that in Figure B2(a) but that the scale may be different. In the particular case of MINITAB this is because the

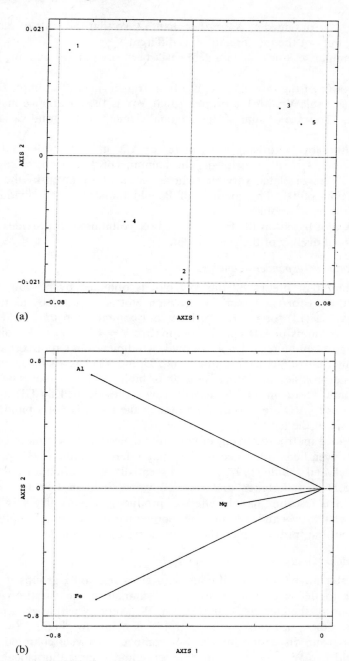

Figure B2: Biplot of the data matrix Y

scores, **P**, are obtained as **XV** rather than **YV** so that the scale of the components, but not the interpretation, is different.

To summarise how matrix algebra has been used in the running example:

(i) A table of the raw data, **X**, has been transformed in a simple fashion to another table **Y** with the property that **Y'Y** is the covariance matrix of the data. The trace or sum of the diagonal elements is the total variance in the data.

(ii) Either by factorising **Y** as in (6) or **Y'Y** as in (7) other tables **V**, **P** = **YV** and **D**^{2} can be defined. The columns of **V** are coefficients that define new uncorrelated variables whose variances are given by the diagonal elements of **D**^{2}. The columns of **P** hold the scores of individuals on each of the new variables.

(iii) A plot based on the first two or three columns of **P** provides a low-dimensional display of the original data.

Biplots and correspondence analysis

In the biplot of Figure B2 the row plot is obtained by plotting the p_{i1} against the p_{i2} for $i = 1 \ldots n$. The column plot is obtained by plotting v_{j1} against v_{j2} in (13) for $j = 1 \ldots p$. Row and column scores are held in the first two columns of **UD** and **V**. The matrix **Y** = **UDV'** and the joint row column plot can be regarded as a visual two-dimensional approximation to **Y**. An alternative factorisation of **Y** into **U** and **VD** has a similar interpretation and provides an alternative form of biplot with row scores held in **U** and column scores in **VD**. Plots with row scores defined by **UD** and column scores by **VD** are also useful although the result is not a biplot in the strict sense of the term.

A detailed matrix treatment of correspondence analysis is more complex and not given here; interested readers may refer to Jolliffe (1986, pp. 201–4) or Pack and Jolliffe (1992, p. 366). Essentially a PCA of data transformed as in (5.1) is undertaken and rescaled values held in **UD** and **VD** are plotted. The rescaling has the effect of producing a symmetry between the way in which rows and columns are represented that does not apply to the biplot, but the spirit in which the methods are used is similar.

Discriminant analysis and cluster analysis

If a cluster analysis of continuous data gives rise to G groups it is then possible to define two matrices which summarise the variation in and separation of the clusters. One of these, **W**, averages within-group dispersion; the other, **B**, treats group centroids as points to obtain a (weighted) between-group dispersion matrix. For compact and well-separated groups we would like **W** to be small in some sense and **B** large. Statistical modelling approaches to cluster analysis try to form groups, for given G, such that these conditions are satisfied. Since 'small' and 'large' can be construed in different ways when dealing with matrices, this has led to several different approaches (Section 8.8).

Minimising the trace of $\mathbf{W}$ converts $\mathbf{W}$ to a single 'size' measure that should be as small as possible. If it is allowed that a determinant can be thought of as a measure of 'size', then minimisation of $|\mathbf{W}|$ is in the same spirit. Thinking of $\mathbf{BW}^{-1}$ as 'division' of $\mathbf{B}$ by $\mathbf{W}$ leads to the idea of making it as large as possible; one approach to doing this is to maximise its trace.

In discriminant analysis the groups are given, and $\mathbf{BW}^{-1}$ can be thought of as summarising information on group compactness and separation. New variables are defined – equivalently to space being 'stretched' – that maximise between- to within-group scatter on the new scale. The eigenvectors of $\mathbf{BW}^{-1}$ provide the coefficients of the new variables in a manner analogous to the way in which elements of $\mathbf{V}$ define principal components. Details are given in the references cited in Section 9.1.

APPENDIX C
Computational Considerations

Introduction

What is available in statistical computer packages, their price, and the capabilities and price of hardware changes so rapidly that any attempt at a summary would soon be hopelessly outdated. The present appendix attempts little more than to note the packages and procedures used in the text, where these are not obvious, including some MINITAB macros for some of the more complex calculations.

My current practice is to use MINITAB for exploratory univariate and bivariate analyses, PCA and correspondence analysis (using my own macros). The only package I have used for cluster analysis is CLUSTAN; for discriminant analysis a combination of SPSS-X and MINITAB have proved useful. STATGRAPHICS is particularly nice for its graphical facilities and can, in principle, be used for all the methods here apart from correspondence analysis. In practice some of the output is of limited value (cluster analysis) and it is limited in the size of data set that can be handled (though this will change). I became aware of the MV-ARCH package in the course of writing the book and have been impressed by its capabilities as well as being in sympathy with the philosophy that underpins it, though it has not been used for any of the illustrative material. Fletcher and Lock (1991, pp. 145–76) give summary details of many other general-purpose packages and of packages designed for and by archaeologists.

In what follows, summary details are given on a chapter-by-chapter basis of the packages used, with additional material and listing of macros where this seems useful.

Chapter 2

MINITAB on a mainframe was used for most of the analyses; it is quick and easy to use and data manipulation is easier than in STATGRAPHICS. The latter package was used for the star plot of Section 2.11 and box-plots of a better graphical quality can be obtained if desired.

In MINITAB box-plots for comparative purposes are obtained as follows. Assume the data of Table A1, excluding the initial identifying column, are read into MINITAB into columns c1–c12. The plot for Fe, for example, is then obtained from

> **boxp c2;**
> **by c12.**

Figure 2.4 uses **lplot c2 c5 c12**; probability plots, such as that for Al in Figure 2.6 are obtained by **nscor c1 c21** followed by **plot c1 c21**.

Chapter 3

MINITAB was used for all analyses except Figure 3.9; again, it is very convenient for exploratory use as it is easy and quick to detect and delete outliers; select subsets of the data, etc. An analysis of the covariance matrix, as in Figure 3.5, is obtained by

> **pca c1-c11;**
> **cova;**
> **score c21-c31;**
> **coef c41-c51.**

Then **lplot c22 c21 c12** gives the (labelled) object plot of Figure 3.5(a) and **plot c42 c41** gives the variable plot of Figure 3.5(b). The variable labels here and elsewhere were added in for publication purposes; more usually I would create a column of the numbers 1, 2, . . . , 11 – call it c99 – and use **lplot c42 c41 c99** for the purpose of identification.

Deleting the **cova** sub-command leads to an analysis of the correlation matrix which is the default if only the command **pca c1-c11** is used. Default output from this last command is that given in Table 3.1. For Figure 3.9 the Multiple X-Y plot facility in STATGRAPHICS was used.

Chapter 4

This chapter illustrates a number of plots obtained within STATGRAPHICS. Scatterplots are straightforward, and if the data set is not too large PCA plots of the kind discussed in the previous section, and a biplot of the first kind (Section 4.4) can be obtained. To get Figure 4.1, a biplot of the second kind, was less straightforward particularly for the h-plot of Figure 4.1(b). What was done, because of the size of the data set, was to use MINITAB to save the scores and coefficients; divide the former by the square root of the appropriate eigenvalue and multiply the latter by the square root of the eigenvalue (see Section 4.4). The output of this was then saved and read into STATGRAPHICS.

Construction of the variable plot as shown in Figure 4.1(b) is not obvious and I am grateful to my colleagues Neville Davies and John Naylor for their assistance with this. Two columns consisting of the first two sets of modified coefficient vectors (loadings) need to be created and 'interleaved' with zeroes so that the general form is $(0 \, c_1 \, 0 \, c_2 \ldots 0 c_p)$. Call these variables V1 and V2. Then the figure, apart from the circle, is obtained from the Multiple X-Y plot facility in STATGRAPHICS with V1 and V2 as the X and Y variables. The 'plot options' and 'graphic options' facilities then need to be used to get the scales right in printing.

To get the circle, go into Multiple X-Y plotting; tap F8 followed by 'exec' to get the interactive calculator screen; type

x1 GETS -1,((COUNT 200)-100)/100
y1 GETS SQRT(1-x1*x1)

where the x1 line evaluates 201 points for the x axis of a unit circle. Tap ESC to return to the plotting input window and enter x1 and y1 for the X-vector and first vector respectively. The resultant circle can be tidied up using plot and graphic options, saved, and superimposed on the h-plot using the Splitscreen/Overlay facility. Obviously some knowledge of STATGRAPHICS is needed for all this; while a little tedious, the above construction is not obvious and is potentially useful.

Figure 4.2 was obtained using the X-Y-Z plotting facility and playing around with the plot and graphics options, experimenting in particular with the facility that allows the angle of view to be altered.

For some of the analyses described MINITAB macros were used. To illustrate the general procedure, construction of a macro for evaluating the singular-value decomposition of a matrix (Appendix B; Section 4.4) is given below.

store 'svd'

\# This macro obtains the singular value decomposition of a matrix
\# X via extraction of the eigenvalues and eigenvectors of X'X.
\#
\# The matrix X must be set up as m1 before use. The SVD of X has
\# the form X = UDV'. The macro calculates m15 = X'X ,then C90 contains
\# the eigenvalues whose square roots (C91) are the diagonal elements
\# of D. The eigenvectors m14 correspond to V and U is obtained as
\# m12.

trans m1 m15
mult m15 m1 m15
eigen m15 c90 m14
let c91 = sqrt(c90)
diag c91 m11
inve m11 m13
mult m1 m14 m10
mult m10 m13 m12
end

Comments following \# can be edited in after typing in the 'business' part if required. The above is written for use with a micro version of MINITAB where there is a limit of 15 on the number of matrix structures allowed and 99 on the columns. If an eigenvalue is effectively zero, the macro will produce an error message that can be ignored. For illustration, to obtain the SVD of the matrix Y in Appendix B, assume that the data is in c1-c3 and that the macro has previously been stored. Then

    copy c1-c3 m1
    exec 'svd'

is all that is needed. **V** ($= $ m14), **U** ($=$ m12) and the diagonal elements of **D** (c91) can be printed out as required.

To analyse fully compositional data using Aitchison's approach, the following set of macros can be used:

    store 'aitch.mtb'
    noecho

\# Macro aitch.mtb transforms the data so that Aitchison's approach
\# to principal component analysis can be performed. The macro calls
\# logtran.mtb and logcent.mtb. Errors will be reported if there are
\# zeroes in the data.

\# Use the command let k1 $=$ c where c is the number of columns
\# before executing this macro. After executing it use the PCA directive
\# with the COVA subcommand to get the analysis recommended by Aitchison.

    let k2 = 1
    exec 'logtran' k1
    rsum c1-ck1 c99
    let c99 = c99/k1
    let k2 = 1
    exec 'logcent' k1
    end

    store 'logtran.mtb'
    noecho
\# Macro LOGTRAN.MTB called by aitch.mtb
    let ck2 = loge(ck2)
    let k2 = k2 + 1
    end

    store 'logcent.mtb'
    noecho
\# Macro LOGCENT.MTB called by aitch.mtb
    let ck2 = ck2-c99
    let k2 = k2 + 1
    end

To illustrate how this might be used suppose data similar to the eleven elements of Table A1 are read into c1-c11. Then

```
rsum c1-c11 c12
let c12 = 100 - c12
let k1 = 12
exec 'aitch'
pca c1-c12;
cova.
```

will carry out the transformation and PCA recommended by Aitchison after the first two lines calculate the value for silica as 100 less the sum of the other elements. Note than zero values in the data are not permitted which means that the method cannot be applied directly to Table A1.

To compare two analyses (Section 4.7) Sibson's coefficient as defined in (4.13) can be obtained with the following macro.

```
store 'sibson.mtb'
noecho
```

\# To use this macro SVD.MTB must already have been stored. This
\# version is designed to calculate Sibson's coefficient for
\# comparing 2 two-dimensional configurations though it is easily
\# modified for higher dimensions. The co-ordinates for the first
\# configuration must be in c1-c2 and for the second in c3-c4.

```
center c1-c2 c1-c2;
location;
scale 1.
center c3-c4 c3-c4;
location;
scale 1.
copy c1-c2 m10
copy c3-c4 m9
trans m10 m8
trans m9 m7
mult m8 m10 m6
mult m7 m9 m5
mult m8 m9 m1
exec 'svd'
diag m6 c92
diag m5 c93
sum c91 k1
sum c92 k2
sum c93 k3
let k4 = 1-(k1*k1/(k2*k3))
prin k4
```

```
            erase m1-m15
            end
```

Chapters 5–6

When I first started using correspondence analysis the method was unavailable in general-purpose statistical packages I had access to and I was ignorant of other software. Accordingly I wrote a set of MINITAB macros based on algorithms given by Underhill and Peisach (1985) to carry out analyses of the kind reported in Chapters 5 and 6. These were written for MINITAB on a mainframe, making full use of the storage facilities (twenty matrix structures and 999 columns), and are not easily converted for micro use. Additionally they are not things of beauty and would appal a professional programmer. Anyone seriously interested in using correspondence analysis is advised to get one of the packages referenced in Fletcher and Lock (1991); however, a version for MINITAB that works for up to forty columns of data (subject to the caveats that follow) and provides the row and column plots is given below.

```
            store 'corran.mtb'
            noecho
```
This macro carries out a correspondence analysis of an nxp data
matrix. The data must be read in as p columns and p declared as
k1 before executing the macro.
The effect of the first 14 lines is to convert the original data
matrix Y to the form X'X . The macro 'SVD' then obtains the
singular value decomposition for X via a principal component
analysis of X'X .

```
            copy c1-ck1 m1
            let k2 = 40 + k1
            let k3 = 1
            exec 'colsums' k1
            rsum c1-ck1 c94
            let c95 = 1/sqrt(c93)
            let c96 = 1/sqrt(c94)
            diag c95 m2
            diag c96 m3
            mult m3 m1 m4
            mult m4 m2 m1
            name c90 'Inertias'
            exec 'svd'
```
The 2nd and 3rd columns of m6 and m7 as determined below
hold the coordinates for the plots for columns and rows
respectively. c90 contains the eigenvalues (inertias)
ignoring the first which should be 1 and is 'trivial'.

```
            diag c90 m8
            mult m2 m14 m6
            mult m6 m8 m6
            mult m3 m12 m7
            mult m7 m8 m7
            set c97
            1:k1
            end
            copy c90 c90;
            use c97 = 2:k1.
            parsum c90 c91
            let c91 = c91/c91(k1−1)
            name c91 'CumPerc.'
            print C90
            print c91
            copy m14 c41-ck2
            print c42-c47
            copy m6 c41-ck2
            copy c42 c98
            copy c43 c99
            copy m7 c41-ck2
            note Column Plot
            lplot c99 c98 c97
            note Row Plot
            plot c43 c42
            end

            store 'colsums.mtb'
            let k21 = sum(ck3)
            stack c93 k21 c93
            let k3 = k3 + 1
            end
```

For analysis of an n by p table this requires that $n > p$ to be sure of working; if $n < p$ and $n < 40$ then enter the data as a p by n table. Zero eigenvalues can give rise to an error message at the start of an analysis which can be ignored. The line **lplot c99 c98 c97** produces a labelled variable plot which is most useful if $p < 27$ (so that letters are not repeated); **plot c99 c98** could also be used though more work is then needed to identify variables.

A correspondence analysis of the data in Table A3, read into c1-c5 is obtained from

```
            let k1 = 5
            exec 'corran'
```

Computational Considerations

Chapters 7–8

All reported analyses were carried out using the CLUSTAN 3.2 package (Wishart, 1987). Over the years this has probably been the most widely used cluster analysis package in archaeology, and specific procedures will not be described in detail here. Of features that are not widely reported in the archaeological literature, some of the material in Section 8.3 on determining the number of clusters uses the RULES command. Cluster comparison, using the co-phenetic correlation coefficient (Section 8.4), is effected using the COMPARE command. The procedures described in Section 8.8 use the INVARIANT command.

That I have had no practical experience of spatial k-means clustering or of Whallon's unconstrained clustering method is acknowledged in the text (Sections 7.3; 8.6). Fortunately Blankholm's (1991) text and the other references given in Section 8.6 provide numerous examples and practical advice. Blankholm (and Fletcher and Lock, 1991, pp. 172–3) give details of a package, ARCOSPACE, that may be used to effect calculations, etc.

Chapters 9–10

To carry out a discriminant analysis in MINITAB on the data in Table A1, assuming the element data is stored in columns c1-c11 and site information in c12, a possible set of commands is

disc c12 c1-c11;
xval.

The **xval** sub-command produces results for 'cross validation' (Section 10.7). If the sub-command **quad** is used a quadratic discriminant analysis is carried out (Section 10.5). MINITAB does not provide graphical output or other of the paraphernalia often associated with discriminant analysis and in general is thus less useful than SPSS-X. Its main advantage is that it allows cross-validation which SPSS-X does not. The BMDP package allows cross-validation but I have no experience of it.

To generate some of the illustrative material scattered throughout Sections 9 and 10 the following SPSS-X commands were used.

data list free/c1 to c22 clus
begin data
[include data here]
end data
discriminant groups = clus(1,3)/
variables = c1 to c22/
method = direct/
statistics all

The variables are the elements in Tables A1 and A2; 'clus' is a variable showing which of three groups' objects are assigned to by a cluster analysis. The data can, alternatively, be read in from a previously-created data file. The **method = direct** command is not strictly necessary; if, however, '**direct**' is replaced by an appropriate command such as '**wilks**', a stepwise discriminant analysis will be obtained. This is an example of the 'kitchen sink' approach where everything is included in the statistics requested. Much of the copious output is typically of limited use to man or beast and can be considerably reduced by selecting only those options you intend to use.

APPENDIX D
MV-ARCH and MV-NUTSHELLX

Examples throughout the text have mostly been based on use of the MINITAB, STATGRAPHICS, SPSS-X and CLUSTAN packages. This reflects a decision made at an early stage of writing the book. The packages are ones available to me (on a mainframe computer in the case of all but STATGRAPHICS) and with which I am familiar. Potential users of the methodologies described in the book may not be in such a fortunate position and might, ideally, want to use a single package, purpose designed for archaeologists, and preferably inexpensive.

An ideal package for such users is the MV-ARCH package, developed by Professor Richard Wright when he was teaching at the University of Sydney, which came to my attention in the course of writing the book. It allows many of the analyses discussed in the book to be carried out rapidly and easily, as well as possessing attractive features not discussed in the book. Another package, MV-NUTSHELL, that contains a subset of MV-ARCH has also been developed. The MV-NUTSHELL package concentrates on the techniques of PCA, correspondence analysis and cluster analysis.

In order to illustrate the potential of these packages Professor Wright has kindly agreed to make available to readers of this book a free demonstration copy of MV-NUTSHELL called MV-NUTSHELLX that includes most of the data sets in Appendix A.

MV-NUTSHELLX is a DOS disk that does principal components, correspondence and cluster analysis. You get scattergrams and dendrograms on screen. On disk are Tables A1 to A6 from Appendix A which can be analysed. Analysis of small trial data sets of your own is also possible (restricted to a maximum of 7 variables and 20 objects for this free demonstration copy).

For a free copy of MV-NUTSHELLX **you should** *write directly* **to Professor Wright at**

> MV-ARCH,
> 72 Campbell St,
> BALMAIN,
> AUSTRALIA, 2041

enclosing a label with your name and address.

(This offer is open until the end of 1994.)

Bibliography

The bibliography is organised into five parts: general references that occur throughout the text, and four subject bibliographies of applications to archaeological problems of PCA, correspondence analysis, cluster analysis and discriminant analysis. The subject bibliographies are subdivided into sections on the analysis of chemical compositions of artefacts; typological or morphological analysis; assemblage comparison, and spatial analysis. In the sections on PCA and correspondence analysis there are additional subsections on factor analysis and review/expository articles respectively.

If a reference in the text is clearly to an archaeological application of a technique, it should be possible to locate it in the relevant subject bibliography. References to non-archaeological works, or those having a wider significance than simply the application of a single technique are contained in the general references. A few papers on specific subjects that are referenced outside the chapters that deal with that subject are, for convenience, repeated in the general references.

To locate a specific subject reference it will sometimes be necessary to look through the various subsections. It is hoped that this will not be too inconvenient and is balanced by the provision of a resource that will enable the rapid location of bodies of material of particular interest to readers. Not all of the references in the subject bibliography are cited in the text; this is particularly the case for the sections on cluster analysis and PCA.

General References

Adams W. Y. (1988) Archaeological classification: theory versus practice. *Antiquity* 62, 40–56.

Adams W. Y. and Adams E. W. (1991) *Archaeological Typology and Practical Reality: A Dialectical Approach to Artifact Classification and Sorting*. Cambridge: Cambridge University Press.

Aitchison J. A. (1983) Principal component analysis of compositional data. *Biometrika* 70, 57–65.

Aitchison J. A. (1984) The statistical analysis of geochemical compositions, *Journal of Mathematical Geology* 16, 617–36.

Aitchison J. A. (1986) *The Statistical Analysis of Compositional Data*. London: Chapman and Hall.

Aldenderfer M. S. (1982) Methods of cluster validation for archaeology. *World Archaeology* 14, 61–72.

Bibliography

Aldenderfer M. S. (ed.) (1987) *Quantitative Research in Archaeology: Progress and Prospects*. Newbury Park, CA: Sage.

Aldenderfer M. S. and Blashfield R. K. (1984) *Cluster Analysis*. Beverly Hills, CA: Sage.

Barcelo J. A. (1991) Some theoretical consequences of the use of advanced statistics in archaeology, in Waldren W. H., Ensenyat J. A. and Kennard R. C. (eds.), *2nd Deya International Conference on Prehistory: Vol. II Archaeological Technology and Theory*, 267–99. BAR International Series 574. Oxford: Tempus Reparatum.

Barclay K., Biddle M. and Orton C. (1990) The chronological and spatial distribution of the objects, in M. Biddle (ed.), *Object and Economy in Medieval Winchester (i)*, 42–73. Oxford: Clarendon Press.

Barnett V. (ed.) (1981) *Interpreting Multivariate Data*. Chichester: John Wiley and Sons.

Barnett V. (1982) *Comparative Statistical Inference, 2nd Edition*. Chichester: John Wiley and Sons.

Barnett V. and Lewis T. (1984) *Outliers in Statistical Data, 2nd Edition*. Chichester: John Wiley and Sons.

Barral L. and Simone S. (1977) Elements d'analyse des données. *Bulletin du Musée d'Anthropologie Préhistorique de Monaco 21*, 5–92.

Baxter M. J. (1991) An empirical study of principal component and correspondence analysis of glass compositions. *Archaeometry 33*, 29–41.

Baxter M. J. and Heyworth M. P. (1991) Comparing correlation matrices: with applications in the study of artefacts and their chemical compositions, in Pernicka E. and Wagner G. A. (eds.), *Archaeometry '90*, 355–64. Basel: Birkhäuser Verlag.

Baxter M. J., Cool H. E. M. and Heyworth M. P. (1990) Principal component and correspondence analysis of compositional data: some similarities. *Journal of Applied Statistics 17*, 229–35.

Benfer R. A. and Benfer A. N. (1981) Automatic classification of inspectional categories: multivariate theories of archaeological data. *American Antiquity 46*, 381–96.

Benzécri J.-P. (1973) *L'analyse des Données, Volume 2*. Paris: Dunod.

Bieber Jr. A. M., Brooks D. W., Harbottle G. and Sayre E. V. (1976) Application of multivariate techniques to analytical data on Aegean ceramics. *Archaeometry 18*, 59–74.

Binford L. R. (1987) Were there elephant hunters at Torralba?, in Nitecki M. H. and Nitecki D. V. (eds.), *The Evolution of Human Hunting*, 47–105. New York: Plenum Press.

Binford L. R. (1989) Technology of early man: an organizational approach to the Oldowan, in Binford L. R., *Debating Archaeology*, 437–63. Sandiago, CA: Academic Press.

Birks H. J. B. and Gordon A. D. (1985) *Numerical Methods in Quaternary Pollen Analysis*. London: Academic Press.

Bishop R. L. and Neff H. (1989) Compositional data analysis in archaeology, in Allen R. O. (ed.), *Archaeological Chemistry IV*, 57–86. Washington, D.C.: American Chemical Society.

Blankholm H. P. (1991) *Intrasite Spatial Analysis in Theory and Practice*. Aarhus: Aarhus University Press.

Bølviken E., Helskog E., Helskog K., Holm-Olsen I. M., Solheim L. and Bertelsen R. (1982) Correspondence analysis: an alternative to principal components. *World Archaeology 14*, 41–60.

Bookstein F. L. (1986) Size and shape spaces for landmark data in two dimensions (with discussion). *Statistical Science 1*, 181–242.

Box G. E. P. (1949) A general distribution theory for a class of likelihood criteria. *Biometrika 36*, 317–46.

Braun D. P. (1981) A critique of some recent North American mortuary studies. *American Antiquity 46*, 398–416.

Broadbent S. (1991) Simulating the ley hunter (with discussion). *Journal of the Royal Statistical Society B 53*, 285–339.

Brown J. A. (1987) Quantitative burial analyses as interassemblage comparisons, in Aldenderfer M. S. (ed.) *Quantitative Research in Archaeology: Progress and Prospects*, 294–308. Newbury Park, CA: Sage.

Buck C. E. and Litton C. D. (1991) A computational Bayes approach to some common archaeological problems, in Lockyear K. and Rahtz S. P. Q. (eds.) *Computer Applications and Quantitative Methods in Archaeology 1990*, 93–9. BAR International Series 565, Oxford: Tempus Reparatum.

Buck C. E., Kenworthy J. B., Litton C. D. and Smith A. F. M. (1991) Combining archaeological and radiocarbon information: a Bayesian approach to calibration. *Antiquity 65*, 808–21.

Budd P., Chapman B., Jackson C., Janaway R. and Ottoway B. (eds.) (1991) *Archaeological Sciences 1989*. Oxford: Oxbow Books.

Carr C. (ed.) (1985) *For Concordance in Archaeological Analysis*. Kansas City: Westport Publishers.

Carr C. (1987) Removing discordance from quantitative analysis, in Aldenderfer M. S. (ed.) *Quantitative Research in Archaeology: Progress and Prospects*, 185–243. Newbury Park, CA: Sage.

Chatfield C. and Collins A. J. (1980) *Introduction to Multivariate Analysis*. London: Chapman and Hall.

Christenson A. L. and Read D. W. (1977) Numerical taxonomy, R-mode factor analysis and archaeological classification. *American Antiquity 42*, 163–79.

Christie O. H. J., Brenna J. A. and Straume E. (1979) Multivariate classification of Roman glasses found in Norway. *Archaeometry 21*, 233–41.

Cooley W. W. amd Lohnes P. R. (1971) *Multivariate Data Analysis*. New York: John Wiley and Sons.

Cormack R. M. (1971) A review of classification. *Journal of the Royal Statistical Society A 134*, 321–67.

Corsten L. C. A. and Gabriel K. R. (1976) Graphical exploration in comparing variance matrices. *Biometrics 32*, 851–63.

Cowgill G. L. (1977) Review of *Mathematics and Computers in Archaeology* (J. E. Doran and F. R. Hodson). *American Antiquity 42*, 126–9.

Cowgill G. L. (1990a) Why Pearson's r is not a good similarity coefficient for comparing collections. *American Antiquity 55*, 512–21.

Cowgill G. L. (1990b) Artifact classification and archaeological purposes, in Voorrips A. (ed.) *Mathematics and Information Science in Archaeology: A Flexible Framework*, 61–78. Bonn: Holos Verlag.

Daniel C. and Wood F. S. (1980) *Fitting Equations to Data*. Chichester: John Wiley and Sons.

Digby P. G. N. (1985) Graphical displays for classification, in Voorrips A. and Loving S. H. (eds.), *PACT 11*, Strasbourg: Council of Europe.

Digby P. G. N and Kempton R. A. (1987) *Multivariate Analysis of Ecological Communities*. London: Chapman and Hall.

Djindjian F. (1985a) Seriation and toposeriation by correspondence analysis. *PACT 11*, 119–35.

Doran J. E. and Hodson F. R. (1975) *Mathematics and Computers in Archaeology*. Edinburgh: Edinburgh University Press.

Draper N. R. and Smith H. (1981) *Applied Regression Analysis, Second Edition*. Chichester: John Wiley and Sons.

Duncan R. J., Hodson F. R., Orton C. R., Tyers P. A. and Vekaria A. (1988) *Data Analysis for Archaeologists: The Institute of Archaeology Programs*. London: Institute of Archaeology.

Ehrenberg A. S. C. (1975) *Data Reduction*. Chichester: John Wiley and Sons.

Ehrenberg A. S. C. (1977) Rudiments of numeracy (with discussion). *Journal of the Royal Statistical Society A* 140, 277–97.

Englestad E. (1988) Pit-houses in Arctic Norway – an investigation of their typology using multiple correspondence analysis, in Madsen T. (ed.), *Multivariate Archaeology*, 71–84. Aarhus: Aarhus University Press.

Everitt B. S. (1980) *Cluster Analysis* 2nd ed. London: Heinemann Educational Books.

Everitt B. S. and Dunn G. (1991) *Applied Multivariate Data Analysis*. London: Edward Arnold.

Fatti L. P., Hawkins D. M. and Raath E. L. (1982) Discriminant analysis, in Hawkins D. M. (ed.), *Topics in Applied Multivariate Analysis*, 1–71. Cambridge: Cambridge University Press.

Fisher R. A. (1936) The use of multiple measurements in taxonomic problems. *Annals of Eugenics* 7, 376–86.

Fletcher M. and Lock G. R. (1991) *Digging Numbers: Elementary Statistics for Archaeologists*. Oxford: Oxford University Committee for Archaeology.

Flury B. and Riedwyl H. (1988) *Multivariate Statistics: A Practical Approach*. London: Chapman and Hall.

Foy D. (1985) Essai de typologie des verres medievaux d'après les fouilles provençales et languedociennes. *Journal of Glass Studies* 27, 18–71.

Freeman P. R. (1976) A Bayesian approach to the megalithic yard. *Journal of the Royal Statistical Society A* 139, 20–55.

Friedman J. H. and Tukey J. W. (1974) A projection pursuit algorithm for exploratory data analysis. *IEEE Transactions in Computing* 23, 881–9.

Gabriel K. R. (1971) The biplot graphical display of matrices with applications to principal components analysis. *Biometrika* 58, 453–67.

Gifi A. (1990) *Nonlinear Multivariate Analysis*. Chichester: John Wiley and Sons.

Goodall C. (1991) Procrustes methods in the statistical analysis of shape (with discussion). *Journal of the Royal Statistical Society B* 53, 285–339.

Gordon A. D. (1981) *Classification*. London: Chapman and Hall.

Gower J. C. (1967) A comparison of some methods of cluster analysis, *Biometrics* 23, 623–28.

Gower J. C. (1971a) A general coefficient of similarity and some of its properties. *Biometrics* 27, 857–74.

Gower J. C. (1971b) Statistical methods of comparing different multivariate analyses of the same data, in Hodson F. R., Kendall D. G. and Tautu P. (eds.), *Mathematics in the Archaeological and Historical Sciences*, 138–49. Edinburgh, Edinburgh University Press.

Gower J. C. (1984) Multivariate analysis: ordination, multidimensional scaling and allied topics, in Lloyd E. (ed.) *Handbook of Applicable Mathematics VI B: Statistics*, 727–81. Chichester: John Wiley and Sons.

Gower J. C. and Digby P. G. N. (1981) Expressing complex relationships in two dimensions, in Barnett V. (ed.) *Interpreting Multivariate Data*, 83–118. Chichester: John Wiley and Sons.

Greenacre M. J. (1981) Practical correspondence analysis, in Barnett V. (ed.) *Interpreting Multivariate Data*, 119–46. Chichester: John Wiley and Sons.

Greenacre M. J. (1984) *Theory and Applications of Correspondence Analysis*. London: Academic Press.
Greenacre M. J. (1988) Correspondence analysis of multivariate categorical data by weighted least squares. *Biometrika 75*, 457–67.
Greenacre M. J. (1990) Some limitations of multiple correspondence analysis. *Computational Statistics Quarterly 3*, 249–56.
Greenacre M. J. and Hastie T. J. (1987) The geometric interpretation of correspondence analysis. *Journal of the American Statistical Association 82*, 437–47.
Haggett P. (1991) Classics in human geography revisited: author's response. *Progress in Human Geography 15*, 302.
Hand D. J. (1981) *Discrimination and Classification*. Chichester: John Wiley and Sons.
Harbottle G. (1976) Activation analysis in archaeology. *Radiochemistry 3*, 33–72.
Harbottle G. (1982) Chemical characterization in archaeology, in Ericson J. E. and Earle T. K. (eds.) *Contexts for Prehistoric Exchange*. New York: Academic Press, 13–51.
Harbottle G. (1991) The efficiencies and error rates of Euclidean and Mahalanobis searches in hypergeometries of archaeological ceramic compositions, in Pernicka E. and Wagner G. A. (eds.), *Archaeometry '90*, 413–24. Basel: Birkhauser Verlag.
Healy M. J. R. (1986) *Matrices for Statistics*. Oxford: Clarendon Press.
Henderson J. (1985) The raw materials of early glass production. *Oxford Journal of Archaeology 4*, 267–91.
Henderson J. (1989) The scientific analysis of ancient glass and its archaeological interpretation, in Henderson J. (ed.), *Scientific Analysis in Archaeology*, 30–62. Oxford: Oxford University Committee for Archaeology.
Hill M. O. (1974) Correspondence analysis: a neglected multivariate method. *Applied Statistics 23*, 340–54.
Hill M. O. and Gauch H. G. (1980) Detrended correspondence analysis, an improved ordination technique. *Vegetatio 42*, 31–43.
Hills M. (1969) On looking at large correlation matrices. *Biometrika 56*, 249–53.
Hodder I. and Orton C. (1976) *Spatial Analysis in Archaeology*. Cambridge: Cambridge University Press.
Hodson F. R. (1980) Cultures as types? Some elements of classification theory. *Institute of Archaeology Bulletin 17*, 1–10.
Hodson F. R. (1990) *Hallstatt: The Ramsauer Graves*. Bonn: Dr Rudolf Habelt GmbH.
Hodson F. R., Sneath P. H. A. and Doran J. E. (1966) Some experiments in the numerical analysis of archaeological data. *Biometrika 53*, 311–24.
Hodson F. R., Kendall D. G. and Tautu P. (eds.) (1971) *Mathematics in the Archaeological and Historical Sciences*. Edinburgh: Edinburgh University Press.
Jackson C. M., Hunter J. R., Warren S. E. and Cool H. E. M. (1991) The analysis of blue-green glass and glassy waste from two Romano-British glass working sites, in Pernicka E. and Wagner G. A. (eds.), *Archaeometry '90*, 295–304. Basel: Birkhäuser Verlag.
Jackson J. E. (1991) *A User's Guide to Principal Components*. Chichester: John Wiley and Sons.
Jolliffe I. T. (1986) *Principal Component Analysis*. New York, Springer-Verlag.
Jolliffe I. T. (1989) Rotation of ill-defined principal components. *Applied Statistics 38*, 139–47.
Jones M. C. and Sibson R. (1987) What is projection pursuit? (with discussion). *Journal of the Royal Statistical Society B*, 1–36.
Jones R. R. (1986) *Greek and Cypriot Pottery*. Athens: British School at Athens.

Kaiser H. F. (1958) The varimax criterion for analytic rotation in factor analysis. *Psychometrika 23*, 187–200

Kendall D. G. (1971) Seriation from abundance matrices, in Hodson F. R., Kendall D. G. and Tautu P. (eds.), *Mathematics in the Archaeological and Historical Sciences*, 215–22. Edinburgh, Edinburgh University Press.

Kendall D. G. (1974) Hunting quanta. *Philosophical Transactions of the Royal Society of London A 276*, 229–64.

Kendall D. G. (1989) A survey of the statistical theory of shape (with discussion). *Statistical Science 4*, 87–120.

Kintigh K. W. (1990) Intra-site spatial analysis in archaeology, in Voorrips A. (ed.), *Mathematics and Information Science in Archaeology: A Flexible Framework*, 165–200. Bonn: Holos-Verlag.

Klecka W. R. (1980) *Discriminant Analysis*. Newbury Park, CA: Sage.

Krzanowski W. J. (1988) *Principles of Multivariate Analysis*. Oxford: Clarendon Press.

Kvamme K. L. (1990) The fundamental principles and practice of archaeological modelling, in Voorrips A. (ed.) *Mathematics and Information Science in Archaeology: A Flexible Framework*. Bonn: Holos Verlag, 257–95.

Lachenbruch P. A. (1975) *Discriminant Analysis*. New York: Hafner Press.

Lawley D. N. and Maxwell A. E. (1971) *Factor Analysis as a Statistical Method*, 2nd ed. London: Butterworths.

Laxton R. R. (1990) Methods of chronological ordering, in Voorrips A. and Ottoway B. (eds.), *New Tools from Mathematical Archaeology*, 37–44. Warsaw: Scientific Information Centre of the Polish Academy of Sciences.

Laxton R. R. and Restorick J. (1989) Seriation by similarity and consistency, in Rahtz S. P. Q. and Richards J. (eds.), *Computer Applications and Quantitative Methods in Archaeology 1989*, 215–25. BAR International Series 548. Oxford: British Archaeological Reports.

Lebart L., Morineau A. and Warwick K. M. (1984) *Multivariate Descriptive Statistical Analysis*. Chichester: John Wiley and Sons.

Leese M. N. and Needham S. P. (1986) Frequency table analysis: examples from Early Bronze Age axe decoration. *Journal of Archaeological Science 13*, 1–12.

Leese M. N., Hughes M. J. and Stopford J. (1989) The chemical composition of tiles from Bordesley: a case study in data treatment, in Rahtz S. P. Q. and Richards J. (eds.), *Computer Applications and Quantitative Methods in Archaeology 1989*, 241–9. BAR International Series 548. Oxford: British Archaeological Reports.

Lewis R. (1986) The analysis of contingency tables in archaeology, in Schiffer M. (ed.) *Advances in Archaeological Method and Theory 9*, 277–310. New York: Academic Press.

Litton C. D. and Leese M. N. (1991) Some statistical problems arising in radiocarbon calibration, in Lockyear K. and Rahtz S. P. Q. (eds.) *Computer Applications and Quantitative Methods in Archaeology 1990*, 93–9. BAR International Series 565, Oxford: Tempus Reparatum.

Lock G. (1991) An introduction to statistics for archaeologists, in Ross S., Moffett J. and Henderson J. (eds.), *Computing for Archaeologists*, 57–95. Oxford: Oxford University Committee for Archaeology.

Lockyear K. and Rahtz S. P. Q. (eds.) (1991) *Computer Applications and Quantitative Methods in Archaeology 1990*. BAR International Series 565. Oxford: Tempus Reparatum.

Lukesh S. S. and Howe S. (1978) Protoapennine vs. Subapennine: mathematical distinction between two ceramic phases. *Journal of Field Archaeology 5*, 339–47.

Madsen T. (ed.) (1988a) *Multivariate Archaeology*. Aarhus: Aarhus University Press.

Madsen T. (1988b) Multivariate statistics and archaeology, in Madsen T. (ed.), *Multivariate Archaeology*, 7–27. Aarhus: Aarhus University Press.

Manly B. F. J. (1986) *Multivariate Statistical Methods: A Primer*. London: Chapman and Hall.

Mardia K. V., Kent J. T. and Bibby J. M. (1979) *Multivariate Analysis*. London: Academic Press.

Mellars P. (1976) Settlement patterns and industrial variability in the British Mesolithic, in Sieveking G. de G., Longworth I. H. and Wilson K. E. (eds.), *Problems in Economic and Social Archaeology*, 375–99. London: Duckworth.

Mojena R. (1977) Hierarchical grouping methods and stopping rules: an evaluation. *Computer Journal 20*, 359–63.

Mommsen H., Krauser A. and Weber J. (1988) A method for grouping pottery by chemical composition. *Archaeometry 30*, 47–57.

Myers A. (1987) All shot to pieces? Inter-assemblage variability, lithic analysis and Mesolithic assemblage 'types'; some preliminary observations, in Brown A. G. and Edmonds M. R. (eds.), *Lithic Analysis and Later British Prehistory*, 137–53. BAR British Series 162. Oxford: British Archaeological Reports.

Naylor J. C. and Smith A. F. M. (1988) An archaeological inference problem. *Journal of the American Statistical Association 83*, 588–95.

Nielsen K. H. (1988) Correspondence analysis applied to hoards and graves of the Germanic Iron Age, in Madsen T. (ed.), *Multivariate Archaeology*, 37–54. Aarhus: Aarhus University Press.

Nishisato S. (1980) *Analysis of Categorical Data: Dual Scaling and Its Applications*. Toronto: University of Toronto Press.

Norusis M. J. (1985) *SPSS-X: Advanced Statistics Guide*. Chicago: SPSS Inc.

Orton C. R. (1980) *Mathematics in Archaeology*. London: Collins.

Orton C. R. (1988) Review of *Quantitative Research in Archaeology* (ed. M. S. Aldenderfer, 1987). *Antiquity 62*, 597–98.

Orton C. R. (1992) Quantitative methods in the 1990s, in Lock G. and Moffett J. (eds.), *Computer Applications and Quantitative Methods in Archaeology 1991*, 137–40. BAR International Series 577. Oxford: Tempus Reparatum.

Pack P. and Jolliffe I. T. (1992) Influence in correspondence analysis. *Applied Statistics 41*, 365–80.

Petrie W. M. F. (1899) Sequences in prehistoric remains. *Journal Anthropological Institute 29*, 295–301.

Pitts M. W. (1979) Hides and antlers: a new look at the gatherer-hunter site at Star Carr, North Yorkshire, England. *World Archaeology 11*, 32–44.

Pollard A. M. (1986) Data analysis, in Jones R. R. (ed.) *Greek and Cypriot Pottery: A Review of Scientific Studies*, 56–83. Athens: British School at Athens.

Rauret G., Casassas E., Rius F. X. and Munoz M. (1987) Cluster analysis applied to spectrochemical data of European Medieval stained glass. *Archaeometry 29*, 240–9.

Rawlings J. O. (1988) *Applied Regression Analysis: A Research Tool*. Belmont, CA: Wadsworth and Brooks/Cole.

Read D. W. (1982) Toward a theory of archaeological classification, in Whallon R. and Brown J. A. (eds.), *Essays in Archaeological Typology*, 56–92. Evanston: Centre for American Archaeology Press.

Rice P. (1987) *Pottery Analysis: A Source Book*. Chicago: University of Chicago Press.

Richards J. D. (1987) *The Significance of Form and Decoration of Anglo-Saxon Cremation Urns*. BAR British Series 166. Oxford: British Archaeological Reports.

Robertson-MacKay M. E. (1980) A 'Head and Hooves' burial beneath a round barrow with other Neolithic and Bronze Age sites on Hemp Knoll, near Avebury, Wiltshire. *Proceedings of the Prehistoric Society 46*, 123–76.

Ruggles C. L. N. (1986) You can't have one without the other? IT and Bayesian statistics and their possible impact within archaeology. *Science and Archaeology 28*, 8–15.
Ruggles C. L. N. and Rahtz S. P. Q. (eds.) (1988) *Computer and Quantitative Methods in Archaeology 1987*. Oxford: BAR International Series 393.
Scheps S. (1982) Statistical blight. *American Antiquity 47*, 836–51.
Scollar I. (1988) The Bonn Archaeological Seriation and Clustering Package, Version 3.1 for IBM/MS DOS compatible PCs. *Archaeological Computing Newsletter 16*, 23–5.
Scott A., Whittaker J., Green M. and Hillson S. (1991) Graphical modelling of archaeological data, in Lockyear K. and Rahtz S. P. Q. (eds.) *Computer Applications and Quantitative Methods in Archaeology 1990*, 111–16. BAR International Series 565. Oxford: Tempus Reparatum.
Seber G. A. F. (1984) *Multivariate Observations*. Chichester: John Wiley and Sons.
Shennan S. (1988) *Quantifying Archaeology*. Edinburgh: Edinburgh University Press.
Sibson R. (1978) Studies in the robustness of multi-dimensional scaling: Procrustes statistics. *Journal of the Royal Statistical Society B 40*, 234–8.
Slater E. A. and Tate J. O. (1988) *Science and Archaeology – Glasgow 1987*. BAR British Series 196. Oxford: British Archaeological Reports.
Sneath P. H. A. (1975) Review of *Mathematics and Computers in Archaeology* (J. E. Doran and F. R. Hodson, 1975). *Archaeological Journal 132*, 373–4.
Sneath P. H. A. and Sokal R. R. (1973) *Numerical Taxonomy*. San Francisco: W. H. Freeman and Co..
Solomon H. (1971) Cluster analysis, in Hodson F. R., Kendall D. G. and Tautu P. (eds.), *Mathematics in the Archaeological and Historical Sciences*. Edinburgh: Edinburgh University Press, 62–81.
Spaulding A. C. (1982) Structure in archaeological data: nominal variables, in Whallon R. and Brown J. A. (eds.), *Essays in Archaeological Typology*, 1–20. Evanston: Centre for American Archaeology Press.
Speth J. D. and Johnson G. A. (1976) Problems in the use of correlation for the investigation of tool-kits and activity areas, in Cleland C. E. (ed.), *Cultural Change and Continuity*, 35–57. New York: Academic Press.
Teil H. (1975) Correspondence factor analysis: an outline of its method. *Mathematical Geology 7*, 3–12.
Thomas D. H. (1978) The awful truth about statistics in archaeology. *American Antiquity 43*, 231–44.
Topping P. G. and MacKenzie A. B. (1988) A test of the use of neutron activation analysis for clay source characterisation. *Archaeometry 30*, 92–101.
Tukey J. W. (1977) *Exploratory Data Analysis*. Reading, Mass: Addison-Wesley.
Underhill L. G. and Peisach M. (1985) Correspondence analysis and its application to multielemental trace analysis. *Journal of Trace and Microprobe Techniques 3*, 41–65.
Valenchon F. (1982) The use of correspondence analysis in geochemistry. *Mathematical Geology 14*, 331–42.
van der Heijden P. G. M., de Falguerolles A. and de Leeuw J. (1989) A combined approach to contingency table analysis using correspondence analysis and log-linear analysis (with discussion). *Applied Statistics 38*, 249–92.
Vierra R. K. and Carlson D. L. (1981) Factor analysis, random data, and patterned results. *American Antiquity 46*, 272–83.
Vitali V. (1991a) Review of *Quantifying Archaeology* (S. Shennan, 1988). *Geoarchaeology 6*, 297–8.
Vitali V. (1991b) Formal methods for the analysis of archaeological data: data analysis *vs* expert systems, in Lockyear K. and Rahtz S. P. Q. (eds.), *Computer*

Applications and Quantitative Methods in Archaeology 1990, 207–209. BAR International Series 565. Oxford: Tempus Reparatum.

Vitali V. and Franklin U. M. (1986a) New approaches to the characterization and classification of ceramics on the basis of their elemental composition. *Journal of Archaeological Science 13*, 61–170.

Vitali V. and Franklin U. M. (1986b) An approach to the use of packaged statistical programs for cluster, classification, and discriminant analysis of trace element data. *Geoarchaeology 1*, 195–201.

Voorrips A. (ed.) (1990a) *Mathematics and Information Science in Archaeology: A Flexible Framework*. Bonn: Holos Verlag.

Voorrips A. (1990b) The evolution of a flexible framework for archaeological analysis, in Voorrips A. (ed.) *Mathematics and Information Science in Archaeology: A Flexible Framework*, 1–6. Bonn: Holos Verlag.

Whallon R. (1982) Variables and dimensions: the critical step in quantitative typology, in Whallon R. and Brown J. A. (eds.), *Essays in Archaeological Typology*, 127–61. Evanston: Centre for American Archaeology Press.

Whallon R. (1984) Unconstrained clustering for the analysis of spatial distributions in archaeology, in Hietala H. (ed.), *Intrasite Spatial Analysis in Archaeology*, 242–77. Cambridge: Cambridge University Press.

Whallon R. (1987) Simple statistics, in Aldenderfer M. S. (ed.) *Quantitative Research in Archaeology: Progress and Prospects*, 135–50. Newbury Park, CA: Sage.

Whallon R. and Brown J. A. (eds.) (1982) *Essays in Archaeological Typology*. Evanston: Centre for American Archaeology Press.

Wilkinson E. M. (1971) Archaeological seriation and the travelling salesman problem, in Hodson F. R., Kendall D. G. and Tautu P. (eds.), *Mathematics in the Archaeological and Historical Sciences*, 276–84. Edinburgh, Edinburgh University Press.

Williams I., Limp W. F. and Bruier F. L. (1990) Using geographic information systems and exploratory data analysis for archaeological site classification and analysis, in Allen K. M. S., Green S. W. and Zubrow E. B. W. (eds.), *Interpreting Space: GIS and Archaeology*, 239–73. London: Taylor and Francis.

Wilson A. L. (1978) Elemental analysis of pottery in the study of its provenance: a review. *Journal of Archaeological Science 5*, 219–36.

Wishart D. (1987) CLUSTAN *User Manual*. St Andrews: University of St Andrews.

Wright R. (1989) *Doing Multivariate Archaeology and Prehistory: Handling Large Data Sets with MV-ARCH*. Sydney: Dept. of Anthropology, University of Sydney.

References – Principal Component Analysis in Archaeology

(a) Analyses of chemical compositions of artefacts

Aitchison S. and Ottoway B. (1986) Neutron activation analysis of Bavarian Altheim pottery and of local clay sources, in Olin J. S. and Blackman M. J. (eds.), *Proceedings of the 24th Archaeometry Symposium*, 321–6. Washington D. C.: Smithsonian Institution Press.

Arnold D. E., Neff H. and Bishop R. L. (1991) Compositional analysis and 'sources' of pottery: an ethnoarchaeological approach. *American Anthropologist 93*, 70–90.

Barrett J., Bradley R., Cleal R. and Pike H. (1978) Characterisation of Deverel-Rimbury pottery from Cranbourne Chase. *Proceedings of the Prehistoric Society 44*, 135–42.

Baxter M. J. (1989) The multivariate analysis of compositional data in archaeology: a methodological note. *Archaeometry 31*, 45–53.

Bibliography

Baxter M. J. (1991) An empirical study of principal component and correspondence analysis of glass compositions. *Archaeometry* 33, 29–41.

Baxter M. J. (1992a) Archaeological uses of the biplot – a neglected technique?, in Lock G. and Moffett J. (eds.), *Computer Applications and Quantitative Methods in Archaeology 1991*, 141–8. BAR International Series 577. Oxford: Tempus Reparatum.

Baxter M. J. (1992b) Statistical analysis of chemical compositional data and the comparison of analyses. *Archaeometry* 34, 267–77.

Baxter M. J., Cool H. E. M. and Heyworth M. P. (1990) Principal component and correspondence analysis of compositional data: some similarities. *Journal of Applied Statistics* 17, 229–35.

Baxter M. J. and Heyworth M. P. (1989) Principal component analysis of compositional data in archaeology, in Rahtz S. P. Q. and Richards J. (eds.), *Computer Applications and Quantitative Methods in Archaeology 1989*, 227–40. BAR International Series 548. Oxford: British Archaeological Reports.

Baxter M. J. and Heyworth M. P. (1991) Comparing correlation matrices: with applications in the study of artefacts and their chemical compositions, in Pernicka E. and Wagner G. A. (eds.), *Archaeometry '90*, 355–64. Basel: Birkhäuser Verlag.

Berthoud T., Besenval R., Cesbron F., Cleuziou S., Pechoux M., Francuix J. and Lisak-Hours J. (1979) The early Iranian metallurgy: analytical studies of copper ores from Iran. *Archaeo-Physika* 10, 68–74.

Betts I. M. (1986) Analytical analysis and manufacturing techniques of Anglo-Saxon tiles. *Medieval Ceramics* 10, 37–42.

Bishop R. L. and Neff H. (1989) Compositional data analysis in archaeology, in Allen R. O. (ed.), *Archaeological Chemistry IV*, 57–86. Washington, D.C.: American Chemical Society.

Cabral J. M. P., Gouveia M. A., Alarcao A. M. and Alarcao J. (1983) Neutron activation analysis of fine grey pottery from Conimbriga, Santa Olaia and Tavarede, Portugal. *Journal of Archaeological Science* 10, 61–70.

Calamiotou M., Fillipakis S. E., Jones R. E. and Kassals D. (1984) X-ray and spectrographic analyses of terracotta figurines from Myrina: an attempt to characterise workshops. *Journal of Archaeological Science* 11, 103–17.

Capannesi G., Seccaroni C., Sedda A. F., Majerini V. and Musco S. (1991) Classification of fifteenth to nineteenth century mortars from Gabii using instrumental neutron activation analysis. *Archaeometry* 3, 255–66.

Christensen L. H., Conradsen K. and Nielsen S. (1985) The Selbjerg project: investigation of Neolithic pottery by means of neutron activation analysis. *ISKOS* 5, 401–10.

Christie O. H. J., Brenna J. A. and Straume E. (1979) Multivariate classification of Roman glasses found in Norway. *Archaeometry* 21, 233–41.

Cox G. A. and Pollard A. M. (1981) The multivariate analysis of data relating to the durability of medieval window glass. *Revue d'Archéométrie* 5, 119–28.

Cracknell S. (1982) An analysis of pottery from Elgin and north-east Scotland. *Medieval Ceramics* 6, 51–65.

Culbert T. P. and Schwalbe L. A. (1987) X-ray fluorescence survey of Tikal ceramics. *Journal of Archaeological Science* 14, 635–57.

D'Altroy T. N. and Bishop R. L. (1990) The provincial organisation of Inka ceramic production. *American Antiquity* 55, 120–37.

De Palol P., Gurt J. M., Tuset F., Planas C., Buxeda J., Cau M. A. and Alcobé X. (1991) Clunia: producer and receiver of Hispanic terra sigillata, in Budd P., Chapman B., Jackson C., Janaway R. and Ottoway B. (eds.), *Archaeological Sciences 1989*, 83–94. Oxford: Oxbow Books.

Devereux D. F., Jones R. F. and Warren S. E. (1983) X-ray fluorescence analysis of Verulamium region ware, in Aspinall A. and Warren S. E. (eds.), *The Proceedings of the 22nd Symposium on Archaeometry*, 333–42. Bradford: University of Bradford.

Evans J. (1989), Neutron activation analysis and Romano-British pottery studies, in Henderson J. (ed.), *Scientific Analysis in Archaeology*, 136–62. Oxford: Oxford University Committee for Archaeology.

Germann K., Gruben G., Knoll H., Vallio V. and Winkler F. J. (1988) Provenance characteristics of cycladic (Paros and Naxos) marbles – a multivariate geologic approach, in Herz N. and Waelkens M. (eds.), *Classical Marble: Geochemistry, Technology, Trade*, 251–62. Dordrecht: Kluwer Academic Publishers.

Gilmore G. R. (1981) The use of mathematical clustering techniques for the analysis of coin composition data. *Revue d'Archéométrie* 5, 97–107.

Glascock M. D., Elam J. E. and Cobean R. H. (1988) Differentiation of obsidian sources in Mesoamerica, in Farquhar R. M., Hancock R. G. V. and Pavlish L. A. (eds.), *Proceedings of the 26th International Archaeometry Symposium*, 245–51. Toronto: University of Toronto.

Goad S. I. and Noakes J. (1978) Prehistoric copper artifacts in the eastern United States, in Carter G. F. (ed.), *Archaeological Chemistry II*, 335–46. Washington, D.C.: American Chemical Society.

Green R. C. and Bird J. R. (1989) Fergusson Island obsidian from the D'Entrecasteaux group in a Lapita site of the Reef Santa Cruz group. *New Zealand Journal of Archaeology* 11, 87–99.

Hart F. A., Storey J. M. V., Adams S. J., Symonds R. P. and Walsh J. N. (1987) An analytical study using inductively coupled plasma (ICP) spectrometry, of Samian and colour-coated wares from the Roman town at Colchester together with related continental Samian wares. *Journal of Archaeological Science* 14, 577–98.

Hatcher H., Hedges R. E. M., Pollard A. M. and Kenrick P. M. (1980) Analysis of Hellenistic and Roman fine pottery from Benghazi. *Archaeometry* 22, 133–51.

Heyworth M. P., Baxter M. J. and Cool H. E. M. (1990) *Compositional Analysis of Roman Glass from Colchester, Essex*. Ancient Monuments Laboratory Report 53/90. London: English Heritage.

Hughes M. J. (1991) Provenance studies of Spanish Medieval tin-glazed pottery by neutron activation analysis, in Budd P., Chapman B., Jackson C., Janaway R. and Ottoway B. (eds.), *Archaeological Sciences 1989*, 54–68. Oxford: Oxbow Books.

Hughes M. J. and Hall J. A. (1979) X-ray fluorescence analysis of late Roman and Sassanian silver plate. *Journal of Archaeological Science* 6, 321–41.

Knapp A. B. (1989) Complexity and collapse in the North Jordan Valley: archaeometry and society in the middle-late Bronze Age. *Israel Exploration Journal* 3–4, 129–48.

Knapp A. B., Duerden P., Wright R. V. S. and Grave P. (1988) Ceramic production and social change: archaeometric analysis of Bronze Age pottery from Jordan. *Journal of Mediterranean Archaeology* 1, 57–113.

Leese M. N. (1981) Gamma plotting, its application to archaeological plotting. *Revue d'Archéométrie Supplement*, 177–84.

Leese M. N., Hughes M. J. and Stopford J. (1989) The chemical composition of tiles from Bordesley: a case study in data treatment, in Rahtz S. P. Q. and Richards J. (eds.), *Computer Applications and Quantitative Methods in Archaeology 1989*, 241–9. BAR International Series 548. Oxford: British Archaeological Reports.

Luedtke B. E. (1978) Chert sources and trace element analysis. *American Antiquity* 43, 413–23.

MacSween A., Warren S. E. and Tate J. (1989) Neutron activation analysis of coarse pottery from the Hebridean Islands of Scotland, in Maniatis Y. (ed.), *Archaeometry*, 551–7. Amsterdam: Elsevier.

Mello E., Meloni S., Monna D. and Oddone M. (1988) A computer based pattern recognition approach to the provenance study of Mediterranean marbles through trace element analysis, in Herz N. and Waelkens M. (eds.), *Classical Marble: Geochemistry, Technology, Trade*, 283–91. Dordrecht: Kluwer Academic Publishers.

Mertz R., Melson W. and Levenbach G. (1979) Exploratory data analysis of Mycenean ceramic compositions and provenances. *Archaeo-Physika 10*, 580–90.

Mirti P., Zelano V., Aruga R., Ferrara E. and Appolonia L. (1990) Roman pottery from Augusta Praetoria (Aosta, Italy): a provenance study. *Archaeometry 32*, 163–75.

Mirti P., Aruga R., Zelano V. and Appolonia L. (1991) Roman pottery from Augusta Praetoria: Provenance investigation by atomic absorption spectroscopy and multivariate analysis of the analytical data, in Pernicka E. and Wagner G. A. (eds.), *Archaeometry '90*, 475–84. Basel: Birkhauser Verlag

Mynors H. S. (1983) An examination of Mesopotamium ceramics using petrographic and neutron activation analysis, in Aspinall A. and Warren S. E. (eds.) *The Proceedings of the 22nd Symposium on Archaeometry*, 377–87. Bradford: University of Bradford.

Neff H., Bishop R. L. and Arnold D. E. (1988) Reconstructing ceramic production from ceramic compositional data: an example from Guatemala. *Journal of Field Archaeology 15*, 339–48.

Neff H., Bishop R. L. and Arnold D. E. (1990) A re-examination of the compositional affiliations of formative period whiteware from highland Guatemala. *Ancient Mesoamerica 1*, 171–80.

Neff H., Bishop R. L. and Sayre E. V. (1988) A simulation approach to the problem of tempering in compositional studies of archaeological ceramics. *Journal of Archaeological Science 15*, 159–72.

Neff H., Bishop R. L. and Sayre E. V. (1989) More observations on the problem of tempering in compositional studies of archaeological ceramics. *Journal of Archaeological Science 16*, 57–69.

Pike H. H. M. and Fulford M. G. (1983) Neutron activation analysis of black-glazed pottery from Carthage. *Archaeometry 25*, 77–86.

Poirier J. and Barrandon J. N. (1983) Non destructive analysis of Roman and Byzantine gold coins by proton saturation, in Aspinall A. and Warren S. E. (eds.), *The Proceedings of the 22nd Symposium on Archaeometry*, 235–44. Bradford: University of Bradford.

Pollard A. M., Hatcher H. and Symonds R. P. (1981) Provenance studies of Rhenish pottery by comparison with terra sigillata. *Revue d'Archaéométrie 5*, 177–85.

Porat N., Yellin J., Heller-Kallai L. and Halicz L. (1991) Correlation between petrography, NAA and ICP analyses: application to Early Bronze Age Egyptian pottery from Canaan. *Geoarchaeology 6*, 133–49.

Prudencio M. I., Cabral J. M. P.and Tavares A. (1989) Identification of clay sources used for Conimbriga and Santa Olaia pottery making, in Maniatis Y. (ed.), *Archaeometry*, 503–14. Amsterdam: Elsevier.

Rauret G., Casassas E., Rius F. X. and Munoz M. (1987) Cluster analysis applied to spectrochemical data of European medieval stained glass. *Archaeometry 29*, 240–9.

Salter C. J. (1982) The relevance of chemical provenance studies to Celtic ironwork in Britain. *Institute of Archaeology Bulletin 19*, 73–81.

Sanderson D. C. W and Hunter J. R. (1981) Major element glass type specification for

Roman, post-Roman and mediaeval glasses. *Revue d'Archaéométrie Supplement*, 255–64.

Schwabe R. and Slussallek K. (1981) Application of the cluster analysis on element concentrations of archaeological bronzes, ceramics and glass. *Revue d'Archéométrie 5*, 109–17.

Shaw T. (1969) Further spectrographic analysis of Nigerian bronzes. *Archaeometry 11*, 85–93.

Sheridan A. (1989) Pottery production in Neolithic Ireland: a petrological and chemical study, in Henderson J. (ed.) *Scientific Analysis in Archaeology*, 112–35. Oxford: Oxford University Committee for Archaeology.

Spoerry P. S. (1990) Ceramic production in Medieval Dorset and the surrounding region. *Medieval Ceramics 14*, 3–17.

Stopford J., Hughes M. J. and Leese M. N. (1991) A scientific study of Medieval tiles from Bordesley Abbey, near Redditch (Hereford and Worcester). *Oxford Journal of Archaeology 10*, 349–60.

Storey J. M. V., Symonds R. P., Hart F. A., Smith D. M. and Walsh J. N. (1989) A chemical investigation of 'Colchester' samian by means of inductively-coupled plasma emission spectrometry. *Journal of Roman Pottery Studies 2*, 33–44.

Thuesen I., Heydorn K. and Gurozdz R. (1982) Investigation of 5000 year old pottery from Mesopotomia by instrumental neutron activation analysis. *PACT 7(II)*, 375–81.

Tubb A., Parker A. J. and Nickless G. (1980) The analysis of Romano British pottery by atomic absorption spectrophotometry. *Archaeometry 22*, 153–71.

Welinder S. and Griffin W. L. (1984) Raw material sources and an exchange network of the earliest farming society in central Sweden. *World Archaeology 16*, 174–85.

Williams J. Ll. W., Jenkins D. A. and Livens R. G. (1974) An analytical study of the composition of Roman coarse wares from the fort of Bryn y Gefeiliau (Caer Llugwy) in Snowdonia. *Journal of Archaeological Science 1*, 47–67.

Williams-Thorpe O., Warren S. E. and Nandris J. O. (1984) The distribution and provenance of archaeological obsidian in Central and Eastern Europe. *Journal of Archaeological Science 11*, 183–212.

Wisseman S. V., Hopke P. K. and Schindler-Kaudelka E. (1987) Multi-elemental and multivariate analysis of Italian terra-sigillata in the World Heritage Museum, University of Illinois at Urbana-Champaign, *Archaeomaterials 1*, 101–17.

(b) Typological/morphological studies

Aldenderfer M. S. (1982) Methods of cluster validation for archaeology. *World Archaeology 14*, 61–72.

Alvey R. C. and Laxton R. R. (1974) Analysis of some Nottingham clay pipes, *Science and Archaeology 13*, 3–12.

Alvey R. C. and Laxton R. R. (1977) Further analysis of some Nottingham clay pipes. *Science and Archaeology 19*, 20–9.

Alvey R. C., Laxton R. R. and Paechter G. F. (1979) Statistical analysis of some Nottingham clay tobacco pipes, in Davey P. (ed.), *The Archaeology of the Clay Tobacco Pipe*, 229–53. BAR British Series 63. Oxford: British Archaeological Reports.

Blanc J.-J. (1987) Discrimination des phases d'éboulements dans quelques cavernes préhistoriques. *Bulletin Du Musée d'Anthropologie Préhistorique de Monaco 87*, 27–41.

Brugal J.-P. (1984–5) Le Bos primigenius Boj., 1827 du Pléistocène moyen des grottes de Lusel-Viel (Hérault). *Bulletin du Musée d'Anthropologie Préhistorique de Monaco 28*, 5–17.

Chappell S. (1987) *Stone Axe Morphology and Distribution in Neolithic Britain*. BAR British Series 177. Oxford: British Archaeological Reports.
Dive J. (1986) Applications archaeologiques de l'analyse de grands tableaux. *Bulletin du Musée d'Anthropologie Préhistorique de Monaco 30*, 27–41.
Fraser D. (1983) *Land and Society in Neolithic Orkney*. BAR British Series 117. Oxford: British Archaeological Reports.
Gendel P. A. (1987) Socio-stylistic analysis of lithic artefacts from the Mesolithic of northwestern Europe, in Rowley-Conwy P., Zvelebil M. and Blankholm H. P. (eds.), *Mesolithic Northwest Europe: Recent Trends*, 65–73. Sheffield: University Dept. of Archaeology and Prehistory.
Green H. S. (1980) *The Flint Arrowheads of the British Isles*. BAR British Series 75. Oxford: British Archaeological Reports.
Habgood P. J. (1986) The origin of the Australians: a multivariate approach. *Archaeology in Oceania 21*, 130–37.
Hinout J. (1984) Les outils et armatures-standards mesolithiques dans le Bassin Parisien par l'analyse des données. *Revue Archéologique de Picardie 1–2*, 9–30.
Hinout J. (1985) Le gisement épipaléolithique de la Muette 1 commune du Vieux-Moulin (Oise). *Bulletin de la Société Préhistorique Française 82*, 377–88.
Hinout J. (1991) Le gisement mésolithique de Bonneuil-en Valois (Oise), lieu-dit: Lieu-Restauré. *Bulletin de la Société Préhistorique Française 88*, 178–86.
Hodder I. and Lane P. (1982) A contextual examination of Neolithic axe distribution in Britain, in Ericson J. E. and Earle T. K. (eds.), *Contexts for Prehistoric Exchange*, 213–35. New York: Academic Press.
Hodson F. R. (1970) Cluster analysis and archaeology: some new developments and applications. *World Archaeology 1*, 299–320.
Impey O. R. and Pollard A. M. (1985) A multivariate metrical study of ceramics made by three potters. *Oxford Journal of Archaeology 4*, 157–64.
Larsen C. U. (1988) A morphological study of biconical Late Bronze Age urns, in Madsen T. (ed.), 133–8. *Multivariate Archaeology*, Aarhus: Aarhus University Press.
Law G. (1984) Shell points of Maori two-piece fish hooks from northern New Zealand. *New Zealand Journal of Archaeology 6*, 5–21.
Madsen T. (1988b) Multivariate statistics and archaeology, in Madsen T. (ed.), *Multivariate Archaeology*, 7–27. Aarhus: Aarhus University Press.
O'Hare G. B. (1990) A preliminary study of polished stone artefacts in prehistoric southern Italy. *Proceedings of the Prehistoric Society 56*, 123–52.
Richards J. D. (1982) Anglo-Saxon pot shapes: cognitive investigations. *Science and Archaeology 24*, 33–46.
Richards J. D. (1987) *The Significance of Form and Decoration of Anglo-Saxon Cremation Urns*. BAR British Series 166. Oxford: British Archaeological Reports.
Shennan S. and Wilcock J. D. (1975) Shape and style variation in Central German bell beakers. *Science and Archaeology 15*, 32–7.
Siromoney G., Bagavandas M. and Govindaraju S. (1980) An application of component analysis to the study of South Indian sculptures. *Computers and the Humanities 14*, 29–38.
Speiser J.-M. (1989) Informatique et céramique. L'exemple de Pergame. *Bulletin de Correspondance Héllénique Supplement 18*, 291–302.
Van Peer P. (1991) Interassemblage variability and Levallois style: the case of the North African Middle Palaeolithic. *Journal of Anthropological Archaeology 10*, 107–151.

(c) Assemblage comparison

Azoury I and Hodson F. R. (1973) Comparing Palaeolithic assemblages: Ksar Akil, a case study. *World Archaeology 4*, 292–306.

Bietti A. and Burani A. (1985) The late upper palaeolithic in continental Italy: old classifications, new data and new perspectives, in Malone C. and Stoddart S. (eds.), *Papers in Italian Archaeology IV: Part (i) The Human Landscape*. BAR International Series 243, 7–27. Oxford: British Archaeological Reports.

Bietti A., Burani A. and Zanello L. (1985) Interactive pattern recognition in prehistoric archaeology: some applications. *PACT 11*, 205–8.

Binford L. R. (1987) Were there elephant hunters at Torralba?, in Nitecki M. H. and Nitecki D. V. (eds.), *The Evolution of Human Hunting*, 47–105. New York: Plenum Press.

Binford L. R. (1989) Technology of early man: an organizational approach to the Oldowan, in Binford L. R., *Debating Archaeology*, 437–63. Sandiago, CA: Academic Press.

Callow P. (1986a) An overview of the industrial succession, in Callow P. and Cornford J. M. (eds.), *La Cotte de St. Brelade 1961–1978*, 219–30. Norwich: Geo Books.

Callow P. (1986b) The La Cotte industries and the European lower and middle Palaeolithic, in Callow P. and Cornford J. M. (eds.), *La Cotte de St. Brelade 1961–1978*, 377–88. Norwich: Geo Books.

Callow P. and Webb R. E. (1981) The application of multivariate statistical techniques to middle palaeolithic assemblages from southwestern France. *Revue d'Archéométrie 5*, 129–38.

Castelletti L and Zimmerman A. (1985) Seriation for a spatial factor. An interpretation of the distribution of wood species in a settlement as a hypothesis of the use of the settlements neighbourhood. *PACT 11*, 111–18.

Close A. E. (1977) *The Identification of Style in Lithic Artefacts from North-East Africa*. Memoires de l'Institut d'Egypte.

Close A. E. (1978) The identification of style in lithic artefacts. *World Archaeology 10*, 223–37.

Fall P. L., Kelso G. and Markgraf V. (1981) Palaeoenvironmental reconstruction at Canyon del Muerto, Arizona, based on principal component analysis. *Journal of Archaeological Science 8*, 297–307.

Fisher A. R. (1985) Winklebury Hillfort: A study of artefact distributions from subsoil features. *Proceedings of the Prehistoric Society 51*, 167–80.

Gob A. (1988) Multivariate analysis of lithic industries: the influence of typology, in Ruggles C. L. N. and Rahtz S. P. Q. (eds.), *Computer and Quantitative Methods in Archaeology 1987*, 15–23. BAR International Series 393. Oxford: British Archaeological Reports.

Gowlett J. A. J. (1986) Culture and conceptualisation: the Oldowan-Acheulian gradient, in Bailey G. N. and Callow P. (eds.), *Stone Age Prehistory*, 243–60. Cambridge: Cambridge University Press.

Gowlett J. A. J. (1988) A case of developed Oldowan in the Acheulean? *World Archaeology 20*, 13–26.

Hodson F. R. (1969) Searching for structure within multivariate archaeological data. *World Archaeology 1*, 90–105.

Hutcheson J. C. C. and Callow P. (1986) The flint debitage and cores, in Callow P. and Cornford J. M. (eds.), *La Cotte de St. Brelade 1961–1978*, 231–49. Norwich: Geo Books.

Isaac G. Ll. (1977) *Olorgesailie*. Chicago: University of Chicago Press.

Jones G. (1987) A statistical approach to the archaeological identification of crop processing. *Journal of Archaeological Science 14*, 311–23.

Leese M. N. (1983) The statistical treatment of grain size data from pottery, in Aspinall A. and Warren S. E. (eds.), *The Proceedings of the 22nd Symposium on Archaeometry*, 47–55. University of Bradford.

Marguerie D. amd Walter Ph. (1986) Approches informatiques de la palynoarchéologie: exemples armoricains. *Bulletin de la Société Préhistorique Française 83*, 345–52.

Muñiz A. M. (1988) On the use of butchering as a palaeocultural index: proposal of a new methodology for the study of bone fracture from archaeological sites. *Archaeo Zoologia 2*, 111–50.

Pitts M. W. (1978a) Towards an understanding of flint industries in post-glacial England. *Institute of Archaeology Bulletin 15*, 179–97.

Pitts M. W. (1978b) On the shape of waste flakes as an index of technological change in lithic industries. *Journal of Archaeological Science 5*, 17–37.

Pitts M. W. (1979) Hides and antlers: a new look at the gatherer-hunter site at Star Carr, North Yorkshire, England. *World Archaeology 11*, 32–44.

Pitts M. W. and Jacobi R. M. (1979) Some aspects of change in flaked stone industries of the Mesolithic and Neolithic in Southern Britain. *Journal of Archaeological Science 6*, 163–77.

Prosch-Danielsen L. and Simonsen A. (1988) Principal component analysis of pollen, charcoal and soil phosphate data as a tool in prehistoric land-use investigation at Forsandmoen, southwest Norway. *Norwegian Archaeological Review 21*, 84–102.

Ringrose T. J. (1988a) Correspondence analysis as an exploratory technique for stratigraphic abundance data, in Ruggles C. L. N. and Rahtz S. P. Q. (eds.), *Computer and Quantitative Methods in Archaeology 1987*, 3–14. BAR International Series 393. Oxford: British Archaeological Reports.

Ringrose T. J. (1988b) Exploratory multivariate analysis of stratigraphic data: Armstrong's data from Pin-Hole cave re-examined, in Slater E. A. and Tate J. O. (eds.), *Science and Archaeology – Glasgow 1987*, 521–39. BAR British Series 196(ii). Oxford: British Archaeological Reports.

Ryan N. S. (1982) Characterising fourth century coin loss: an application of principal component analysis to archaeological time series data. *Science and Archaeology 24*, 25–32.

Ryan N. S. (1988) *Fourth Century Coin Finds from Roman Britain*. BAR British Series 183. Oxford: British Archaeological Reports.

Tomber R. (1988) Multivariate statistics and assemblage comparison, in Ruggles C. L. N. and Rahtz S. P. Q. (eds.), *Computer and Quantitative Methods in Archaeology 1987*, 29–38. BAR International Series 393. Oxford: British Archaeological Reports.

(d) Spatial

Cribb R. and Minnegal M. (1989) Spatial analysis on a dugong consumption site at Princess Charlotte Bay, North Queensland. *Archaeology in Oceania 24*, 1–12.

Whallon R. (1984) Unconstrained clustering for the analysis of spatial distributions in archaeology, in Hietala H. (ed.), *Intrasite Spatial Analysis in Archaeology*, 242–77. Cambridge: Cambridge University Press.

(e) Factor analysis

Bartel B. (1981) Cultural associations and mechanisms of change in anthropomorphic figurines during the Neolithic in the eastern Mediterranean basin. *World Archaeology 13*, 73–86.

Beneke N. (1990) The Krabbe collection of Icelandic horses and its significance for archaeozoological research. *Journal of Archaeological Science 17*, 161–85.

Bettinger R. L. (1979) Multivariate statistical analysis of a regional subsistence-settlement model for Owens Valley. *American Antiquity 44*, 455–70.

Binford L. R. (1972) Contemporary model building: paradigms and the current state of Palaeolithic research, in Clarke D. L. (ed.), *Models in Archaeology*, 109–66. London: Methuen.

Binford L. R. and Binford S. R. (1966) A preliminary analysis of functional variability in the Mousterian of Levallois facies. *American Anthropologist 68*, 238–95.

Buikstra J. E., Frankenberg S., Lambert J. B. and Xue L. (1989) Multiple elements: multiple expectations, in Price T. D. (ed.), *The Chemistry of Prehistoric Human Bone*, 155–210. Cambridge: Cambridge University Press.

Ciolek-Torrello R. (1985) A typology of room function at Grasshopper Pueblo, Arizona. *Journal of Field Archaeology 12*, 41–63.

Draper N. (1985) Back to the drawing board: a simplified approach to assemblage variability in the early palaeolithic. *World Archaeology 17*, 3–19.

Findlow F. J. and Bolognese M. (1982) Regional modelling of obsidian procurement in the American southwest, in Ericson J. E. and Earle T. K. (eds.), *Contexts for Prehistoric Exchange*, 53–81. New York: Academic Press.

Frankel D. (1974) Inter-site relationships in the Middle Bronze Age of Cyprus. *World Archaeology 6*, 109–208.

Hagstrum M. B. (1985) Measuring prehistoric ceramic craft specialization: a test case in the American southwest. *Journal of Field Archaeology 12*, 65–75.

Hoffman C. M. (1985) Projectile point maintenance and typology: assessment with factor analysis and canonical correlation, in Carr C. (ed.), *For Concordance in Archaeological Analysis*, 45–86. Kansas City: Westport Publishers.

Lambert J. B., Xue L. and Buikstra J. E. (1991) Inorganic analysis of excavated human bone after surface removal. *Journal of Archaeological Science 18*, 363–83.

McClellan T. L. (1979) Chronology of the 'Philistine' burials at Tell el-Far'ah (South). *Journal of Field Archaeology 6*, 57–73.

O'Shea J. M. (1984) *Mortuary Variability: An Archaeological Investigation*. New York: Academic Press.

Simek J. F. (1987) Spatial order and behavioural change in the French Palaeolithic. *Antiquity 61*, 25–40.

Read D. W. (1985) The substance of archaeological analysis and the mold of statistical method: enlightenment out of discordance, in Carr C. (ed.), *For Concordance in Archaeological Analysis*, 45–86. Kansas City: Westport Publishers.

Rigaud J.-P. and Simek J. F. (1991) Interpreting spatial patterns at the Grotte XV: a multiple method approach, in Kroll E. and Price T. D. (eds.), *The Interpretation of Archaeological Spatial Patterning*, 197–220. New York: Plenum Press.

Tainter J. A. (1975) Social inference and mortuary practices: an experiment in numerical classification. *World Archaeology 7*, 1–16.

Vierra R. K. and Carlson D. L. (1981) Factor analysis, random data, and patterned results. *American Antiquity 46*, 272–83.

Walker M. J. (1985) *Characterising Local Southeastern Spanish Populations of 3000–1500 BC*. BAR International Series 263. Oxford: British Archaeological Reports.

Williams M. W. (1986) Sub-surface patterning at Puerto-Real: a 16th century town on Haiti's north coast. *Journal of Field Archaeology 13*, 283–96.

Wynn T. and Tierson F. (1990) Regional comparison of the shapes of later Acheulean handaxes. *American Anthropologist* 92, 73-84.

References – Correspondence Analysis in Archaeology

(a) Analysis of chemical compositions of artefacts

Baxter M. J. (1991) An empirical study of principal component and correspondence analysis of glass compositions. *Archaeometry* 33, 29-41.

Baxter M. J. (1992b) Statistical analysis of chemical compositional data and the comparison of analyses. *Archaeometry* 34, 267-77.

Baxter M. J. and Heyworth M. P. (1989) Principal components analysis of compositional data in archaeology, in Rahtz S. P. Q. and Richards J. (eds.), *Computer Applications and Quantitative Methods in Archaeology 1989*, 227-40. BAR International Series 548. Oxford: British Archaeological Reports.

Baxter M. J., Cool H. E. M. and Heyworth M. P. (1990) Principal component and correspondence analysis of compositional data: some similarities. *Journal of Applied Statistics* 17, 229-35.

Gihwala D., Jacobson L., Peisach M. and Pineda C. A. (1984) Determining the origin of 18th and 19th century pottery and glasses using PIXE and PIPPS. *Nuclear Instruments and Methods in Physics Research B3*, 408-11.

Gratuze B. and Barrandon J.-N. (1990) Islamic glass weights and stamps: analysis using nuclear techniques. *Archaeometry* 32, 155-162.

Heyworth M. P., Baxter M. J. and Cool H. E. M. (1990) *Compositional Analysis of Roman Glass from Colchester, Essex*. Ancient Monuments Laboratory Report 53/90. London: English Heritage.

Jacobson L., Loubser J. H. N., Peisach M., Pineda C. A. and Van der Westhuizen W. (1991) PIXE analysis of pre-European pottery from the Northern Transvaal and its relevance to the distribution of ceramic styles, social interaction and change. *The South African Archaeological Bulletin* 153, 19-24.

Peisach M., Jacobson L., Boulle G. J., Gihwala D. and Underhill L. G. (1982) Multivariate analysis of trace elements determined in archaeological materials and its use for characterisation. *Journal of Radioanalytical Chemistry* 69, 349-64.

(b) Typological/morphological analysis

Barral L. and Simone S. (1981) Handaxes classifications. *Bulletin du Musée d'Anthropologie Préhistorique de Monaco* 25, 5-17.

Barral L. and Simone S. (1990) Calculs et graphs aux Merveilles (Tendes, Alpes-Maritimes). *Bulletin du Musée d'Anthropologie Préhistorique de Monaco* 33, 99-111.

Barral L., Bussière J.-F. and Simone S, (1988) Partition des produits de débitage. *Bulletin du Musée d'Anthropologie Préhistorique de Monaco* 31, 83-90.

Blanc J.-J. (1987) Discrimination des phases d'éboulements dans quelques cavernes préhistoriques. *Bulletin du Musée d'Anthropologie Préhistorique de Monaco* 30, 27-41.

Boutin P., Tallur B. and Chollet A. (1977) Essai d'application des techniques de l'analyse des données aux pointes à dos des niveaux aziliens de Rochereil. *Bulletin de la Société Préhistorique Française* 74, 362-75.

Decormeille A. and Hinout J. (1982) Mise en évidence des différentes cultures Mésolithiques dans le Bassin Parisien par l'analyse des données. *Bulletin de la Société Préhistorique Française* 79, 81-8.

Diaz-Andreu M. and Fernandez-Miranda M. (1991) Cuevas sepulcrales pretalayoticas de Mallorca: una esayo de clasificacion y analysis, in Waldren W. H.,

Ensenyat J. A. and Kennard R. C. (eds.), *2nd Deya International Conference on Prehistory: Vol. II Archaeological Technology and Theory*, 79–114. BAR International Series 574. Oxford: Tempus Reparatum.

Dive J. (1986) Applications archaéologiques de l'analyse de grands tableaux. *Bulletin du Musée d'Anthropologie Préhistorique de Monaco 29*, 45–64.

Djindjian F. (1977) Burin de Noailles, burin sur troncature et sur cassure: statistique descriptive appliquée à l'analyse typologique. *Bulletin de la Société Préhistorique Française 74*, 145–54.

Djindjian F. and de Croisset E. (1976a) Un essai de reconnaissance de formes sur une serie de deux cents bifaces Mousteriéns de Tabaterie (Dordogne) par l'analyse des données. *Cahier du Centre de Recherches Préhistoriques de Paris 5*, 39–60.

Djindjian F. and de Croisset E. (1976b) Etude typométrique d'une série de deux cents bifaces Mousteriéns de Tabaterie (Dordogne) par l'analyse des données. *IXth Congress UISPP, Thèmes Spécialisés*, 38–50. Nice: UISPP.

Englestad E. (1988) Pit-houses in Arctic Norway – an investigation of their typology using multiple correspondence analysis, in Madsen T. (ed.), *Multivariate Archaeology*, 71–84. Aarhus: Aarhus University Press.

Gaillard C., Raju D. R., Misra V. N. and Rajaguru S. N. (1986) Handaxe assemblages from the Didirana region, Thar Desert, India: a metrical analysis. *Proceedings of the Prehistoric Society 52*, 189–214.

Holm-Olsen I. M. (1985) Farm mounds and land registers in Helgoy, North Norway: an investigation of trends in site location by correspondence analysis. *American Archaeology 5*, 27–34.

Mohen J.-P. (1980) L'Age du fer en Aquitaine. *Mémoires de la Société Préhistorique Française 14*, 1–562.

Tuffreau A. and Bouchet J.-P. (1985) Le gisement Achéuleen de la Vallée du Muid à Gouzeancourt (Nord). *Bulletin de la Société Préhistorique Française 82*, 291–306.

Verjux C. and Rousseau D.-D. (1986) La retouche Quina: une mise au point. *Bulletin de la Société Préhistorique Française 83*, 404–15.

Welte A. C. and Lambert G. (1986) Analyse de données sur les chevaux gravés Magdaleniéns de Fontalès (Tarn-et-Garonne), de la collection Darasse du Muséum d'Histoire Naturelle de Toulouse. *Bulletin de la Société Préhistorique Française 83*, 335–44.

(c) Assemblage comparison

Annable R. (1987) *The Later Prehistory of Northern England*. BAR British Series 160. Oxford: British Archaeological Reports.

Avery G. and Underhill L. G. (1986) Seasonal exploitation of seabirds by late Holocene coastal foragers: analysis of modern and archaeological data from the Western Cape, South Africa. *Journal of Archaeological Science 13*, 339–60.

Barclay K., Biddle M. and Orton C. (1990) The chronological and spatial distribution of the objects, in Biddle M. (ed.), *Object and Economy in Medieval Winchester (i)*, 42–73. Oxford: Clarendon Press.

Barral L. and Simone S. (1989) Que sont les merveilles? *Bulletin du Musée d'Anthropologie Préhistorique de Monaco 32*, 109–59.

Bayliss A. and Orton C. (1988) Seriation with parallel series – an historical example, in Rahtz S. P. Q. (ed.), *Computer and Quantitative Methods in Archaeology 1988*, 161–76. BAR International Series 446(i). Oxford: British Archaeological Reports.

Bech J.-H. (1988) Correspondence analysis and pottery chronology, in Madsen T. (ed.), *Multivariate Archaeology*, 29–35. Aarhus: Aarhus University Press.

Beck C. and Shennan S. (1991) *Amber in Prehistoric Britain*. Oxford: Oxbow Books.

Bertelsen R. (1988a) The finds pattern of archaeological excavations: correspondence analysis as explorative tool, in Ruggles C. L. N. and Rahtz S. P. Q. (eds.), *Computers and Quantitative Methods in Archaeology 1987*, 25–8. BAR International Series 393. Oxford: British Archaeological Reports.

Bertelsen R. (1988b) Find pattern of multi-stratified sites: correspondence analysis as an explorative tool, in Madsen T. (ed.), *Multivariate Archaeology*, 85–90. Aarhus: Aarhus University Press.

Bosselin B. and Djindjian F (1988) Un essai de structuration de Magdalénien Français à partir de l'outillage lithique. *Bulletin de la Société Préhistorique Française 85*, 304–31.

Delporte H., Maziere G. and Djindjian F. (1977) L'Aurignacien de la Ferrassie: observations preliminaires à la suite de fouilles récentes. *Bulletin de la Société Préhistorique Française 74*, 343–61.

Denys C. (1985) Palaeoenvironmental and palaeobiological significance of fossil rodent assemblages of Laetoli (Pliocene, Tanzania). *Palaeogeography, Palaeoclimatology, Palaeoecology 52*, 77–97.

Djindjian F. (1985a) Seriation and toposeriation by correspondence analysis. *PACT 11*, 119–35.

Djindjian F. (1985b) Typologie et culture: l'exemple de l'Aurignacien, in Otte M. (ed.), *La Signification Culturelle des Industries Lithiques*, 338–73. BAR International Series 239. Oxford: British Archaeological Reports.

Djindjian F. (1986) Recherches sur l'Aurignacien du Périgord à partir des données nouvelle de La Ferrassie. *L'Anthropologie 90*, 89–106.

Djindjian F. (1988a) Les rapports entre les industries Magdaleniennes, Cresswelliennes et Hambourgiennes du Nord de l'Europe, in Otte M. (ed.), *De la Loire à l'Oder*, 683–705. BAR International Series 444(i). Oxford: British Archaeological Reports.

Gebauer A. B. (1988) Stylistic variation in the pottery of the Funnel Beaker Culture, in Madsen T. (ed.), *Multivariate Archaeology*, 91–117. Aarhus: Aarhus University Press.

Gillis C. (1990) Minoan conical cups: form, function and significance. *Studies in Mediterranean Archaeology 86*, 1–200.

Girard M. and Hinout J. (1990) Essai de chronologie de sites mesolithique du Bassin Parisien par l'analyse pollinique. *Bulletin de la Société Préhistorique Française 87*, 113–16.

Gob A. (1988) Multivariate analysis of lithic industries: the influence of typologies, in Ruggles C. L. N. and Rahtz S. P. Q. (eds.), *Computers and Quantitative Methods in Archaeology 1987*, 15–23. BAR International Series 393. Oxford: British Archaeological Reports.

Gorecki P. P., Horton D. R., Stern N. and Wright R. V. S. (1984) Coexistence of humans and megafauna in Australia: improved stratified evidence. *Archaeology in Oceania 19*, 117–19.

Grönlund E., Simola H. and Vimonen-Simola P. (1990) Early agriculture in the eastern Finnish Lake District. *Norwegian Archaeological Review 23*, 79–85.

Gruel K., Lleres J. and Wideman F. (1981) Etude de seriation du tresor de Trebry et chronologie analytique. *Revue d'Archaéométrie 5*, 145–58.

Hamard D. (1984) La répartition du matériel chasséen de 'Canneville' a Saint-Maximin (Oise). *Revue Archaeologique de Picardie 1–2*, 167–71.

Hill M. O. (1974) Correspondence analysis: a neglected multivariate method. *Applied Statistics 23*, 340–54.

Højlund F. (1988) Chronological and functional differences in Arabian Bronze Age pottery: a case study in correspondence analysis, in Madsen T. (ed.), *Multivariate Archaeology*, 55–60. Aarhus: Aarhus University Press.

Holm-Olsen I. M. (with Solheim I.) (1981) The Helgoy project: Economy and settlement pattern 1350–1600 AD, based on evidence from farm mounds. *Norwegian Archaeological Review 14*, 86–101.

Holm-Olsen I. M. (1988) The archaeological survey of North Norway: an evaluation using correspondence analysis, in Madsen T. (ed.), *Multivariate Archaeology*, 61–9. Aarhus: Aarhus University Press.

Hours P. F. (1976) L'Epipaléolithique au Liban; resultats acquis en 1975. *IXth Congress UISPP, Colloque III*. Nice: UISPP.

Jensen H. T. (1986) Unretouched blades in the late mesolithic of south Scandinavia. A functional study. *Oxford Journal of Archaeology 5*, 19–33.

Jones G. E. M. (1991) Numerical analysis in archaeobotany, in van Zeist W., Wasylikowa K. and Behre K.-E. (eds.), *Progress in Old World Palaeoethnobotany*, 63–80. Rotterdam: Balkema.

Jørgensen A. N. (1992) Weapon sets in Gotlandic grave finds from 530–800 AD: a chronological analysis, in Jørgensen L. (ed.), *Chronological Studies of Anglo-Saxon England, Lombard Italy and Vendel Period Switzerland*, 5–34. Copenhagen: Institute of Prehistoric and Classical Archaeology, University of Copenhagen.

Jørgensen L. (1992b) AD568 – A chronological analysis of Lombard graves in Italy, in Jørgensen L. (ed.), *Chronological Studies of Anglo-Saxon England, Lombard Italy and Vendel Period Switzerland*, 94–112. Copenhagen: Institute of Prehistoric and Classical Archaeology, University of Copenhagen.

Lange A. G. (1990) *De Hordern Near Wijk Bij Duurstere*. Amersfoort: ROB.

Laxton R. (1990) Methods of chronological ordering, in Voorrips A. and Ottoway B. (eds.), *New Tools from Mathematical Archaeology*, 37–44. Warsaw: Scientific Information Centre of the Polish Academy of Sciences.

Madsen T. (1988b) Multivariate statistics and archaeology, in Madsen T. (ed.), *Multivariate Archaeology*, 7–27. Aarhus: Aarhus University Press.

Madsen T. (1989) Seriation and multivariate statistics, in Rahtz S. P. Q. and Richards J. (eds.), *Computer Applications and Quantitative Methods in Archaeology 1989*, 205–14. BAR International Series 548. Oxford: British Archaeological Reports.

Mohen J.-P. and Bergougnan D. (1984) Le camp néolithique de chez Reine, à Semussac (Charente-Maritime). *Gallia Préhistoire 27*, 7–40.

Muñiz A. M. (1988) On the use of butchering as a paleocultural index: proposal of a new methodology for the study of bone fractures from archaeological data. *Archaeo Zoologia 2*, 111–50.

Nielsen K. H. (1986) Zur chronologie der jüngeren Germanischen eisezeit auf Bornholm. *Acta Archaeologica 57*, 47–86.

Nielsen K. H. (1988) Correspondence analysis applied to hoards and graves of the Germanic Iron Age, in Madsen T. (ed.), *Multivariate Archaeology*, 37–54. Aarhus: Aarhus University Press.

Ollendorf A. L. (1987) Archaeological implications of a phytolith study at Tel Mique (Ekron), Israel. *Journal of Field Archaeology 14*, 453–63.

Orton C. R. and Tyers P. A. (1990) Statistical analysis of ceramic assemblages. *Archeologia E Calcolatori 1*, 81–110.

Orton C. R. and Tyers P. A. (1991) A technique for reducing the size of sparse contingency tables, in Lockyear K. and Rahtz S. P. Q. (eds.), *Computer Applications and Quantitative Methods in Archaeology 1990*, 121–6. BAR International Series 565. Oxford: Tempus Reparatum.

Palm M and Pind J. (1992) Anglian English women's graves in the fifth to seventh centuries AD – a chronological analysis, in Jørgensen L. (ed.), *Chronological Studies of Anglo-Saxon England, Lombard Italy and Vendel Period Switzerland*, 50–80.

Copenhagen: Institute of Prehistoric and Classical Archaeology, University of Copenhagen.
Powers A. H. (1988) Phytoliths: animal, vegetable *and* mineral, in Slater E. A. and Tate J. O. (eds.), *Science and Archaeology – Glasgow 1987*, 459–72. BAR British Series 196(ii). Oxford: British Archaeological Reports.
Powers A. H., Padmore J. and Gilbertson D. D. (1989) Studies of the late prehistoric and modern opal phytoliths from coastal sand dunes and machair in northwest Britain. *Journal of Archaeological Science 16*, 27–45.
Ringrose T. J. (1988a) Correspondence analysis as an exploratory technique for stratigraphic abundance data, in Ruggles C. L. N. and Rahtz S. P. Q. (eds.), *Computers and Quantitative Methods in Archaeology 1987*, 3–14. BAR International Series 393. Oxford: British Archaeological Reports.
Ringrose T. J. (1988b) Exploratory multivariate analysis of stratigraphic data: Armstrong's data from Pin-Hole cave re-examined, in Slater E. A. and Tate J. O. (eds.), *Science and Archaeology – Glasgow 1987*, 521–39. BAR British Series 196(ii). Oxford: British Archaeological Reports.
Ringrose T. J. (1992) Bootstrapping and correspondence analysis in archaeology. *Journal of Archaeological Science 19*, 615–29.
Sauvet G. and Sauvet S. (1979) Fonction sémiologique de l'art pariétal animalier Franco-Cantabrique. *Bulletin de la Société Préhistorique Française 76*, 340–54.
Slachmuylder J.-L. (1985) Seriation by correspondence analysis for mesolithic assemblages. *PACT 11*, 137–48.
Sognnes K. (1987) Rock art and settlement pattern in the Bronze age. Example from Stordel, Trondelog, Norway. *Norwegian Archaeological Review 20*, 110–19.
Stilborg O. (1992) A chronological analysis of Anglo-Saxon men's graves in England, in Jørgensen L. (ed.), *Chronological Studies of Anglo-Saxon England, Lombard Italy and Vendel Period Switzerland*, 35–49. Copenhagen: Institute of Prehistoric and Classical Archaeology, University of Copenhagen.
Stopford J., Hughes M. J. and Leese M. N. (1991) A scientific study of Medieval tiles from Bordesley Abbey, near Redditch (Hereford and Worcester). *Oxford Journal of Archaeology 10*, 349–60.
Van der Veen M. (1987) The plant remains, in Heslop D. H., *The Excavation of an Iron Age settlement at Thorpe Thewles, Cleveland, 1980–82*, 93–7. London: CBA.
Vandkilde H. (1988) A late neolithic hoard with objects of bronze and gold from Skeldah, Central Jutland. *Journal of Danish Archaeology 7*, 115–35.
Voruz J.-L. (1984) *Outillages Osseux et Dynamisme Industriel dans le Néolithique Jurassien*. Lausanne: Cahiers D'Archéologie Romande.

(d) Spatial analysis

Blankholm H. P. (1991) *Intrasite Spatial Analysis in Theory and Practice*. Aarhus: Aarhus University Press.
Chataigner C. and Plateaux M. (1986) Analyse spatiale des habitats Rubanés et informatique. *Bulletin de la Société Préhistorique Française 83*, 319–24.
Djindjian F. (1988b) Improvements in intra-site spatial analysis techniques, in Rahtz S. P. Q. (ed.), *Computer and Quantitative Methods in Archaeology 1988*, 95–106. BAR International Series 446(i). Oxford: British Archaeological Reports.
Johnson I. (1984) Cell frequency recording and analysis of artefact distributions, in Hietala H. (ed.), *Intrasite Spatial Analysis in Archaeology*, 75–96. Cambridge: Cambridge University Press.

(e) Expository and review articles and edited collections

Barral L. and Simone S. (1977) Eléments d'analyse des données. *Bulletin du Musée d'Anthropologie Préhistorique de Monaco 21*, 5–92.

Bølviken E., Helskog E., Helskog K., Holm-Olsen I. M., Solheim L. and Bertelsen R. (1982) Correspondence analysis: an alternative to principal components. *World Archaeology 14*, 41–60.

Djindjian F. (1989) Fifteen years of contributions of the French school of data analysis, in Rahtz S. P. Q. and Richards J. (eds.), *Computer Applications and Quantitative Methods in Archaeology 1989*, 193–204. BAR International Series 548. Oxford: British Archaeological Reports.

Djindjian F. (1990a) Ordering and structuring in archaeology, in Voorrips A. (ed.), *Mathematics and Information Science in Archaeology: A Flexible Framework*, 79–92. Bonn: Holos-Verlag.

Djindjian F. (1990b) Data analysis in archaeology. *Science and Archaeology 32*, 57–62.

Djindjian F. (1990c) A select bibliography of French data analysis applications in archaeology. *Science and Archaeology 32*, 63–8.

Djindjian F. and Vigneron E. (1980) L'analyse des données au service de l'archaeologie préhistorique. *Bulletin de la Société Préhistorique Française 77*, 177–81.

Ihm P. (1990) Stochastic models and data analysis in archaeology, in Voorrips A. (ed.), *Mathematics and Information Science in Archaeology: A Flexible Framework*, 115–34. Bonn: Holos-Verlag.

Jørgensen L. (ed.) (1992a) *Chronological Studies of Anglo-Saxon England, Lombard Italy and Vendel Period Switzerland*. Copenhagen: Institute of Prehistoric and Classical Archaeology, University of Copenhagen.

Leredde H. and Djindjian F. (1980) Traitement automatique des données en archéologie. *Les Dossiers de l'Archéologie 42*, 52–69.

Madsen T. (ed.) (1988a) *Multivariate Archaeology*. Aarhus: Aarhus University Press.

Wright R. (1985) Detecting pattern in tabled archaeological data by principal components and correspondence analysis: programs in BASIC for portable microcomputers. *Science and Archaeology 27*, 35–8.

References – Cluster Analysis in Archaeology

(a) Analyses of chemical compositions of artefacts

Aitchison S. and Ottoway B. S. (1986) Neutron activation analysis of Bavarian Altheim pottery and of local clay sources, in Olin J. S. and Blackman M. J. (eds.), *Proceedings of the 24th International Archaeometry Symposium*, 321–6. Washington D.C.: Smithsonian Institution Press.

Alvey R. C. and Laxton R. R. (1978) A note on the chemical analysis of some Nottingham clay tobacco pipes. *Archaeometry 20*, 189–96.

'Amr K. (1987) *The Pottery from Petra: A Neutron Activation Analysis Study*. BAR International Series 324. Oxford: British Archaeological Reports.

Attas M., Yaffe L. and Fossey J. M. (1977) Neutron activation analysis of Early Bronze Age pottery from Lake Vouliageni, Perakhora, Central Greece. *Archaeometry 19*, 33–43.

Barrett J., Bradley R, Cleal R. and Pike H. (1978) Characterization of Deverel-Rimbury pottery from Cranbourne Chase. *Proceedings of the Prehistoric Society 44*, 135–142.

Baxter M. J. (1989) The multivariate analysis of compositional data in archaeology: a methodological note. *Archaeometry 31*, 45–53.

Baxter M. J. (1992b) Statistical analysis of chemical compositional data and the comparison of analyses. *Archaeometry 34*, 267–77.

Bell M. A. and Ottoway B. S. (1988) Neutron activation analysis of late neolithic pottery from Czechoslovakia and Bavaria, in Slater E. A. and Tate J. O. (eds.), *Science and Archaeology – Glasgow 1987*, 83–94. BAR British Series 196(i). Oxford: British Archaeological Reports.

Bello M. A. and Martin A. (1992) Microchemical characterization of building stone from Seville cathedral, Spain. *Archaeometry 34*, 21–9.

Bieber Jr. A. M., Brooks D. W., Harbottle G. and Sayre E. V. (1976) Application of multivariate techniques to analytical data on Aegean ceramics. *Archaeometry 18*, 59–74.

Birgül O., Dikšić M. and Yaffe L. (1979) X-ray fluorescence analysis of Turkish clays and pottery, Archaeometry. *Archaeometry 21*, 203–18.

Blackman M. J. and Nagle C. (1983) Characterization of Dorset paleo-eskimo nephritic jade artifacts from Central Labrador, Canada, in Aspinall A. and Warren S. E. (eds.), *Proceedings of the 22nd Symposium on Archaeometry*, 411–19. Bradford: University of Bradford.

Blackman M. J. (1984) Provenance studies of Middle Eastern obsidian from sites in highland Iran, in Lambert J. (ed.), *Archaeological Chemistry III*, 19–50. Washington, D.C.: American Chemical Society.

Blasius E., Wagner H., Braun H., Krumbholz R. and Schwartz B. (1983) Tile fragments as characteristic evidence for ancient Roman settlements: scientific investigations. *Archaeometry 25*, 165–78.

Bower N. W., Faciszewski S., Renwick S. and Peckham S. (1986) A preliminary analysis of Rio-Grande glazes of the Classic Period using scanning electron microscopy with X-ray fluorescence. *Journal of Field Archaeology 13*, 307–15.

Bower N., Radamacher D., Shimotake T., Anselmi T. and Peckham S. (1988) A pattern recognition and scanning electron microscope with X-ray fluorescence analysis of glaze-paint Anasazi pottery from the American Southwest, in Farquhar R. M., Hancock R. G. V. and Pavlish L. A. (eds.), *Proceedings of the 26th International Archaeometry Symposium*, 204–9. Toronto: University of Toronto.

Burmester A. (1983a) Historical coating materials – East Asian Lacquer, in Aspinall A. and Warren S. E. (eds.), *Proceedings of the 22nd Symposium on Archaeometry*, 184–93. Bradford: University of Bradford.

Burmester A. (1983b) Far eastern lacquers: classification by pyrolisis mass spectrometry. *Archaeometry 25*, 45–58.

Cabral J. M. P., Gouveia M. A., Alarcão A. M. and Alarcão J. (1983) Neutron activation analysis of fine grey pottery from Conimbriga, Santa Olaia and Tavarede, Portugal. *Journal of Archaeological Science 10*, 61–70.

Cabral J. M. P., Prudencio M. I., Gouveia M. A. and Arnaud J. M. (1988) Chemical and mineralogical characterization of pre-beaker and beaker pottery from Ferreira do Alentejo (Beja, Portugal), in Farquhar R. M., Hancock R. G. V. and Pavlish L. A. (eds.), *Proceedings of the 26th International Archaeometry Symposium*, 172–8. Toronto: University of Toronto.

Calamioutou M., Fillipakis S. E., Jones R. E. and Kassals D. (1984) X-ray and spectrographic analyses of terracotta figurines from Myrina: an attempt to characterise workshops. *Journal of Archaeological Science 11*, 103–17.

Camalich-Massieu M. D., Martin-Sosas D., Casasus-Latorre L. and Gonzalez-Quintero P. (1989) Pottery of the group of Purchena (Almeira, Spain): a cluster analysis, Maniatis Y. (ed.), *Archaeometry*, 603–12. Amsterdam: Elsevier.

Capannesi G., Seccaroni C., Sedda A. F., Majerini V. and Musco S. (1991)

Classification of fifteenth to nineteenth century mortars from Gabii using instrumental neutron activation analysis. *Archaeometry 33*, 255–66.

Christie O. H. J., Brenna J. A. and Straume E. (1979) Multivariate classification of Roman glasses found in Norway. *Archaeometry 21*, 233–41.

Christensen L. H., Conradsen K and Nielsen S. (1985) The Selbjerg project: investigation of neolithic pottery by means of neutron activation analysis. *ISKOS 5*, 401–10.

Cox G. A. and Gillies K. J. S. (1986) The X-ray fluorescence analysis of medieval durable blue soda glass from York Minster. *Archaeometry 28*, 57–68.

Cox G. A. and Pollard A. M. (1981) The multivariate analysis of data relating to the durability of medieval window glass. *Revue d'Archéométrie 5*, 119–28.

Cracknell S. (1982) An analysis of pottery from Elgin and north-east Scotland. *Medieval Ceramics 6*, 51–65.

D'Altroy T. N. and Bishop R. L. (1990) The provincial organisation of Inka ceramic production. *American Antiquity 55*, 120–38.

De Palol P., Gurt J. M., Tuset F., Planas C., Buxeda J., Cau M. A. and Alcobé X. (1991) Clunia: producer and receiver of Hispanic terra sigillata, in Budd P., Chapman B., Jackson C., Janaway R. and Ottoway B. (eds.), *Archaeological Sciences 1989*, 83–94. Oxford: Oxbow Books.

Desbat A. and Picon M. (1986) Les importations d'amphores de Méditerranée Orientale à Lyon (fin du Ier siècle avant J.-C. et Ier siècle après). *Bulletin de Correspondance Hellénique Supplement 13*, 637–48.

Devereux D. F., Jones R. F. and Warren S. E. (1983) X-ray fluorescence analysis of Verulamium region ware, in Aspinall A. and Warren S. E. (eds.), *Proceedings of the 22nd Symposium on Archaeometry*, 333–42. Bradford: University of Bradford.

Djingova R. and Kuleff I. (1992) An archaeometric study of Medieval glass from the first Bulgarian capital, Pliska (ninth to tenth century AD). *Archaeometry 34*, 53–61.

Elam J. M., Glascock M. D. and Slane K. W. (1988) A re-examination of the provenance of Eastern Sigillata A, in Farquhar R. M., Hancock R. G. V. and Pavlish L. A. (eds.), *Proceedings of the 26th International Archaeometry Symposium*, 179–83. Toronto: University of Toronto.

Elam J. M., Glascock M. D. and Finstein L. (1991) The implications of obsidian artifact proveniences from Jalieza, Oaxaca, Mexico, in Pernicka E. and Wagner G. A. (eds.), *Archaeometry '90*, 365–74. Basel: Birkhäuser Verlag.

Empereur J.-Y. and Picon M. (1986) A la recherche des fours d'amphores. *Bulletin de Correspondance Hellénique Supplement 13*, 103–26.

Empereur J.-Y. and Picon M. (1988) The production of Aegean amphorae: field and laboratory studies, in Jones R. E. and Catling H. W. (eds.), *New Aspects of Archaeological Science in Greece*, 33–8. Athens: British School at Athens.

Esse D. and Hopke P. K. (1986) Levantine trade in the early Bronze Age, in Olin J. S. and Blackman M. J. (eds.), *Proceedings of the 24th International Archaeometry Symposium*, 327–39. Washington, D. C.: Smithsonian Institution Press.

Evans J. (1989) Neutron activation analysis and Romano-British pottery studies, in Henderson J. (ed.), *Scientific Analysis in Archaeology*, 136–62. Oxford: Oxford University Committee for Archaeology.

Fontes P., Laubenheimer F., Leblanc J., Bonnefoy F., Gruel K. and Widemann F. (1981) Nouvelles données analytiques et hypologiques sur les ateliers de production d'amphores en Gaule du Sud. *Revue d'Archéométrie Supplement*, 95–111.

Gebhard R., Ihra W., Wagner F. E., Wagner U., Bischof H., Riederer J. and Wippern A. M. (1988) Mössbauer and neutron activation analysis study of ceramic finds from Canapote, Columbia, in Farquhar R. M., Hancock R. G. V. and Pavlish

L. A. (eds.), *Proceedings of the 26th International Archaeometry Symposium*, 196–203. Toronto: University of Toronto.

Gebhard R., Kossack G., Riederer J., Schwabe R. and Wagner U. (1989) Colouration of Celtic glass from Manching, in Maniatis Y. (ed.), *Archaeometry*, 207–15. Amsterdam: Elsevier.

Gilmore G. R. (1981) The use of mathematical clustering techniques for the analysis of coin composition data. *Revue d'Archéométrie 5*, 97–107.

Glascock M. D., Elam J. M. and Cobean R. H. (1988) Differentiation of obsidian sources in Mesoamerica, in Farquhar R. M., Hancock R. G. V. and Pavlish L. A. (eds), *Proceedings of the 26th International Archaeometry Symposium*, 245–51. Toronto: University of Toronto.

Goad S. I. and Noakes J. (1978) Prehistoric copper artifacts in the eastern United States, in Carter G. F. (ed.), *Archaeological Chemistry II*, 335–46. Washington, D.C.: American Chemical Society.

Grimanis A. P., Katsanos A. A., Kilikoglou V., Kourou N., Maniatis Y., Panakleridou D. and Vassilaki-Grimani M. (1989) An interdisciplinary approach to geometric pottery from Naxos: provenance and technological studies, in Maniatis Y. (ed.), *Archaeometry*, 169–75. Amsterdam: Elsevier.

Gunneweg J., Beier Th., Diehl U., Lambrecht D. and Mommsen H. (1991) 'Edomite', 'Negbite' and 'Midianite' pottery from the Negev Desert and Jordan: instrumental neutron activation analysis results. *Archaeometry 32*, 239–53.

Gurt J. M., Tuset F., Buxeda J., Planas C. and Alcobé X. (1991) The study of Hispanic terra sigillata from the kilns of Pla D'Abella (Catalonia): a preliminary analysis, in Budd P., Chapman B., Jackson C., Janaway R. and Ottoway B. (eds.), *Archaeological Sciences 1989*, 36–45. Oxford: Oxbow Books.

Hammond N., Harbottle G. and Gazard T. (1976) Neutron activation and statistical analysis of Maya ceramics and clays from Labaantun, Belize. *Archaeometry 18*, 147–68.

Harbottle G. (1976) Activation analysis in archaeology. *Radiochemistry 3*, 33–72.

Hart F. A., Storey J. M. V., Adams S. J., Symonds R. P. and Walsh J. N. (1987) An analytical study using inductively coupled plasma (ICP) spectrometry, of Samian and colour-coated wares from the Roman town of Colchester together with related continental Samian wares. *Journal of Archaeological Science 14*, 577–98.

Hatch J. W., Michels J. W., Stevenson C. M., Scheetz B. E. and Geidel R. A. (1990) Hopewell obsidian studies: behavioural implications of recent sourcing and dating research. *American Antiquity 55*, 461–79.

Hatcher H., Hedges R. E. M., Pollard A. M. and Kenrick P. M. (1980) Analysis of Hellenistic and Roman fine pottery from Benghazi. *Archaeometry 22*, 133–51.

Herz N. and Doumas C. (1991) Marble sources in the Aegean early Bronze Age, in Pernicka E. and Wagner G. A. (eds.), *Archaeometry '90*, 425–34. Basel: Birkhäuser Verlag.

Heyworth M. P., Hunter J. R., Warren S. E. and Walsh N. (1989) The role of inductively coupled plasma spectrometry in glass provenance studies, in Maniatis Y. (ed.), *Archaeometry*, 661–9. Amsterdam: Elsevier.

Hughes M. J. (1991) Provenance studies of Spanish Medieval tin glazed pottery by neutron activation analysis, in Budd P., Chapman B., Jackson C., Janaway R. and Ottoway B. (eds.), *Archaeological Sciences 1989*, 54–68. Oxford: Oxbow Books.

Hughes M. J. and Vince A. G. (1986) Neutron activation analysis and petrology of Hispano-Maresque pottery, in Olin J. S. and Blackman M. J. (eds.), *Proceedings of the 24th International Archaeometry Symposium*, 353–67. Washington, D.C.: Smithsonian Institution Press.

Hurtado de Mendoza L., and Jester W. A. (1978) Obsidian sources in Guatemala: a regional approach. *American Antiquity 43*, 424–35.
Jones R. E. (1986) *Greek and Cypriot Pottery*. Athens: British School at Athens.
Kilikoglou V., Maniatis Y. and Grimanis A. P. (1988) The effect of purification and firing of clays on trace element provenance studies. *Archaeometry 30*, 37–46.
King R. H., Rupp D. W. and Sorenson L. W. (1986) A multivariate analysis of pottery from southwestern Cyprus using neutron activation analysis data. *Journal of Archaeological Science 13*, 361–74.
Knapp A. B., Duerden P., Wright R. V. S. and Grave P. (1988) Ceramic production and social change: archaeometric analysis of Bronze Age pottery from Jordan. *Journal of Mediterranean Archaeology 1*, 57–113.
Krywonos W., Newton G. W. A., Robinson V. J. and Riley J. A. (1980) Neutron activation analysis of Roman coarse-ware from Cyrenaica. *Archaeometry 22*, 189–96.
Krywonos W., Newton G. W. A., Robinson V. J. and Riley J. A. (1982) Neutron activation analysis of some Roman and Islamic coarse-ware of Western Cyrenaica and Crete. *Journal of Archaeological Science 9*, 63–78.
Kuleff I., Djingova R. and Djingov G. (1985) Provenience study of medieval Bulgarian glasses by NAA and cluster analysis. *Archaeometry 27*, 185–93.
Kuleff I., Djingova R. and Djingov G. (1989) Provenance study of sgraffito ceramics (XIII–XIVc) from Shumen, Varna and Tcherven (North-Eastern Bulgaria), in Maniatis Y. (ed.), *Archaeometry*, 533–42. Amsterdam: Elsevier.
Kuleff I., Djingova R., Alexandrova A. and Angelova Chr. (1991) INAA and AAS of ancient lead anchors found along the Bulgarian Black Sea coast, in Pernicka E. and Wagner G. A. (eds.), *Archaeometry '90*, 199–208. Basel: Birkhäuser Verlag.
Lambert J. B., McLaughlin C. D. and Leonard Jr. A. (1978) X-ray photoelectron spectroscopic analysis of the Mycenean pottery from Megiddo. *Archaeometry 20*, 107–22.
Leese M. N., Hughes M. J. and Cherry J. (1986) A scientific study of N. Midlands medieval tile production. *Oxford Journal of Archaeology 5*, 355–70.
Lemoine C., Walker S. and Picon M. (1982) Archaeological, geochemical and statistical methods in ceramic provenance studies, in Olin J. S. and Franklin A. D. (eds.), *Archaeological Ceramics*, 57–64. Washington, D.C.: Smithsonian Institution Press.
Liddy D. J. (1988) A chemical study of early Iron Age pottery from the North cemetery, Knossos, in Jones R. E. and Catling H. W. (eds.), *New Aspects of Archaeological Science in Greece*, 29–32. Athens: British School at Athens.
Liddy D. J. (1989) A provenance study of decorated pottery from an Iron Age cemetery at Knossos, Crete, in Maniatis Y. (ed.), *Archaeometry*, 559–70. Amsterdam: Elsevier.
Lintzis Y. and McKerrell H. (1979) Some T. L. dates and neutron activation analysis of Greek neolithic pottery. *Archaeo-Physika 10*, 486–98.
MacSween A., Hunter J. R. and Warren S. E. (1988) Analysis of coarse wares from the Orkney Islands, in Slater E. A. and Tate J. O. (eds.), *Science and Archaeology – Glasgow 1987*, 95–106. BAR British Series 196(i). Oxford: British Archaeological Reports.
Matthers J., Liddy D. J., Newton G. W. A., Robinson V. J. and Al-Tawel H. (1983) Black-on-red ware in the Levant: a neutron activation analysis study. *Journal of Archaeological Science 10*, 369–82.
McGovern P. E. (1989) Ancient ceramic technology and stylistic change: contrasting studies from Southwest and Southeast Asia, in Henderson J. (ed.), *Scientific*

Analysis in Archaeology, 63–81. Oxford: Oxford University Committee for Archaeology.

McGovern P. E., Harbottle G. and Wnuk C. (1982) Late bronze age pottery fabrics from the Baq'ah Valley, Jordan: composition and origin. *MASCA Journal 2*, 8–12.

Mertz R., Melson W. and Levenbach G. (1979) Exploratory data analysis of Mycenean ceramic compositions and provenances. *Archaeo-Physika 10*, 580–90.

Michels J. W. (1982) Bulk element composition versus trace element composition in the reconstruction of an obsidian source system. *Journal of Archaeological Science 9*, 113–23.

Mirti P., Zelano V., Aruga R., Ferrara E. and Appolonia L. (1990) Roman pottery from Augusta Praetoria (Aosta, Italy): a provenance study. *Archaeometry 32*, 163–75.

Mirti P., Aruga R., Zelano V. and Appolonia L. (1991) Roman pottery from Augusta Praetoria: provenance investigation by atomic absorption spectroscopy and multivariate analysis of the analyticaal data, in Pernicka E. and Wagner G. A. (eds.), *Archaeometry '90*, 475–84. Basel: Birkhäuser Verlag.

Mommsen H., Krauser A. and Weber J. (1988) A method for grouping pottery by chemical composition. *Archaeometry 30*, 47–57.

Mommsen H., Lewandowski E., Weber J. and Podzweit Ch. (1988) Neutron activation analysis of Mycenean pottery from the Argolid: the search for reference groups, in Farquhar R. M., Hancock R. G. V. and Pavlish L. A. (eds.), *Proceedings of the 26th International Archaeometry Symposium*, 165–71. Toronto: University of Toronto.

Mommsen H., Krauser A., Weber J. and Podzuweit Ch. (1989) Classification of Mycenaen pottery from Kastonas by neutron activation analysis, in Maniatis Y. (ed.), *Archaeometry*, 515–23. Amsterdam: Elsevier.

Mynors H. S. (1983) An examination of mesoamerican ceramics using petrographic and neutron activation analysis, in Aspinall A. and Warren S. E. (eds.), *Proceedings of the 22nd Symposium on Archaeometry*, 377–87. Bradford: University of Bradford.

Neff H., Bishop R. L. and Sayre E. V. (1989) More observations on the problem of tempering in compositional studies of archaeological ceramics. *Journal of Archaeological Science 16*, 57–69.

Newman J. R. and Neilsen R. L. (1985) Initial notes on the X-ray fluorescence sourcing of northern New-Mexico obsidian. *Journal of Field Archaeology 12*, 377–83.

Newman J. R. and Neilsen R. L. (1987) Initial notes on the X-ray fluorescence characterization of the rhyodacite sources of the Taos Plateau, New Mexico. *Archaeometry 29*, 262–74.

Newton G. W. A., Robinson V. J., Oladipo M., Chandratillake M. R. and Whitbread I. K. (1988) Clay sources and Corinthian amphorae, in Slater E. A. and Tate J. O. (eds.), *Science and Archaeology – Glasgow 1987*, 59–82. BAR British Series 196(i). Oxford: British Archaeological Reports.

Olcese G. (1991) Roman coarse ceramics from Albintimilium (Ventimigliaa, Italy): an example of archaeometric and archaeological studies, in Pernicka E. and Wagner G. A. (eds.), *Archaeometry '90*, 495–504. Basel: Birkhäuser Verlag.

Ottoway B. (1974) Cluster analysis of impurity patterns in Armorico-British daggers. *Archaeometry 16*, 221–31.

Ottoway B. (1979) Interpretation of prehistoric metal artifacts with the aid of cluster analysis. *Archaeo-Physika 10*, 597–607.

Ottoway B. (1981) Mixed data classification in archaeology. *Revue d'Archéométrie 5*, 139–44.

Pernicka E., Begemann F., Schmitt-Streker S. and Grimanis A. P. (1990) On the composition and provenance of metal artefacts from Poliochni on Lemnos. *Oxford Journal of Archaeology 9*, 263–98.

Picon M. (1984) Le traitement des données d'analyse. *PACT 10*, 379–99.

Picon M. and Garlan Y. (1986) Recherches sur l'implementation des ateliers amphoriques à Thasos et analyse de la parte des amphores Thasienes. *Bulletin de Correspondance Hellénique Supplement XIII*, 287–309.

Pollard A. M. (1983) A critical study of multivariate methods as applied to provenance data, in Aspinall A. and Warren S. E. (eds.), *Proceedings of the 22nd Symposium on Archaeometry*, 56–66. Bradford: University of Bradford.

Pollard A. M. (1986) Data analysis, in Jones R. R. (ed), *Greek and Cypriot Pottery: A Review of Scientific Studies*, 56–83. Athens: British School at Athens.

Pollard A. M., Hatcher H. and Symonds R. P. (1981) Provenance studies of Rhenish pottery by comparison with terra-sigillata. *Revue d'Archéométrie 5*, 177–85.

Pollard A. M., Hatcher H. and Symonds R. P. (1983) Provenance studies of 'Rhenish' wares – a concluding report, in Aspinall A. and Warren S. E. (eds.), *Proceedings of the 22nd Symposium on Archaeometry*, 343–54. Bradford: University of Bradford.

Pollard A. M. and Hatcher H. (1986) The chemical analysis of oriental ceramic body compositions: Part 2 – greenwares. *Journal of Archaeological Science 13*, 261–87.

Prag A. J. N. W., Schweizer F., Williams J. Ll. W. and Schubiger P. (1974) Hellenistic glazed wares from Athens and Southern Italy: analytical techniques and implications. *Archaeometry 16*, 153–87.

Prudencio M. I., Cabral J. M. P. and Tavares A. (1989) Identification of clay sources used for Conimbriga and Santa Olaia pottery making, in Maniatis Y. (ed.), *Archaeometry*, 503–14. Amsterdam: Elsevier.

Rapp G. Jr., Allert J. and Henrickson E. (1984) Trace element discrimination of discrete sources of native copper, in Lambert J. (ed.), *Archaeological Chemistry III*, 273–93. Washington, D.C.: American Chemical Society.

Rauret G., Casassas E., Rius F. X. and Muñoz M. (1987) Cluster analysis applied to spectrochemical data of European Medieval stained glass. *Archaeometry 29*, 240–9.

Rehman F., Robinson V. J., Newton G. W. A. and Shennan S. J. (1991) Neutron activation analysis and the bell beaker folk, in Budd P., Chapman B., Jackson C., Janaway R. and Ottoway B. (eds.), *Archaeological Sciences 1989*, 95–103. Oxford: Oxbow Books.

Rice P. M. (1977) Whiteware pottery production in the Valley of Guatemala: specialisation and resource utilization. *Journal of Field Archaeology 4*, 221–33.

Rice P. M. and Saffer M. E. (1982) Cluster analysis of mixed-level data: pottery provenience as an example. *Journal of Archaeological Science 9*, 395–409.

Roos P., Moens L., de Rudder J., de Paepe P., van Hende J. and Waelkens M. (1988) Chemical, isotopic and petrographic characterization of ancient white marble quarries, in Farquhar R. M., Hancock R. G. V. and Pavlish L. A. (eds.), *Proceedings of the 26th International Archaeometry Symposium*, 220–6. Toronto: University of Toronto.

Salazar R., Wagner U., Wagner F. E., Korschinsky W., Zahn M., Riederer J. and Kaufmann-Doig F. (1986) Mössbauer spectroscopy and neutron activation analysis of recent finds from Chaurin, in Olin J. S. and Blackman M. J. (eds.), *Proceedings of the 24th International Archaeometry Symposium*, 143–151. Washington D. C.: Smithsonian Institution Press.

Schneider G. (1989) A technological study of North-Mesopotamian stone ware. *World Archaeology 21*, 30–50.

Schneider G., Knoll H., Gallis C. J. and Demoule J.-P. (1991) Production and distribution of coarse and fine pottery in Neolithic Thessaly, Greece, in Pernicka E. and Wagner G. A. (eds.), *Archaeometry '90*, 513–22. Basel: Birkhäuser Verlag.

Schubert P. (1986) Petrographic modal analysis – a necessary complement to chemical analysis of ceramic coarse wear. *Archaeometry 28*, 163–78.

Schwabe R. and Slusallek K. (1981) Application of the cluster analysis on element concentrations of archaeological bronzes ceramics and glass. *Revue d'Archéométrie 5*, 109–17.

Sheridan A. (1989) Pottery production in Neolithic Ireland: a petrological and chemical study, in Henderson J. (ed.), *Scientific Analysis in Archaeology*, 112–35. Oxford: Oxford University Committee for Archaeology.

Stevenson C. M. and McCurry M. O. (1990) Chemical characterization and hydration rate development for New Mexican obsidian sources. *Geo-Archaeology 5*, 149–70.

Stewart J. D., Fralick P., Hancock R. G. V., Kelley J. H. and Garrett E. M., (1990) Petrographic analysis and INAA geochemistry of prehistoric ceramics from Robinson Pueblo, New Mexico. *Journal of Archaeological Science 17*, 601–25.

Stopford J., Hughes M. J. and Leese M. N. (1991) A scientific study of Medieval tiles from Bordesley Abbey, near Redditch (Hereford and Worcester). *Oxford Journal of Archaeology 10*, 349–60.

Storey J. M. V., Symonds R. P., Hart F. A., Smith D. M. and Walsh J. N. (1989) A chemical investigation of 'Colchester' samian by means of inductively coupled plasma emission spectrometry. *Journal of Roman Pottery Studies 2*, 33–43.

Tobey M. H., Nielsen E. O. and Rowe M. W. (1986) Elemental analysis of Etruscan ceramics from Murlo, Italy, in Olin J. S. and Blackman M. J. (eds.), *Proceedings of the 24th International Archaeometry Symposium*, 115–27. Washington, D.C.: Smithsonian Institution Press.

Tobey M. H., Shafer H. J. and Rowe M. W. (1986) Trace element investigations of Mayan chert from Belize, in Olin J. S. and Blackman M. J. (eds.), *Proceedings of the 24th International Archaeometry Symposium*, 431–59. Washington, D.C.: Smithsonian Institution Press.

Topping P. G. (1986) Neutron activation analysis of later prehistoric pottery from the Western Isles of Scotland. *Proceedings of the Prehistoric Society 52*, 105–29.

Topping P. G. and MacKenzie A. B. (1988) A test of the use of neutron activation analysis for clay source characterization. *Archaeometry 30*, 92–101.

Tubb A., Parker A. J. and Nickless G. (1980) The analysis of Romano British pottery by atomic absorption spectrophotometry. *Archaeometry 22*, 153–71.

Wagner U., Brandis S. von., Ulbert U., Wagner F. E., Muller-Karpe H., Riederer J. and Tellenbach M. (1988) First results of a Mössbauer and neutron activation analysis study of recent ceramic finds from Montegrande, Peru, in Farquhar R. M., Hancock R. G. V. and Pavlish L. A. (eds.), *Proceedings of the 26th International Archaeometry Symposium*, 35–42. Toronto: University of Toronto.

Wagner U., Brandis S. von., Mathicorena B., Salazar R., Schwabe R., Riederer J. and Wagner F. E. (1989) Mössbauer studies of ceramics from the Inca period, in Maniatis Y. (ed.), *Archaeometry*, 159–68. Amsterdam: Elsevier.

White S. R., Warren S. E. and Jones R. E. (1983) The provenance of Bronze Age pottery from Thessaly in Eastern Greece, in Aspinall A. and Warren S. E. (eds.), *Proceedings of the 22nd Symposium on Archaeometry*, 323–32. Bradford: University of Bradford.

Widemann F. (1982) Why is archaeometry so boring for archaeologists?, in Olin J. S. and Franklin A. D. (eds.), *Archaeological Ceramics*, 29–36. Washington, D.C.: Smithsonian Institution Press.

Widemann F. (1985) Neutron activation analysis of archaeological ceramics and related problems. *ISKOS 5*, 391–400.

Wisseman S. U., Hopke P. K. and Schindler-Kaudelka E. (1987) Multi-elemental and multivariate analysis of Italian terra-sigillata in the World Heritage Museum, University of Illinois at Urbana-Champaign. *Archaeomaterials 1*, 101–17.

Wolff S. R., Liddy D. J., Newton G. W. A., Robinson V. J. and Smith R. J. (1986) Classical and Hellenistic black glaze ware in the Mediterranean: a study by epithermal neutron activation analysis. *Journal of Archaeological Science 13*, 245–59.

Wyttenbach A. and Schubiger P. A. (1973) Trace element content of Roman lead by neutron activation analysis. *Archaeometry 15*, 199–207.

(b) Typological/morphological studies

Aldenderfer M. S. (1982) Methods of cluster validation for archaeology. *World Archaeology 14*, 61–72.

Allsworth-Jones P. and Wilcock J. (1974) A computer assisted study of European Palaeolithic 'leafpoints'. *Science and Archaeology 11*, 25–46.

Alvey R. C. and Laxton R. R. (1974) Analysis of some Nottingham clay pipes. *Science and Archaeology 13*, 3–12.

Alvey R. C. and Laxton R. R. (1977) Further analysis of some Nottingham clay pipes. *Science and Archaeology 19*, 20–9.

Amson D. (1986) Lapita pottery of the Bismarck Archipelago and its affinities. *Archaeology in Oceania 21*, 157–65.

Buck C. E. and Litton C. D. (1991) A computational Bayes approach to some common archaeological problems, in Lockyear K. and Rahtz S. P. Q. (eds.), *Computer Applications and Quantitative Methods in Archaeology 1990*, 93–9. BAR International Series 565. Oxford: Tempus Reparatum Reports.

Cavalloro R. and Shimada I. (1988) Some thoughts on Sican marked adobes and labor organization. *American Antiquity 53*, 75–101.

Celoria F. S. and Wilcock J. D. (1975) A computer assisted classification of British neolithic axes and a comparison with some Mexican and Guatemalan axes. *Science and Archaeology 16*, 11–29.

Chappell S. (1987) *Stone Axe Morphology and Distribution in Neolithic Britain*. BAR British Series 177. Oxford: British Archaeological Reports.

Christenson A. L. and Read D. W. (1977) Numerical taxonomy, R-mode factor analysis and archaeological classification. *American Antiquity 42*, 163–79.

Ciolek-Torrello R. (1984) An alternative model of room function from Grasshopper Pueblo, Arizona, in Hietala H. (ed.), *Intrasite Spatial Analysis in Archaeology*, 127–53. Cambridge: Cambridge University Press.

Ciolek-Torrello R. (1985) A typology of room function at Grasshopper Pueblo, Arizona. *Journal of Field Archaeology 12*, 41–63.

Diaz-Andreu M. and Fernandez-Miranda M. (1991) Cuevas sepulcrales pretalayoticas de Mallorca: una esayo de clasificacion y analysis, in Waldren W. H., Ensenyat J. A. and Kennard R. C. (eds.) *2nd Deya International Conference on Prehistory: Vol. II Archaeological Technology and Theory*, 79–114. BAR International Series 574. Oxford: Tempus Reparatum.

Evans J. G. and Smith I. F. with Darvill T., Grigson C. and Pitts M. W., (1983) Excavations at Cherhill, North Wiltshire, 1967. *Proceedings of the Prehistoric Society 49*, 43–117.

Freedman L. (1985) Human skeletal remains from Mossgiel, NSW. *Archaeology in Oceania 20*, 21–31.

Freij H. (1988) Some attempts to relate ancient land use to soil properties by means of statistics, in Madsen T. (ed.), *Multivariate Archaeology*, 139–44. Aarhus: Aarhus University Press.

Gaillard C., Raju D. R., Misra V. N. and Rajaguru S. N. (1986) Handaxe assemblages from the Didirana region, Thar Desert, India: a metrical analysis. *Proceedings of the Prehistoric Society* 52, 189–214.

Guenode A. and Hensard A. (1983) Typologie d'amphores Romains par une methode logique de classification. *Computers and the Humanities* 17, 185–98.

Guralnick E. (1976) The proportions of some archaic Greek sculptured figures: a computer analysis. *Computers and the Humanities* 10, 153–69.

Habgood P. J. (1986) The origins of the Australians: a multivariate approach. *Archaeology in Oceania* 21, 130–7.

Hodson F. R. (1970) Cluster analysis and archaeology: some new developments and applications. *World Archaeology* 1, 299–320.

Hodson F. R. (1971) Numerical typology and prehistoric archaeology, in Hodson F. R., Kendall D. G. and Tautu P. (eds.), *Mathematics in the Archaeological and Historical Sciences*, 30–45. Edinburgh: Edinburgh University Press.

Hodson F. R., Sneath P. H. A. and Doran J. E. (1966) Some experiments in the numerical analysis of archaeological data. *Biometrika* 53, 311–24.

Impey O. R. and Pollard A. M. (1985) A multivariate metrical study of ceramics made by three potters. *Oxford Journal of Archaeology* 4, 157–64.

Kaplan M. F. (1980) The origin and distribution of Tell El Yahudiyeh ware. *Studies in Mediterranean Archaeology* 62, 1–336.

Kaplan M. F., Harbottle G. and Sayre E. V. (1982) Multi-disciplinary analysis of Tell El Yahudiyeh ware. *Archaeometry* 24, 127–42.

Kaplan M. F., Harbottle G. and Sayre E. V. (1984) Tell El Yahudiyeh ware: a re-evaluation, in Rice P. M. (ed.), *Pots and Potters*, 227–41. Los Angeles: Institute of Archaeology, UCLA.

Larsen C. U. (1988) A morphological study of biconical late Bronze Age urns, in Madsen T. (ed.), *Multivariate Archaeology*, 133–8. Aarhus: Aarhus University Press.

Law G. (1984) Shell points of Maori two-piece fish hooks from northern New Zealand. *New Zealand Journal of Archaeology* 6, 5–21.

Magne M. P. R. and Klassen M. A. (1991) A multivariate study of rock art anthropomorphs at Writing-on Stone, Southern Alberta. *American Antiquity* 56, 389–418.

Perry C. M. and Davidson D. A. (1987) A spatial analysis of chambered cairns on the Island of Arran, Scotland. *Geoarchaeology* 2, 121–30.

Phillip G. and Ottoway B. S. (1983) Mixed data cluster analysis: an illustration using Cypriot hooked-tang weapons. *Archaeometry* 25, 119–33.

Redman C. L. (1978) Multivariate artifact analysis: a basis for multidimensional interpretation, in Redman C. L., Burman M. J., Curtin E. V., Langhorne W. T. Jr., Versaggi N. M. and Wanser J. C. (eds.), *Social Archaeology: Beyond Subsistence and Dating*, 159–92. New York: Academic Press.

Richards L. C. and Telfer P. J. (1979) The use of dental characters in the assessment of genetic distance in Australia. *Archaeology in Oceania* 14, 184–94.

Shennan S. J. and Wilcock J. D. (1975) Shape and style variation in central German bell beakers: a computer assisted study. *Science and Archaeology* 15, 17–31.

Smith C. A. (1974) A morphological analysis of late prehistoric and Roman-British settlements in north west Wales. *Proceedings of the Prehistoric Society* 40, 157–69.

Stahl A. M. (1978) A numerical taxonomy of Merovingian coins. *Computers and the Humanities* 12, 201–14.

Tyldesley J. A., Johnson J. S. and Snape S. R. (1985) 'Shape' in archaeological artifacts: two case studies using a new analytical method. *Oxford Journal of Archaeology* 4, 19–30.

Verjux C. and Rousseau D.-D. (1986) La retouche Quina: une mise au point. *Bulletin de la Société Préhistorique Française 83*, 404–15.

Voruz J.-L. (1984) *Outillages Osseux et Dynamisme Industriel dans le Néolithique Jurassien*. Lausanne: Cahiers d'Archaéologie Romande.

Whallon R. (1972) A new approach to pottery typology. *American Antiquity 37*, 13–33.

Whallon R. (1990) Defining structure in clustering dendrograms with multi-level clustering, in Voorrips A. and Ottoway B. (eds.), *New Tools from Mathematical Archaeology*, 1–13. Warsaw: Scientific Information Centre of the Polish Academy of Sciences.

Williams I., Limp W. F. and Bruier F. L. (1990) Using geographic information systems and exploratory data analysis for archaeological site classification and analysis, in Allen K. M. S., Green S. W. and Zubrow E. B. W. (eds.), *Interpreting Space: GIS and Archaeology*, 239–73. London: Taylor and Francis.

(c) Assemblage comparison

Bard K. A. (1989) The evolution of social complexity in Predynastic Egypt: an analysis of the Naqada cemeteries. *Journal of Mediterranean Archaeology 2*, 223–48.

Bayliss A. and Orton C. (1988) Seriation with parallel series – an historical example, in Rahtz S. P. Q. (ed.), *Computer and Quantitative Methods in Archaeology 1988*, 161–76. BAR International Series 446(i). Oxford: British Archaeological Reports.

Braun D. P. (1981) A critique of some recent North American mortuary studies. *American Antiquity 46*, 398–416.

Burton J. (1980) Making sense of waste flakes: new methods for investigating the technology and economics behind chipped stone assemblages. *Journal of Archaeological Science 7*, 131–48.

Callow P. (1986c) A comparison of British and French Acheulian bifaces, in Collcutt S. N. (ed.), *The Palaeolithic of Britain and Its Nearest Neighbours*, 3–7. Sheffield: Dept. of Archaeology and Prehistory, University of Sheffield.

Close A. E. (1977) *The Identification of Style in Lithic Artefacts from North-East Africa*. Memoires de l'Institut d'Egypte.

Close A. E. (1979) The identification of style in lithic artefacts. *World Archaeology 10*, 223–37.

Cowgill G. L., Altschul J. H. and Sload R. S. (1984) Spatial analysis of Teotihuacan: a Mesoamerican metropolis, in Hietala H. (ed.), *Intrasite Spatial Analysis in Archaeology*, 154–95. Cambridge: Cambridge University Press.

Croes D. R. (1989) Prehistoric ethnicity on the Northwest coast of North America: an evaluation of style in basketry and lithics. *Journal of Anthropological Archaeology 8*, 101–30.

Fisher A. R. (1985) Winklebury Hillfort: A study of artefact distributions from subsoil features. *Proceedings of the Prehistoric Society 51*, 167–80.

Frankel D. (1975) Inter-site relationships in the middle bronze age of Cyprus. *World Archaeology 6*, 190–208.

Gebhard R., Grosse G. and Wagner U. (1991) Waste in open settlements: the evidence from the Celtic oppidum of Manching, in Pernicka E. and Wagner G. A. (eds.), *Archaeometry '90*, 385–94. Basel: Birkhäuser Verlag.

Gillis C. (1990) Minoan conical cups: form, function and significance. *Studies in Mediterranean Archaeology 86*, 1–200.

Gowlett J. A. J. (1986) Culture and conceptualisation: the Oldowan-Acheulian gradient, in Bailey G. N. and Callow P. (eds.), *Stone Age Prehistory*, 243–60. Cambridge: Cambridge University Press.

Gowlett J. A. J. (1988) A case of developed Oldowan in the Acheulean. *World Archaeology 20*, 13–26.

Hemingway M. F. (1980) *The Initial Magdalenian in France.* BAR International Series 90. Oxford: British Archaeological Reports.

Hodson F. R. (1969) Searching for structure within multivariate archaeological data. *World Archaeology 1*, 90–105.

Hodson F. R. (1977) Quantifying Hallstatt: some initial results. *American Antiquity 42*, 394–412.

Hodson F. R. (1990) *Halstatt: The Ramsaeur Graves.* Bonn: Dr. Rudolf Habelt GmbH.

Hutcheson J. C. C. and Callow P. (1986) The flint debitage and cores, in Callow P. and Cornford J. M. (eds.), *La Cotte de St. Brelade 1961–1978*, 231–49. Norwich: Geo Books.

Hynes R. A. and Chase A. K. (1982) Plants, sites and domiculture: aboriginal influence upon plant communities in Cape York Peninsula. *Archaeology in Oceania 17*, 38–50.

Jones G. (1987) A statistical approach to the archaeological identification of crop processing. *Journal of Archaeological Science 14*, 311–23.

Jones R. (1980) Computers and cemeteries: opportunities and limitations, in Rahtz P., Dickinson T. and Watts L. (eds.), *Anglo-Saxon Cemeteries 1979*, 179–95. BAR British Series 82. Oxford: British Archaeological Reports.

King T. F. (1978) Don't that beat the band? Non-egalitarian political organization in prehistoric central California, in Redman C. L., Burman M. J., Curtin E. V., Langhorne W. T. Jr., Versaggi N. M. and Wanser J. C. (eds.), *Social Archaeology: Beyond Subsistence and Dating*, 225–48. New York: Academic Press.

Lock G. R. (1983) Computer assisted seriation of the pits at Danebury Hillfort. *Science and Archaeology 25*, 3–8.

McManamon F. P. (1982) Prehistoric land use on outer Cape Cod. *Journal of Field Archaeology 9*, 1–20.

McClellan T. L. (1979) Chronology of the 'Philistine' burials at Tell el-Far'ah (South). *Journal of Field Archaeology 6*, 57–73.

Millett M. (1979) An approach to the functional interpretation of pottery, in Millett M. (ed.), *Pottery and the Archaeologist*, 35–48. London: Institute of Archaeology.

Myers A. (1987) All shot to pieces? Inter-assemblage variability, lithic analysis and Mesolithic assemblage 'types'; some preliminary observations, in Brown A. G. and Edmonds M. R. (eds.), *Lithic Analysis and Later British Prehistory*, 137–53. BAR British Series 162. Oxford: British Archaeological Reports.

O'Shea J. M. (1984) *Mortuary Variability: An Archaeological Investigation.* New York: Academic Press.

O'Shea J. M. (1985) Cluster analysis and mortuary patterning: an experimental assessment. *PACT 11*, 91–110.

O'Shea J. M. and Zvelebil M. (1984) Oleneostroviski Mogilnik: reconstructing the social and economic organisation of prehistoric foragers in Northern Russia. *Journal of Anthropological Archaeology 3*, 1–40.

Palumbo G. (1987) 'Egalitarian' or 'Stratified' society? Some notes on mortuary practices and social structure at Jericho in EBiv. *Bulletin of the American Schools of Oriental Research 267*, 43–59.

Pearson R., Lee J.-W. and Koh W. (1989) Social ranking in the Kingdom of Old Silla, Korea: analysis of burials. *Journal of Anthropological Archaeology 8*, 1–50.

Peebles C. S. (1972) Monothetic-divisive analysis of the Moundsville burials: an initial report. *Newsletter of Computer Archaeology 8*, 1–13.

Perry D. W., Buckland P. C. and Snaesdothir M. (1985) The application of numerical techniques to insect assemblages from the site of Storaborg, Iceland. *Journal of Archaeological Science 12*, 335–45.

Pitts M. W. (1978a) Towards an understanding of flint industries in post-glacial England. *Institute of Archaeology Bulletin 15*, 179–97.

Pitts M. W. (1978b) On the shape of waste flakes as an index of technological change in lithic industries. *Journal of Archaeological Science 5*, 17–37.

Pitts M. W. and Jacobi R. M. (1979) Some aspects of change in flaked site industries of the mesolithic and neolithic in southern Britain. *Journal of Archaeological Science 6*, 163–77.

Pyszczyk H. W. (1989) Consumption and ethnicity: an example from the fur trade in Western Canada. *Journal of Anthropological Archaeology 8*, 213–49.

Richards J. (1987) *The Significance of Form and Decoration of Anglo-Saxon Cremation Urns*. BAR British Series 166. Oxford: British Archaeological Reports.

Rothschild N. A. (1979) Mortuary behaviour and social organization at Indian Knoll and Dickson Mounds. *American Antiquity 44*, 658–75.

Sognnes K. (1987) Rock art and settlement pattern in the Bronze Age. Example from Stordal, Trondeleg, Norway. *Norwegian Archaeological Review 20*, 110–19.

Tainter J. A. (1975) Social inference and mortuary practices: an experiment in numerical classification. *World Archaeology 7*, 1–16.

Wood J. J. (1978) Optimal location in settlement space: a model for describing location strategies. *American Antiquity 43*, 258–70.

(d) Spatial analysis

Ammerman A. J., Kintigh K. W. and Simek J. F. (1987) Recent developments in the application of the k-means approach to spatial analysis, in Sieveking G. De G. and Newcomer M. (eds.), *The Human Uses of Flint and Chert*, 210–16. Cambridge: Cambridge University Press.

Blankholm H. P. (1991) *Intrasite Spatial Analysis in Theory and Practice*. Aarhus: Aarhus University Press.

Cribb R. and Minnegal M. (1989) Spatial analysis on a dugong consumption site at Princess Charlotte Bay, North Queensland. *Archaeology in Oceania 24*, 1–12.

Ford S. (1987) Flint scatters and prehistoric settlement patterns in South Oxfordshire and East Berkshire, in Brown A. G. and Edmonds M. R. (eds.), *Lithic Analysis and Later British Prehistory*, 101–35. BAR British Series 162. Oxford: British Archaeological Reports.

Gregg S. A., Kintigh K. W. and Whallon R. (1991) Linking ethno-archaeological interpretation and archaeological data: the sensitivity of spatial analytical methods to post-depositional disturbance, in Kroll E. and Price T. D. (eds.), *The Interpretation of Archaeological Spatial Patterning*, 149–96. New York: Plenum Press.

Kintigh K. W. and Ammerman A. J. (1982) Heuristic approaches to spatial analysis in archaeology. *American Antiquity 47*, 31–63.

Kintigh K. W. (1990) Intra-site spatial analysis in archaeology, in Voorrips A. (ed.), *Mathematics and Information Science in Archaeology: A Flexible Framework*, 165–200. Bonn: Holos-Verlag.

Koetje T. A. (1987) *Spatial Patterns in Magdalenian Open Air Sites from the Isle Valley, Southwestern France*. BAR International Series 346. Oxford: British Archaeological Reports.

Koetje T. A. (1991) Simulated archaeological levels and the analysis of Le Flageolet II, The Dordogne, France. *Journal of Field Archaeology 18*, 187–98.

Ridings R. and Sampson G. G. (1990) Theres no percentage in it: intersite spatial analysis of Bushman (San) pottery decorations. *American Antiquity 55*, 766–80.

Bibliography

Rigaud J.-P. and Simek J. F. (1991) Interpreting spatial patterns at the Grotte XV: a multiple-method approach, in Kroll E. and Price T. D. (eds.), *The Interpretation of Archaeological Spatial Patterning*, 197–220. New York: Plenum Press.

Siegel P. E. and Roe P. G. (1986) Shipibo ethno-archaeology: site formation processes and archaeological interpretation. *World Archaeology 18*, 96–115.

Simek J. M. (1984a) *A K-means Approach to the Analysis of Spatial Structure in Upper Palaeolithic Habitation Sites*. BAR International Series. Oxford: British Archaeological Reports.

Simek J. M. (1984b) Integrating pattern and context in spatial archaeology. *Journal of Archaeological Science 11*, 405–20.

Simek J. M. (1987) Spatial order and behavioural change in the French Paleolithic. *Antiquity 61*, 25–40.

Simek J. M. (1989) Structure and diversity in intra-site spatial analysis, in Leonard R. D. and Jones G. T. (eds.), *Quantifying Diversity in Archaeology*, 59–68. Cambridge: Cambridge University Press.

Simek J. M. and Larick R. R. (1983) The recognition of multiple spatial patterns: a case study from the French upper palaeolithic. *Journal of Archaeological Science 10*, 165–80.

Simek J. F., Ammerman A. J. and Kintigh K. W. (1985) Explorations in heuristic spatial analysis: analyzing the structure of material accumulation over space. *PACT 11*, 229–47.

Van Waarden C. (1989) The granaries of Vumba: structural interpretation of a Khami Period commoner site. *Journal of Anthropological Archaeology 8*, 131–57.

Whallon R. (1984) Unconstrained clustering for the analysis of spatial distributions in archaeology, in Hietala H. (ed.), *Intrasite Spatial Analysis in Archaeology*, 242–77. Cambridge: Cambridge University Press.

References – Discriminant Analysis in Archaeology

(a) Analysis of chemical compositions of artefacts

Adan-Bayewitz D. and Perlman I. (1990) The local trade of Sepphoris in the Roman period. *Israel Exploration Journal 2–3*, 153–72.

Adan-Bayewitz D. and Wieder M. (1992) Ceramics from Roman Galilee: a comparison of several techniques for fabric characterization. *Journal of Field Archaeology 19*, 189–205.

Arnold D. E., Neff H. and Bishop R. L. (1991) Compositional analysis and 'sources' of pottery: An ethnoarchaeological approach. *American Anthropologist 93*, 70–90.

Attas M., Yaffe L. and Fossey J. M. (1977) Neutron activation analysis of early bronze age pottery from Lake Vouliageni, Perakhora, Central Greece. *Archaeometry 19*, 33–43.

Bimson M., La Niece S. and Leese M. N. (1982) The characterisation of mounted garnets. *Archaeometry 24*, 51–8.

Birgül O., Dikšić M. and Yaffe L. (1979) X-ray fluorescence analysis of Turkish clays and pottery. *Archaeometry 21*, 203–18.

Burmester A. (1983b) Far eastern lacquers: classification by pyrolisis mass spectrometry. *Archaeometry 25*, 45–58.

Capannesi G., Seccaroni C., Sedda A. F., Majerini V. and Musco S. (1991) Classification of fifteenth to nineteenth century mortars from Gabii using instrumental neutron activation analysis. *Archaeometry 33*, 255–66.

Carter G. F. and Frurip D. J. (1985) Discriminant analysis of the chemical compositions and physical measurements of 245 Augustan quadrantes. *Archaeometry 27*, 117–26.

Christensen L. H., Conradsen K. and Neilsen S. (1985) The Selbjerg project: investigation of neolithic pottery by means of neutron activation analysis. *ISKOS* 5, 401–10.

Craddock P. T., Cowell M. R., Leese M. N. and Hughes M. J. (1983) The trace element composition of polished flint axes as an indicator of source. *Archaeometry* 25, 135–63.

Culbert T. P. and Schwalbe L. A. (1987) X-ray fluorescence survey of Tikal ceramics. *Journal of Archaeological Science 14*, 635–57.

D'Altroy T. N. D. and Bishop R. L. (1990) The provincial organization of Inka ceramic production. *American Antiquity 55*, 120–38.

De Bruin M., Kotthoven P. J. M., Bakels C. C. and Groen F. C. A. (1972) The use of non-destructive activation analysis and pattern recognition in the study of flint artefacts. *Archaeometry 14*, 55–63.

Djingova R. and Kuleff I. (1992) An archaeometric study of Medieval glass from the first Bulgarian capital, Pliska (ninth to tenth century AD). *Archaeometry 34*, 53–61.

Faust R. D. Jr., Ambler J. R. and Turner L. D. (1989) Trace element analysis of Pueblo II Kayenta Anasazi sherds, in Allen R. O. (ed.), *Archaeological Chemistry IV*, 125–43. Washington, D.C.: American Chemical Society.

Fillieres D., Harbottle G. and Sayre E. V. (1983) Neutron activation study of figurines, pottery, and workshop materials from the Athenian Agora, Greece. *Journal of Field Archaeology 10*, 55–69.

Francalacci P. (1989) Dietary reconstruction at Arene Candide Cave (Liguria, Italy) by means of trace element analysis. *Journal of Archaeological Science 16*, 109–24.

Gale N. H. (1989) Lead isotope analyses applied to provenance studies – a brief review, in Maniatis Y. (ed.), *Archaeometry*, 469–502. Amsterdam: Elsevier.

Gale N. H. (1991) Metals and metallurgy in the Chalcolithic period. *Bulletin of the American Schools of Oriental Research 282–283*, 37–61.

Gale N. H. and Stos-Gale Z. A. (1989) Bronze Age archaeometallurgy of the Mediterranean: the impact of lead isotope studies, in Allen R. O. (ed.), *Archaeological Chemistry IV*, 159–98. Washington, D.C.: American Chemical Society.

Germann K., Gruben G., Knoll H., Vallio V. and Winkler F. J. (1988) Provenance characteristics of cycladic (paros and Naxos) marbles – a multivariate geologic approach, in Herz N. and Waelkens M. (eds.), *Classical Marbles: Geochemistry, Technology, Trade*, 251–62. Dordrecht: Kluwer Academic Publishers.

Glascock M. D., Elam J. M. and Cobean R. H. (1988) Differentiation of obsidian sources in Mesoamerica, in Farquhar R. M., Hancock R. G. V. and Pavlish L. A. (eds.), *Proceedings of the 26th International Archaeometry Symposium*, 245–51. Toronto: University of Toronto.

Hansen B. A., Sorensen M. A., Heydon K., Mejdahl V. and Conradsen K. (1979) Provenance study of medieval, decorated floor-tiles carried out by means of neutron activation analysis. *Archaeo-Physika 10*, 119–40.

Hart F. A., Storey J. M. V., Adams S. J., Symonds R. P. and Walsh J. N. (1987) An analytical study using inductively coupled plasma (ICP) spectrometry, of Samian and colour-coated wares from the Roman town of Colchester together with related continental Samian wares. *Journal of Archaeological Science 14*, 577–98.

Hedges R. E. M. and Salter C. J. (1979) Source determination of iron currency bars through analysis of the slag inclusions. *Archaeometry 21*, 161–75.

Herz N. and Doumas C. (1991) Marble sources in the Aegean early Bronze Age, in Pernicka E. and Wagner G. A. (eds.), *Archaeometry '90*, 425–34. Basel: Birkhauser Verlag.

Henderson J. and Warren S. E. (1983) Analysis of prehistoric lead glass, in Aspinall

A. and Warren S. E. (eds.), *The Proceedings of the 22nd Symposium on Archaeometry*, 168–80. Bradford: University of Bradford.

Heyworth M. P., Hunter J. R., Warren S. E. and Walsh J. N. (1988) The analysis of archaeological materials using inductively coupled plasma spectroscopy, in Slater E. A. and Tate J. O. (eds.), *Science and Archaeology – Glasgow 1988*, 27–40. BAR British Series 196(i). Oxford: British Archaeological Reports.

Heyworth M. P., Hunter J. R., Warren S. E. and Walsh J. N. (1989) The role of inductively coupled plasma spectrometry in glass provenance studies, in Maniatis Y. (ed.), *Archaeometry*, 661–69. Amsterdam: Elsevier.

Hughes M. J. (1991) Provenance studies of Spanish Medieval tin-glazed pottery by neutron activation analysis, in Budd P., Chapman B., Jackson C., Janaway R. and Ottaway B. (eds.), *Archaeological Sciences 1989*, 54–68. Oxford: Oxbow Books.

Hughes M. J. and Hall J. A. (1979) X-ray fluorescence analysis of late Roman and Sassanian silver plate. *Journal of Archaeological Science* 6, 321–41.

Hughes M. J., Leese M. N. and Bailey D. M. (1983) Neutron activation analysis of pottery lamps from Ephesus dating from the archaic to the late Roman period, in Aspinall A. and Warren S. E. (eds.), *The Proceedings of the 22nd Symposium on Archaeometry*, 368–76. Bradford: University of Bradford.

Hughes M. J. and Vince A. G. (1986) Neutron activation analysis and petrology of Hispano-Maresque pottery, in Olin J. S. and Blackman M. J. (eds.), *Proceedings of the 24th International Archaeometry Symposium*, 353–67. Washington, D.C.: Smithsonian Institution Press.

King R. H., Rupp D. W. and Sorenson L. W. (1986) A multivariate analysis of pottery from southwestern Cyprus using neutron activation analysis data. *Journal of Archaeological Science* 13, 361–74.

Kohl P. L., Harbottle G. and Sayre E. V. (1979) Physical and chemical analyses of soft stone vessels from Southwest Asia. *Archaeometry* 21, 131–59.

Kuleff I., Djingova R. and Djingov G. (1989) Provenance study of sgraffito ceramics (XIII–XIVc) from Shumen, Varna and Tcherven (North-Eastern Bulgaria), in Maniatis Y. (ed.), *Archaeometry*, 533–42. Amsterdam: Elsevier.

Leach F. and Manly B. (1982) Minimum Mahalanobis distance functions and lithic source characterisation by multi-element analysis. *New Zealand Journal of Archaeology* 4, 77–109.

Leese M. (1983) Statistical methodology in numismatic studies. *Journal of Archaeological Science* 10, 29–33.

Leese M. (1988) Statistical treatment of stable isotope data, in Herz N. and Waelkens M. (eds.), *Classical Marbles: Geochemistry, Technology, Trade*, 347–54. Dordrecht: Kluwer Academic Publishers.

Leese M. N., Hughes M. J. and Cherry J. (1986) A scientific study of N. Midlands medieval tile production. *Oxford Journal of Archaeology* 5, 355–70.

Longworth G. and Warren S. E. (1979) The application of Mossbauer spectroscopy to the characterization of Western Mediterranean obsidian. *Journal of Archaeological Science* 6, 179–93.

Luedtke B. E. (1979) The identification of sources of chert artifacts. *American Antiquity* 44, 744–57.

Mello E., Meloni S., Monna D. and Oddone M. (1988) A computer based pattern recognition approach to provenance study of Mediterranean marbles through trace element analysis, in Herz N. and Waelkens M. (eds.), *Classical Marbles: Geochemistry, Technology, Trade*, 251–62. Dordrecht: Kluwer Academic Publishers.

Mello E., Monna D. and Oddoe M. (1988) Discriminating sources of Mediterranean marble: a pattern recognition approach. *Archaeometry* 30, 102–8.

Mommsen H., Beier T., Diehl U. and Podzuweit Ch. (1992) Provenance determination of Mycenaean sherds found in Tell el Amarna by neutron activation analysis. *Journal of Archaeological Science 19*, 295–302.

Needham S. P., Leese M. N., Hooke D. R. and Hughes M. J. (1989) Developments in the early Bronze Age metallurgy of southern Britain. *World Archaeology 20*, 383–402.

Neff H. and Bishop R. L. (1988) Plumbate origins and development. *American Antiquity 53*, 502–22.

Neff H., Bishop R. L. and Arnold D. E. (1988) Reconstructing ceramic production from ceramic compositional data: an example from Guatemala. *Journal of Field Archaeology 15*, 339–48.

Newton G. W. A., Robinson V. J., Oladipo M., Chandratillake M. R. and Whitbread I. K. (1988) in Slater E. A. and Tate J. O. (eds.), *Science and Archaeology – Glasgow 1988*, 59–82. BAR British Series 196(i). Oxford: British Archaeological Reports.

Northover P. (1982) The metallurgy of the Wilburton hoards. *Oxford Journal of Archaeology 1*, 69–109.

Pernicka E., Begemann F., Schmitt-Strecker S. and Grimanis A. P. (1990) On the composition and provenance of metal artefacts from Poliochni on Lemnos. *Oxford Journal of Archaeology 9*, 263–98.

Picon M., Carre C., Cordoliani M. L., Vichy M., Hernandez J. A. and Migiard J. L. (1975) Composition of the La Graufesenque, Banassac and Montans terra sigillata. *Archaeometry 17*, 191–9.

Pollard A. M. (1986) Data analysis, in Jones R. R. (ed.), *Greek and Cypriot Pottery: A Review of Scientific Studies*, 56–83. Athens: British School at Athens.

Pollard A. M. and Hatcher H. (1986) The chemical analysis of oriental ceramic body compositions: Part 2 – greenwares. *Journal of Archaeological Science 13*, 261–87.

Salter C. J. (1982) The relevance of chemical provenance studies to Celtic ironwork in Britain. *Institute of Archaeology Bulletin 19*, 73–81.

Sanderson D. C. W, Hunter J. R. and Warren S. E. (1984) Energy dispersive X-ray fluorescence analysis of Ist Millenium AD glass from Britain. *Journal of Archaeological Science 11*, 53–69.

Sieveking G. De G., Bush P., Ferguson J., Craddock P. T., Hughes M. J. and Cowell M. R. (1972) Prehistoric flint mines and their identification as sources of raw material. *Archaeometry 14*, 151–76.

Spoerry P. (1988) Problem-specific provenance – a case study from medieval Wessex, in Slater E. A. and Tate J. O. (eds.), *Science and Archaeology – Glasgow 1988*, 107–19. BAR British Series 196(i). Oxford: British Archaeological Reports.

Spoerry P. (1990) Ceramic production in medieval Dorset and the surrounding region. *Medieval Ceramics 14*, 3–17.

Stevenson C. M. and McCurry M. O. (1990) Chemical characterization and hydration rate development for New Mexican obsidian sources. *Geo-Archaeology 5*, 149–70.

Storey J. M. V., Symonds R. P., Hart F. A., Smith D. M. and Walsh J. N. (1989) A chemical investigation of 'Colchester' samian by means of inductively coupled plasma emission spectrometry. *Journal of Roman Pottery Studies 2*, 33–43.

Stos-Gale Z. A. (1989) Lead isotope studies of metals and the metal trade in the Bronze Age Mediterranean, in Henderson J. (ed.), *Scientific Analysis in Archaeology*, 274–301. Oxford: Oxford University Committee for Archaeology.

Tobey M. H., Nielsen E. O. and Rowe M. W. (1986) Elemental analysis of Etruscan ceramics from Murlo, Italy, in Olin J. S. and Blackman M. J. (eds.), *Proceedings of the 24th International Archaeometry Symposium*, 115–27. Washington, D.C.: Smithsonian Institution Press.

Tobey M. H., Shafer H. J. and Rowe M. W. (1986) Trace element investigations of Mayan chert from Belize, in Olin J. S. and Blackman M. J. (eds.), *Proceedings of the 24th International Archaeometry Symposium*, 353–67. Washington, D.C.: Smithsonian Institution Press.

Vitali V. and Franklin U. M. (1986a) New approaches to the characterization and classification of ceramics on the basis of their elemental composition. *Journal of Archaeological Science 13*, 161–70.

Vitali V., Simmons J. W., Henrickson E. F., Levine L. D. and Hancock R. G. V. (1987) A hierarchic taxonomic procedure for provenance determination: a case study of Chalcolithic ceramics from the Central Zagros. *Journal of Archaeological Science 14*, 423–35.

Walthall J. A., Stow S. H. and Karson M. J. (1980) Copena Galena: source identification and analysis. *American Antiquity 45*, 21–42.

Ward G. K. (1974) A systematic approach to the definition of sources of raw material. *Archaeometry 16*, 41–53.

White S. R., Warren S. E. and Jones R. E. (1983) The provenance of Bronze Age pottery from Thessaly in Eastern Greece, in Aspinall A. and Warren S. E. (eds.), *The Proceedings of the 22nd Symposium on Archaeometry*, 323–32. Bradford: University of Bradford.

Williams-Thorpe O., Warren S. E. and Nandris J. O. (1984) The distribution and provenance of archaeological obsidian in Central and Eastern Europe. *Journal of Archaeological Science 11*, 183–212.

(b) Typological/morphological

Aldenderfer M. S. (1982) Methods of cluster validation for archaeology. *World Archaeology 14*, 61–72.

Bain P. (1985) Geographic and temporal variation in Maori rock drawings in two regions of southern New Zealand. *New Zealand Journal of Archaeology 7*, 39–59.

Bartel B. (1979) A discriminant analysis of Harappan civilization human populations. *Journal of Archaeological Science 6*, 49–61.

Bartel B. (1981) Cultural associations and mechanisms of change in anthropomorphic figurines during the Neolithic in the eastern Mediterranean basin. *World Archaeology 13*, 73–86.

Baxter M. J. (1988) The morphology and evolution of post Medieval wine bottles revisited. *Science and Archaeology 30*, 10–14.

Beneke N. (1987) Studies on early dog remains from Northern Europe. *Journal of Archaeological Science 14*, 31–49.

Beneke N. (1990) The Krabbe collection of Icelandic horses and its significance for archaeozoological research. *Journal of Archaeological Science 17*, 161–85.

Benfer R. A. and Benfer A. N. (1981) Automatic classification of inspectional categories: multivariate theories of archaeological data. *American Antiquity 46*, 381–96.

Boyd C. C. Jr. and Boyd D. C. (1991) A multi-dimensional investigation of biocultural relationships amon three late prehistoric societies in Tennessee. *American Antiquity 56*, 75–88.

Brothwell D. (1972) Palaeodemography and earlier British populations. *World Archaeology 4*, 75–89.

Brothwell D. and Krzanowski W. (1974) Evidence of biological differences between early British populations from neolithic to medieval times, as revealed by eleven commonly available cranial measurements. *Journal of Archaeological Science 1*, 249–69.

Brown P. (1981) Sex determination of Aboriginal crania from the Murray River

Valley: A reassessment of the Larnod and Freedman technique. *Archaeology in Oceania 16*, 53–63.

Brown P. (1987) Pleistocene homogeneity and Holocene size reduction: The Australian human skeletal evidence. *Archaeology in Oceania 22*, 41–67.

Browne S. (1983) Investigations into the evidence for postcranial variation in Bos primigenius (Bojanus) in England and the problem of its differentiation from Bison priscus (Bojanus). *Institute of Archaeology Bulletin 20*, 1–42.

Burton J. (1980) Making sense of waste flakes: new methods for investigating the technology and economics behind chipped stone assemblages. *Journal of Archaeological Science 7*, 131–48.

Butzer K. W. (1981) Cave sediments, Upper Pleistocene stratigraphy and Mousterian facies in Cantabrian Spain. *Journal of Archaeological Science 8*, 133–83.

Collier S. (1989) The influence of economic behaviour and environment upon robusticity of the post-cranial skeleton: A comparison of Australian Aborigines and other populations. *Archaeology in Oceania 24*, 17–30.

De Guio A. (1985) Towards an analytical-mathematical approach to locational strategies: some preliminary steps from fieldwork in the 'Vicentino' (Veneto), in Malone C. and Stoddart S. (eds.), *Papers in Italian Archaeology IV: The Human Landscape*, 153–83. BAR International Series 243(i). Oxford: British Archaeological Reports.

Freedman L. and Lofgren M. (1981) Odontometrics of Western Australian Aborigines. *Archaeology in Oceania 16*, 87–93.

Gowlett J. A. J. (1988) A case of developed Oldowan in the Acheulean? *World Archaeology 20*, 13–26.

Graham I. M. (1970) Discrimination of British lower and middle Palaeolithic handaxe groups using canonical variates. *World Archaeology 1*, 321–37.

Grebinger P. and Adam D. P. (1974) Hard times? Classic period Hohokam cultural development in the Tucson Basin, Arizona. *World Archaeology 6*, 226–41.

Guenode A. and Hensard A. (1983) Typologie d'amphores Romains par une methode logique de classification. *Computers and the Humanities 17*, 185–98.

Higham C. F. W., Kijngam A. and Manly B. F. J. (1980) An analysis of prehistoric canid remains from Thailand. *Journal of Archaeological Science 7*, 149–65.

Higham C. F. W., Kijngam A., Manly B. F. J. and Moore S. J. E. (1981) The Bovid third phalanx and prehistoric ploughing. *Journal of Archaeological Science 8*, 353–65.

Hills M. and Brothwell D. R. (1974) The use of large numbers of variables to measure the shape of a restricted area of bone. *Journal of Archaeological Science 1*, 135–50.

Kirch P. V. (1981) Lapitoid settlements of Futura and Alofi, Western Polynesia. *Archaeology in Oceania 16*, 127–43.

Knutsson K., Dahlquist B. and Knutsson H. (1988) Patterns of tool use: the microwear analysis of the quartz and flint assemblage from the Bjurselet site, Vasterbotten, Northern Sweden, in Beyries S. (ed.), *Industrie Lithiques: Aspectes Archaeologique*, 253–94. BAR International Series 411(i). Oxford: British Archaeological Reports.

Lukesh S. S. and Howe S. (1978) Protoapennine *vs.* Subapennine: mathematical distinction between two ceramic phases. *Journal of Field Archaeology 5*, 339–47.

Martin S. R. (1989) A reconsideration of aboriginal fishing strategies in the Northern Great Lakes region. *American Antiquity 54*, 594–604.

McKern T. W. and Munro E. H. (1959) A statistical technique for classifying human skeletal remains. *American Antiquity 24*, 375–82.

Meltzer D. J. (1981) A study of style and function in a class of tools. *Journal of Field Archaeology 8*, 313–26.

Montet-White A. and Johnson A. E. (1976) Kadar: a late Gravettian site in Northern Bosnia, Yugoslavia. *Journal of Field Archaeology 3*, 407–24.
Morey D. F. (1986) Studies on Amerindian dogs: taxonomic analysis of canid crania from the Northern Plains. *Journal of Archaeological Science 13*, 119–45.
Pearsall D. M. and Piperino D. R. (1990) Antiquity of maize cultivation in Ecuador: Summary and re-evaluation of the evidence. *American Antiquity 55*, 324–37.
Smith M. F. Jr. (1988) Function from whole vessel shape: a method and an application to Anasazi Black Mesa, Arizona. *American Anthropologist 90*, 912–23.
Solberg B. (1986) Automatic versus intuitive and impressionistic classification of Norweigan spearheads from AD 550–1100. *Norwegian Archaeological Review 19*, 77–89.
Walker D. N. and Frison G. C. (1982) Studies on Amerindian dogs 3: prehistoric wolf/dog hybrids from the Northwestern Plains. *Journal of Archaeological Science 9*, 125–72.
Wenban-Smith F. F. (1989) The use of canonical variates for determination of biface manufacturing technology at Boxgrove lower palaeolithic site and the behavioural implications of this technology. *Journal of Archaeological Science 16*, 17–26.
Whittaker J. C. (1987) Individual variation as an approach to economic organization: projectile points at Grasshopper Pueblo, Arizona. *Journal of Field Archaeology 14*, 465–79.
Wynn T. and Tierson F. (1990) Regional comparison of the shapes of later Acheulean handaxes. *American Anthropologist 92*, 73–84.

(c) Assemblage comparison

Bartosiewicz L. (1988) Biometrics at an early Medieval butchering site in Hungary, in Slater E. A. and Tate J. O. (eds.), *Science and Archaeology – Glasgow 1988*, 361–7. BAR British Series 196(ii). Oxford: British Archaeological Reports.
Bettinger R. L. (1979) Multivariate statistical analysis of a regional subsistence-settlement model for Owens Valley. *American Antiquity 44*, 455–70.
Bietti A. and Burani A. (1985) The late upper palaeolithic in continental Italy: old classifications, new data and new perspectives, in *Italian Archaeology IV: Part (ii) Prehistory*. British Archaeological Reports, International Series 243, 7–27.
Callow P. and Webb R. E. (1981) The application of multivariate statistical techniques to middle palaeolithic assemblages from southwestern France. *Revue d'Archeometrie 5*, 129–38.
Cowgill G. L., Altschul J. H. and Sload R. S. (1984) Spatial analysis of Teotihuacan: a Mesoamerican metropolis, in Hietala H. (ed.), *Intrasite Spatial Analysis in Archaeology*, Cambridge: Cambridge University Press.
Jones G. (1987) A statistical approach to the archaeological identification of crop processing. *Journal of Archaeological Science 14*, 311–23.
Pitts M. W. and Jacobi R. M. (1979) Some aspects of change in flaked site industries of the mesolithic and neolithic in southern Britain. *Journal of Archaeological Science 6*, 163–77.
Schlanger S. H. and Orcutt J. D. (1986) Site surface characteristics and functional inferences. *American Antiquity 51*, 296–312.
Stright M. J. (1986) Human occupation of the continental shelf during the late Pleistocene/early Holocene: Methods for site location. *Geo-Archaeology 1*, 347–64.
Tomber R. (1988) Multivariate statistics and assemblage comparison, in Ruggles C. L. N. and Rahtz S. P. Q. (eds.), *Computer and Quantitative Methods in Archaeology 1987*, 29–38. BAR International Series 393. Oxford: British Archaeological Reports.

Index

abundance data
  problems with correlation, 38
artificial abundance data, 121
  seriation of, 119 121–2
assemblage comparisons, 20–2
  cluster analysis and, 182–3
  discriminant analysis and, 215–16
  correspondence analysis and, 101–7, 135–7
  PCA and, 71, 97–9

Bayesian methods, 9, 174, 222–3
bi-modality 50, 210
binary data, 18
  cluster analysis of, 21, 149–52, 157, 172
  discriminant analysis of, 197
Binford L. R., 109
biplots, 57–8, 66–71, 81, 83, 96, 100, 102, 253–4
  correspondence analysis and, 100, 102, 250
Box's M statistic, 199, 209
Bronze Age cup dimensions data, 233–4
  PCA of, 90–4
  discriminant analysis of, 209–10

Carr, C., 220–1
centred data, 46
chemical composition of artefacts, 12–5
  cluster analysis and, 180–1
  discriminant analysis and, 214-15
  PCA and, 70, 90–6
  correspondence analysis and, 127–8, 138
chi-square, 98, 109, 112–13, 126
  see also correspondence analysis; distance and similarity
cluster analysis, 140–84
  association analysis, 151, 157, 173
  average linkage, 141–2, 146, 158
  comparison of, 165–7
  complete linkage, 142, 146, 158
  hierarchical agglomerative methods, 141–7
  k-means, 147–8, 159, 174
  monothetic divisive clustering, 150–2, 157, 160, 172–3
  number of clusters, 161–5

relocation, *see* k-means
single linkage, 141, 146, 158
spatial k-means clustering, *see* spatial data analysis
Ward's method, 142, 146, 158
see also dendrograms
compositional data (summing to 100%) 38, 62, 255–6
  discriminant analysis and, 216
  PCA and, 72–7, 98, 128
  unconstrained spatial clustering and, 171
computer packages
  BMDP, 3–4 200, 202, 208
  Bonn Seriation and Statistics Package, 120
  CLUSTAN, 3–4, 141, 159, 162–3, 174, 178, 181, 259
  DECORANA, 120
  IASTATS, 173
  MINITAB, 4, 28, 30, 34, 54, 70, 78, 175, 202–4, 207–8, 252–8
  MV-ARCH, 3–4, 9, 79, 135, 171, 262
  MV-NUTSHELLX, 261
  SPSS-X, 3–4, 84, 89, 94, 185–6, 188–9, 195, 199–200, 202, 204, 207–8, 259–60
  STATGRAPHICS, 45, 70, 252–4
constellation analysis, 79, 95, 97
co-phenetic correlation, 166, 184
Cormack, R. M., 155–6, 159
correlation, 33–8,
  coefficient of, 33
  among variables in PCA, 50–1, 53
  cluster analysis and, 159, 167–70
  Mahalanobis distance and, 81
  matrix, 56, 65, 67, 91
  negative bias in, 74, 76
correspondence analysis (CA), 100–39
  CA as PCA, 98, 107–9
  CA and chi-squared, 113–4
  CA and cluster analysis, 164, 166, 169
  detrended, 120, 137
  compositional data and, 77, 108
  multiple correspondence analysis, 123–33, 166
  supplementary points, 80
  see also horseshoe effect; inertia; seriation

Index

covariance
  definition of, 33
  matrix 65, 67, 197
Cowgill, G. L., 86
cross-validation, *see* discriminant analysis

Danish Neolithic pottery data, 128
  PCA of, 128–9
  multiple correspondence analysis of, 128–32
data sets and analyses, *see* artificial abundance data; Bronze Age cup dimensions data, Danish Neolithic pottery data; Germanic Iron Age grave assemblage data; Mesolithic assemblage data; Norwegian Early Stone Age assemblage data; Norwegian pit-house morphology data; Roman glass from Norway; Romano-British glass compositional data; Romano-British vessel glass from Winchester; tool types from Ksar Akil
dendrograms, 144, 146, 157
  chaining in, 146
  co-phenetic correlation and, 166
  inspection of, 162
discriminant analysis, 185–218
  canonical discriminant function, 189–90
  cluster analysis and, 165, 181, 196, 204–6
  cross-validation in, 201–4, 211–2, 217, 259
  error rates in, 196
  quadratic, 198–201, 211, 216, 259
  resubstitution method and, 201, 204, 211
  sample size in, 200–1
  stepwise discriminant analysis, 195, 206–9, 213, 217, 260
  three or more groups, 192, 194
  two-groups, 188–91, 196
distance and similarity
  chi-square distance, 108, 110–2
  chi-square statistic, 151
  correlation as a similarity measure, 38
  Euclidean distance, 63–4, 80, 141, 156, 168
  Gower's similarity coefficient, 152–3, 156–7
  information statistic, 151, 157, 173
  Jaccard coefficient, 150, 157, 172, 182
  Mahalanobis distance, 68, 80–2, 168–9, 191, 194, 202–3, 217
  Manhattan (city block) metric, 156
  matching coefficient, 150
  Minkowski metric, 156
Djindjian, F., 123

eigenvalue, 67, 114–15, 246
Everitt, B. S., 156
exploratory data analysis (EDA), 1, 220–1
Euclidean distance, *see* distance and similarity

factor analysis 83, 85–90, 94, 95, 99
  cluster analysis of factors, 169–70

Germanic Iron Age grave assemblage data, 236–7
  correspondence analysis of, 104–7
  similarity coefficients for, 149–54
Gower's similarity coefficient, *see* distance and similarity
Guttman effect, 119

h-plots, 70, 90, 96, 207–8
Haggett, P., 7
Hodson, F. R., 122–3, 182
horseshoe effect (in correspondence analysis), 119–20, 122
Hotelling's T statistic, 194

inertia
  definition of, 114
  decomposition of, 114–18, 138
  multiple correspondence analysis and, 125–6
information statistic, *see* distance and similarity

Jaccard coefficient, *see* distance and similarity
Jolliffe, I. T., 84–5

Kendall, D. G., 118
Krzanowski, W., 5, 90

log-linear models, 9, 138–9
log-normal distribution, *see* normal distribution

Madsen, T., 128
Mahalanobis distance *see under* distance and similarity
matrix
  abundance, 121
  definition of, 240
  determinant of, 243
  incidence, 118–19
  indicator, 124
  Petrie, 118–19
  Q-, 121
  trace of, 174, 242
  transpose of, 241
  see also eigenvalue; singular value decomposition; vectors
matrix algebra, 67, 240–51
maximum likelihood
  cluster analysis and, 174, 181
  factor analysis and, 87–8
Mesolithic assemblage data, 43, 232
  correspondence analysis of, 103–4, 115–18
  PCA of, 61–2
  tabular presentation of, 42–3
minimum spanning tree, 143–4
mixed data, 18–19, 47, 96–7

cluster analysis of, 152–3
PCA of, 65–6
morphology, *see* typology and morphology
mortuary studies, 150–1, 155
multiple correspondence analysis, *see* correspondence analysis
multi-modal data, 30, 40

New Archaeology, 8
Norwegian Early Stone Age assemblage data, 235
  correspondence analysis of, 101–3
Norwegian pit-house morphology data, 238–9
  multiple correspondence analysis of, 124–6
normal distribution,
  assessing normality, 38–40
  desirability of, 30
  discriminant analysis and, 196–7
  log-normal distribution, 40–1, 197
  multivariate normality, 40, 199, 223
  reasons for transformation to, 45, 47
notation, 45–6

Orton, C. R., 8
Ottoway, B., 164
outliers, 30, 40–1, 92
  cluster analysis and, 156, 166, 177–8
  discriminant analysis and, 188, 199, 203
  Gower's coefficient and, 156
  PCA and, 79–80, 96, 175
plots
  box-plots, 30–2
  dot-plots, 28–30, 38, 50, 52–3, 91
  faces, 45
  probability plots, 79
  scatterplots, 32–3
  scree plot, 60–1
  star plots, 44–5, 252
  three-dimensional, 43, 73–5, 95, 99, 111, 130–2
  see also biplots; h-plots
principal components, 248
  cluster analysis of, 169, 179, 183
  form of, 49
  interpretation of, 54–6, 68, 83, 86
  number of, 59–62
  size and shape interpretation, 71–2
  see also rotation (of components)
principal component analysis (PCA), 48–99
  aims of, 48–9
  cluster analysis and, 96, 165, 169, 181
  comparison of analyses, 57–8, 77–9
  correspondence analysis and, 107–9, 113–14
  geometry of, 50–2
  normalisation of coefficients, 50, 67–8, 84–5
  PCA and factor analysis, 85–90

scale dependence of, 49, 87, 89
standardisation of data, 49, 57, 63–5
success of representation, 53, 67
see also correlation; factor analysis; principal components
projection pursuit, 221

quantitative revolution, 7–8, 86

regression analysis, 111
  discriminant analysis and, 206–7
Roman glass from Norway, 142
  cluster analysis of, 144–8, 162–3
  PCA of, 143–4
Romano-British glass compositional data, 28, 228–31
  bivariate analysis of, 32–3
  cluster analysis of, 174–9
  discriminant analysis of, 186–8, 210–14
  multiple correspondence analysis of, 133
  PCA of, 52–62
  univariate analysis of, 28–32
Romano-British vessel glass from Winchester, 50
  discriminant analysis of, 207–8
rotation (of components), 55, 83–5, 97
  factor analysis and, 88
  orthogonal, 84
  oblique, 84, 90
  varimax, 84, 89–90, 99

Seber, G. A. F., 88, 156–7
seriation, 3, 21–2, 104–7, 118–23, 135–7
shape
  of clusters, 155
  correspondence analysis and, 22, 103, 108, 111
  log-ratios interpreted as, 42
  PCA and, 71–2, 97
shape analysis, 20, 222
Shennan S., 2
Sibson's coefficient, 78, 129, 133, 256
similarity, *see* distance and similarity
simple structure, 83
single linkage, *see* cluster analysis
singular value decomposition, 67, 98, 246–7, 254–5
size
  log-products interpreted as, 42
  PCA and, 71–2, 90, 92–3, 97, 103
skewed data, 30, 40, 46, 156, 173, 197
spatial data analysis, 23–4
  cluster analysis and, 155–6, 170–2, 183
  PCA and, 99
  spatial k-means clustering, 148–9
  unconstrained spatial clustering 171–2
standardisation, 38, 45–6
  cluster analysis and, 156

discriminant analysis and, 186–7, 190–1
  PCA and, 57, 63–5
STATGRAPHICS, *see* computer packages
statistical modelling of data, 66
status tables, 173

tool types from Ksar Akil, 74
  three-dimensional plots of, 73–5
transformations
  arc-sin, 46, 76, 98, 183, 216
  centred log-ratio, 46, 73, 95
  linear, 42, 49
  logarithmic, 22, 40–1, 45, 64, 95, 216
  ratio, 41–2, 46, 93, 98, 216
typology and morphology, 15–20
  cluster analysis and, 155, 182–3

discriminant analysis and, 215–6
correspondence analysis and, 137
PCA and, 70–1, 72, 94, 96–7
multiple correspondence analysis and, 126, 137

variance
  between-group, 189, 194
  definition of, 33
  within-group, 189, 193
vectors, 243–6
Vitali, V., 4, 6

Whallon, R., 171, 219
Wilk's lambda, 195–6, 213
Wright, R. V. S., 1, 9, 25, 100